Sustainable Textile Marketing

Sustainable Textile Marketing

Editors

Hafeezullah Memon
Xiaoke Jin
Wei Tian
Chengyan Zhu

MDPI • Basel • Beijing • Wuhan • Barcelona • Belgrade • Manchester • Tokyo • Cluj • Tianjin

Editors
Hafeezullah Memon
Zhejiang Sci-Tech University
China

Xiaoke Jin
Zhejiang Sci-Tech University
China

Wei Tian
Zhejiang Sci-Tech University
China

Chengyan Zhu
Zhejiang Sci-Tech University
China

Editorial Office
MDPI
St. Alban-Anlage 66
4052 Basel, Switzerland

This is a reprint of articles from the Special Issue published online in the open access journal *Sustainability* (ISSN 2071-1050) (available at: https://www.mdpi.com/journal/sustainability/special_issues/textile_marketing).

For citation purposes, cite each article independently as indicated on the article page online and as indicated below:

LastName, A.A.; LastName, B.B.; LastName, C.C. Article Title. *Journal Name* **Year**, *Volume Number*, Page Range.

ISBN 978-3-0365-5525-6 (Hbk)
ISBN 978-3-0365-5526-3 (PDF)

Cover image courtesy of Han Chen

Contents

About the Editors

Hafeezullah Memon

Dr. Hafeezullah Memon CText FTI is an employed foreign expert (professional) at the College of Textile Science and Engineering, Zhejiang Sci-Tech University after completing his postdoc at the same institution. He received his Ph.D. degree in Textile Engineering from Donghua University. Dr. Memon focuses on sustainable textiles and circular economy, dyeing and finishing, vitrimers, and recyclable composites. Dr. Memon has published more than 60 peer-reviewed technical papers in international journals and conferences, and he has been working on more than ten industrial projects. He has been serving as editor and reviewer for more than 40 top journals and books. He is the recipient of several top awards, including Outstanding Foreign Expert and Teacher, CSC Outstanding Award, DHU Outstanding Student Award, NZ Spring International Outstanding Student Award, Excellent Oral Presentation, Excellent Social Award, Fun with Flags-Voluntary Teaching Award, Jing Wei Cultural Ambassador, etc. He is a member of the Textile Society of America, the Society of Wood Science and Technology, the Zhejiang Provincial Key Laboratory of Industrial Textile Materials, and the International Relations Committee of the International Textile and Apparel Association. Moreover, he is a Fellow of the Textile Institute and an active member of the Textile Council, Foundation Committee, and Course Approval Committee. Furthermore, he is a registered textile consultant at The Textile Institute and a registered engineer at Pakistan Engineering Council.

Xiaoke Jin

Dr. Xiaoke Jin received his Ph.D. degree in Zhejiang Sci-Tech University and has been a lecturer at the university. Dr. Jin mainly engaged in the research of spectral imaging technology, textile composition identification, and intelligent textile development. Dr. Jin has published nearly 10 SCI and EI papers, obtained two invention patents, and undertaken or participated in a number of research and development projects; moreover, he has won several provincial and ministerial science and technology awards.

Wei Tian

Dr. Wei Tian is an Associate Professor and postgraduation level mentor. In June 2008, she graduated from the Textile College of Donghua University, majoring in textile materials and textile design, and received a doctor's degree. In June 2008, she entered the College of Materials and Textiles, Zhejiang Sci-Tech University to engage in teaching and research work. She worked as a postdoctoral researcher at Zhejiang University. From December 2016 to December 2017, she was sponsored by China Scholarship Council to visit the University of Manchester in the UK. Her research interests are textile structural composites and functional textiles. She has presided over more than ten projects of Zhejiang Natural Science Foundation and enterprise cooperation. As a backbone member, she has participated in many projects of the National Natural Science Foundation of China, Zhejiang Provincial Department of Science and Technology International Cooperation Project; Economic and Trade Commission of Zhejiang province; and Industry, Education, and Research Project. She has participated in more than 10 provincial and ministerial science and technology progress awards and authorized more than ten patents. She has published more than 30 academic papers as the first author or corresponding author and served as a reviewer for internationally renowned academic journals such as the *Textile Research* Journal.

Chengyan Zhu

Prof. Chengyan Zhu, is doctoral supervisor, director of the national experimental teaching demonstration center of textile engineering, standing director of the international scientific and technological cooperation base of "fiber multidimensional structure preparation and application" of the Ministry of Science and Technology of the People's Republic of China, the person in charge of the construction of the national first-class specialty of textile engineering ("Double 10000" plan), and the person in charge of the provincial teaching team.

He is also the deputy director of the Reinforcement Professional Committee of China Composite Society, the deputy director of the Science and Technology Committee of China Filament Weaving Association, the deputy chairman of the 13th Council of Provincial Textile Engineering Society, and the chairman of Hangzhou Textile Engineering Society.

He is a young and middle-aged discipline leader in provincial colleges and universities, selected as the second level of provincial "151" and winner of the provincial teacher award. His research interests include textile silk science and engineering technology, product performance and R&D, textile structure and composite materials, textile color science, textile design, CAD technology, etc. He has published more than 300 papers and 5 textbooks, and he has been authorized for 20 national invention patents. He has completed more than 60 national, provincial, international, and horizontal cooperation projects, and he has won more than 20 provincial and ministerial science and technology advancement prizes.

sustainability

Editorial

Sustainable Textile Marketing—Editorial

Hafeezullah Memon *, Xiaoke Jin, Wei Tian and Chengyan Zhu

College of Textile Science and Engineering, International Institute of Silk, Zhejiang Sci-Tech University, Hangzhou 310018, China
* Correspondence: hm@zstu.edu.cn

Sustainability has become a vital issue in the textile and garment industry. Over the years, the sector has implemented various initiatives and practices to reduce its environmental impact [1,2]. Today, almost all apparel brands, retailers, and supply chain partners of textiles and garments encourage sustainability throughout the product lifecycle; in addition, sustainability has become an increasingly important theme for consumers [3]. For instance, one recent comparative study about sustainable textile consumption of young millennials in two countries, the US and China, with data collected from 379 Chinese college students and 590 US college students, revealed that sustainability is a large and influential consumer segment for textile and garment consumption [4].

Product Environmental Footprint regulations at an EU level and among the NYC Fashion Policies are the most up-to-date regulations that impact how marketers market sustainable textiles and apparel. The same goes for the recent Higg Index, which refers to Sustainable Apparel Coalition/Life cycle assessment (LCA) issues faced by H&M and how marketers can/should use LCA data in communication. Thus, it has become harder for textile marketers to encourage sustainability in their advertisements to increase the purchase intention of sustainable textile goods. Marketers should set a higher standard and focus on public welfare and related environmental issues of sustainable textiles to persuade consumers to purchase sustainable textile goods by emphasizing that their purchase decision matters a lot with respect to environmental problems. Hong and Kang established a cause–effect relationship between moral philosophy (i.e., idealism and relativism), moral intensity (i.e., magnitude, temporal immediacy, probability, social consensus, and proximity), and purchase behavior toward environmentally sustainable textile and apparel products for the study of female Korean consumers of textile and apparel products dyed with natural and organic dyes for the development of effective marketing communication strategies. It was found that only the idealism of moral philosophy had a significant impact on overall moral intensity. In contrast, moral intensity significantly impacted consumer purchase behavior toward sustainable textile and apparel products [5]. Thus, textile marketers should stress a higher idealism to ensure higher purchases of sustainable textile goods.

Textile marketers should also have good knowledge of textile materials so that they can be ready to answer the queries of customers or consumers, media, and industry experts. There have been concerns about microplastic pollution in the ocean due to microfiber shedding in home laundering, and most consumers believe that only synthetic fibers are prone to microfiber shedding. This microfiber shedding is not limited to synthetic fibers; it is also related to natural fibers [6]. However, textiles' biodegradability might be another concern. Thus, the biodegradability of textile goods may be an additional perk for sustainability. Some biodegradable polymers, including PLA fiber, have been proposed to be used for sustainable textile materials [7]. However, most of the materials in the market have a lower glass transition temperature and poor elasticity, making it challenging to fully replace existing non-biodegradable fibers.

There is increasing interest in making other synthetic fibers biodegradable using chemical or biochemical methods. However, much is remaining to explore the biodegradability mechanisms and their ability to compost without compromising quality during usage. Textile recycling, textile-to-textile recycling, or even garment-to-garment recycling might

Citation: Memon, H.; Jin, X.; Tian, W.; Zhu, C. Sustainable Textile Marketing—Editorial. *Sustainability* **2022**, *14*, 11860. https://doi.org/10.3390/su141911860

Received: 13 September 2022
Accepted: 18 September 2022
Published: 21 September 2022

be another approach for sustainable textile manufacturing. There is increasing interest in the industry to work towards a circular economy and there are various developments taking place, such as pilots, to develop new garment-to-garment recycling technologies. The textile fibers obtained from pre-consumer or post-consumer textile waste might be reused to produce new materials with acceptable properties [8]. An interesting study from Sweden revealed that some companies, e.g., Zhejiang Jiaren, offer fully post-consumer chemical recycling of garments via depolymerization [9].

The second-hand clothing business might be an interesting strategy to reduce textile waste. Recently, there has been an increasing trend for apparel and garment resale sites such as Vestiaire Collective. Moreover, some new startups, e.g., Early Majority and Reflaunt, have recently emerged in the second-hand cloth business, suggesting enormous potential. However, to achieve successful business targets, careful, sustainable textile marketing strategies, including but not limited to customer partnerships, innovative revenue streams, and improved fashion and store appeal, are required [10].

Textiles have remained a significant income source for most countries [11]. Even today, in New York only, there are more than 470 sheep, alpaca, and goat fiber farms and a large number of fiber processing plants for converting raw fibers into final products [12]. Moreover, sustainable textile marketers should have updated knowledge about sustainability trends. The trend of fast fashion in recent decades has emerged as a dominant business model, i.e., Shein; it ensures that each garment is available to customers at a relatively low price due to bulk production, causing health-related, social, and environmental issues throughout the life cycle of textile products. Additionally, it yields higher textile waste to be dumped into landfills. Bick et al. discussed the role of policymakers, consumers, industry, and scientists in ethical consumption and promoting sustainable textile production [13]. On the contrary, slow fashion encourages sustainable fashion design and ensures less textile waste and overall human and environmental impact.

Recently, much research has been made focusing on sustainable textile marketing. Karpova and coworkers studied the views on the impact of the fashion environment by analyzing and interpreting comments made by the readers of the *New York Times* on sustainability, focusing on fashion based on institutional theory [14]. They described how fashion logic is challenged by its moral legitimacy because the values of logical materialism are inconsistent with the importance of sustainability and focus on environmental justice. Multala et al. presented a theoretical microeconomic model of clothing durability standards with which manufacturers may comply by supporting clothing libraries, which can improve overall sustainability by promoting collaborative, sustainable clothing consumption [15]. Wang et al. provided insightful managerial implications to fashion practitioners to formulate product development and marketing strategies by identifying preferred product attributes from the consumer-centric perspective [16]. We received many wonderful publications in this Special Issue of *Sustainability* focusing on the theme of sustainable textile marketing; in addition to these research works presented below, there is yet more to explore in this research area.

List of Contributions

1. Lin, P.-H.; Chen, W.-H. Factors That Influence Consumers' Sustainable Apparel Purchase Intention: The Moderating Effect of Generational Cohorts.
2. Chen, H.; Xu, H.; Zhang, Y.; Jia, L. The Revitalization Design of Regional Ethnic Cultural Capital in a Sustainable Perspective: The Case of Traditional Chinese Garment Yunjian.
3. Memon, H.; Ranathunga, G.M.; Karunaratne, V.M.; Wijayapala, S.; Niles, N. Sustainable Textiles in the Past "Wisdom of the Past: Inherited Weaving Techniques Are the Pillars of Sustainability in the Handloom Textile Sector of Sri Lanka".
4. Fu, Y.; Liang, H.e. Sinicized Exploration of Sustainable Digital Fashion: Chinese Game Players' Intention to Purchase Traditional Costume Skins.

5. Chen, L.; Halepoto, H.; Liu, C.; Yan, X.; Qiu, L. Research on Influencing Mechanism of Fashion Brand Image Value Creation Based on Consumer Value Co-Creation and Experiential Value Perception Theory.
6. Jamnongkan, T.; Intaramongkol, N.; Kanjanaphong, N.; Ponjaroen, K.; Sriwiset, W.; Mongkholrattanasit, R.; Wongwachirakorn, P.; Lin, K.-Y.A.; Huang, C.-F. Study of the Enhancements of Porous Structures of Activated Carbons Produced from Durian Husk Wastes.
7. Halepoto, H.; Gong, T.; Memon, H. A bibliometric analysis of antibacterial textiles.
8. Domingos, M.; Vale, V.T.; Faria, S. Slow Fashion Consumer Behavior: A Literature Review.
9. Hasan, M.M.; Cai, L.; Ji, X.; Ocran, F.M. Eco-Friendly Clothing Market: A Study of Willingness to Purchase Organic Cotton Clothing in Bangladesh.
10. Kim, G.; Kang, P.-G.; Kim, E.; Seo, K. Application of Best Available Techniques to Remove Air and Water Pollutants from Textile Dyeing and Finishing in South Korea.
11. Arshad, M.U.; Zhao, Y.; Hanif, O.; Fatima, F. Evolution of Overall Cotton Production and Its Determinants: Implications for Developing Countries Using Pakistan Case.
12. Chen, L.; Halepoto, H.; Liu, C.; Kumari, N.; Yan, X.; Du, Q.; Memon, H. Relationship Analysis among Apparel Brand Image, Self-Congruity, and Consumers' Purchase Intention.
13. Wang, Y.; Wang, Z.; Zhu, L. A Short Review of Recent Progress in Improving the Fracture Toughness of FRP Composites Using Short Fibers.

Funding: The authors would like to acknowledge funding support from the Research Fund for International Scientists (RFIS-52150410416), National Natural Science Foundation of China.

Conflicts of Interest: The authors declare no conflict of interest.

References

1. Memon, H.; Khoso, N.A.; Memon, S.; Wang, N.N.; Zhu, C.Y. Formulation of Eco-Friendly Inks for Ink-Jet Printing of Polyester and Cotton Blended Fabric. *Key Eng. Mater.* **2016**, *671*, 109–114. [CrossRef]
2. Memon, H.; Khatri, A.; Ali, N.; Memon, S. Dyeing Recipe Optimization for Eco-Friendly Dyeing and Mechanical Property Analysis of Eco-Friendly Dyed Cotton Fabric: Better Fixation, Strength, and Color Yield by Biodegradable Salts. *J. Nat. Fibers* **2016**, *13*, 749–758. [CrossRef]
3. Chen, L.; Qie, K.; Memon, H.; Yesuf, H.M. The Empirical Analysis of Green Innovation for Fashion Brands, Perceived Value and Green Purchase Intention—Mediating and Moderating Effects. *Sustainability* **2021**, *13*, 4238. [CrossRef]
4. Su, J.; Watchravesringkan, K.; Zhou, J.; Gil, M. Sustainable clothing: Perspectives from US and Chinese young Millennials. *Int. J. Retail. Distrib. Manag.* **2019**, *47*, 1141–1162. [CrossRef]
5. Hong, H.; Kang, J.H. The impact of moral philosophy and moral intensity on purchase behavior toward sustainable textile and apparel products. *Fash. Text.* **2019**, *6*, 16. [CrossRef]
6. Vassilenko, E.; Watkins, M.; Chastain, S.; Mertens, J.; Posacka, A.M.; Patankar, S.; Ross, P.S. Domestic laundry and microfiber pollution: Exploring fiber shedding from consumer apparel textiles. *PLoS ONE* **2021**, *16*, e0250346. [CrossRef] [PubMed]
7. Egan, J.; Salmon, S. Strategies and progress in synthetic textile fiber biodegradability. *SN Appl. Sci.* **2021**, *4*, 22. [CrossRef]
8. Memon, H.; Ayele, H.S.; Yesuf, H.M.; Sun, L. Investigation of the Physical Properties of Yarn Produced from Textile Waste by Optimizing Their Proportions. *Sustainability* **2022**, *14*, 9453. [CrossRef]
9. Roos, S.; Sandin, G.; Peters, G.; Spak, B.; Schwarz Bour, L.; Perzon, E.; Jönsson, C. *White Paper on Textile Recycling*; Mistra Future Fashion: Stockholm, Sweden, 2019.
10. Gopalakrishnan, S.; Matthews, D. Collaborative consumption: A business model analysis of second-hand fashion. *J. Fash. Mark. Manag.* **2018**, *22*, 354–368. [CrossRef]
11. Yan, X.; Chen, L.; Memon, H. Introduction. In *Textile and Fashion Education Internationalization: A Promising Discipline from South Asia*; Yan, X., Chen, L., Memon, H., Eds.; Springer Nature: Singapore, 2022; pp. 1–12.
12. Trejo, H.X.; Smith, H.A.; Trejo, N.K.; Lewis, T.L. Made in New York: A Collaborative Model to Encourage Slow Fashion. *Cloth. Text. Res. J.* **2019**, *37*, 139–154. [CrossRef]
13. Bick, R.; Halsey, E.; Ekenga, C.C. The global environmental injustice of fast fashion. *Environ. Health* **2018**, *17*, 92. [CrossRef] [PubMed]
14. Karpova, E.E.; Reddy-Best, K.L.; Bayat, F. The Fashion System's Environmental Impact: Theorizing the Market's Institutional Actors, Actions, Logics, and Norms. *Fash. Theory* **2022**, *1*, 1–22. [CrossRef]
15. Multala, B.; Wagner, J.; Wang, Y. Durability standards and clothing libraries for strengthening sustainable clothing markets. *Ecol. Econ.* **2022**, *194*, 107358. [CrossRef]
16. Wang, L.; Xu, Y.; Lee, H.; Li, A. Preferred product attributes for sustainable outdoor apparel: A conjoint analysis approach. *Sustain. Prod. Consum.* **2022**, *29*, 657–671. [CrossRef]

sustainability

MDPI

Article

A Bibliometric Analysis of Antibacterial Textiles

Habiba Halepoto [1,2], Tao Gong [2,*] and Hafeezullah Memon [3]

1 Engineering Research Center of Digitized Textile and Fashion Technology, Ministry of Education, Donghua University, Shanghai 201620, China
2 College of Information Science and Technology, Donghua University, Shanghai 201620, China
3 International Institute of Silk, College of Textile Science and Engineering, Zhejiang Sci-Tech University, Hangzhou 310018, China
* Correspondence: taogong@dhu.edu.cn

Abstract: Scientists worldwide are always interested in making new discoveries; thus, the scientific literature has been growing exponentially. Keeping abreast of the body of literature at a rapidly advancing pace poses significant challenges to active researchers and society. Although numerous data resources have been made openly available, effectively navigating the vast amount of information with heightened levels of uncertainty remains a significant bottleneck. Here, we searched and analyzed the literature regarding antibacterial treatments in the textile industry. The Web of Science, the most extensive database for scientific literature, was targeted to extract the data. We extracted the raw data from the Web of Science Core Collection. The documents were published from 1998 to 2022, from a total of 878 sources. A total of 3625 documents were found, 2898 of which were articles. A total of 11,192 authors contributed to the topic during this period. We used the Bibliometrix sub-tool Biblioshiny and extracted the basic information about the documents. With an annual growth rate of 26.27%, there was a boost in the literature during the period under study.

Keywords: antibacterial; textiles; bibliometric analysis; database; web of science

Citation: Halepoto, H.; Gong, T.; Memon, H. A Bibliometric Analysis of Antibacterial Textiles. *Sustainability* **2022**, *14*, 11424. https://doi.org/10.3390/su141811424

Academic Editor: Shashi Kant Bhatia

Received: 2 August 2022
Accepted: 5 September 2022
Published: 12 September 2022

1. Introduction

Textiles can be considered a suitable source for the growth of microbes such as fungi, bacteria, and even some protozoans [1]. These microbes affect not only textile materials but also the end user. Besides other effects, microbe growth may generate unwanted odor, discoloration, and contamination and reduce the fabric's mechanical strength [2,3]. The use of antimicrobial textiles that eradicate microorganisms can stop the spread of pathogens via textiles [4]. Antimicrobial agents are applied to textile substrates to create antibacterial textiles, or textile fibers with intrinsic antimicrobial efficiency can be used instead. Depending on the chemistry of the antimicrobial agent and the textile, there are multiple ways to attach antimicrobial chemicals to textiles [5]. Consumers' demand for comfort and well-being and their views toward hygiene have generated a quickly expanding market for antibacterial materials.

Consequently, there has been substantial research in this field in recent years [6]. Because they are hydrophobic, synthetic fibers have been shown to be more resistant to microbial attack, while natural fibers are more susceptible. Thus, some researchers have combined the effect of hydrophobicity and the antibacterial activity of textile materials in their research [7]. Additionally, microbes can obtain nutrients from dust, sweat, and soil [8]. About 85% of the total manufacturing of antimicrobial textiles is made up of lingerie, activewear, socks, and shoe linings. A significant market for antimicrobial fibers has also recently emerged in air filters, outdoor textiles, furniture, and medical textiles [9].

In addition to having antimicrobial properties, an antimicrobial agent must meet a few prerequisites to be successfully applied to textiles and allow for their commercial usage. The fundamental needs of an effective antimicrobial agent for textile substrates can be summarized as follows: they should have a preference for a particular fabric and kind of fiber, be suitable to apply to textile substrates, be capable of eliminating unwanted microorganisms yet keeping favorable microorganisms unaffected, inert to any chemicals that may be used to treat the textile, and so on [5,10]. Furthermore, an infestation of fibers by pathogenic bacteria can cause health problems such as asthma, allergic sensitization, or eczema [11]. Therefore, textiles should be protected to resist bacterial attachment.

Textiles have significant potential to be used for biological and chemical applications, particularly as antibacterial material [12,13]. There are several methods to protect textile materials from microbes. The most general approach to protecting fabrics from bacteria is based on antimicrobial agents in textile finishing, such as silver nanoparticles [14,15]. A wide variety of antibacterial products and chemicals are currently available. The majority of antimicrobial substances used in textile sanitizers have long been found in food preservatives, antiseptics, bandages, and pool cleaners [16]. The attachment of these chemicals to textile surfaces or their interaction with the fiber can significantly diminish their activity and limit the availability of antibacterial agents.

Furthermore, the antibacterial ingredient may be gradually removed by washing and using the textile. Metal salts (for example, silver), quaternary ammonium compounds, halogenated phenols (for example, triclosan), polybiguanide (for example, polyaminopropyl biguanide), chitosan, and N-halamines are the most often used antimicrobial agents for textile applications [17]. In addition, siloxane sulfopropylbetaine has also been proposed for antibacterial cotton fabrics and is claimed to be an eco-friendly product [18].

Bibliometric analysis has recently gained immense fame in research in multiple fields [19–33]. This research method can reveal the distribution patterns of articles published in databases within a given topic, field, institution, and country. The Science Citation Index Expanded from the Web of Science Core Collection of Clarivate Analytics, previously known as Thomson Reuters, is the most valuable and widely used data repository for analyzing scientific achievements across all research fields. Several factors play roles in the popularity of bibliometric studies, such as the advancement, availability, and accessibility of bibliometric software such as Gephi [34], Leximancer [35,36], and VOS viewer [37], and the easy accessibility to scientific databases such as Scopus and Web of Science. More importantly, bibliometric analysis plays a role in handling large volumes of scientific data and, as a result, generates a high research impact. Bibliometric analysis can be used for various reasons, such as to uncover emerging trends in article and journal performance, partnership patterns, and research constituents, and to explore the intellectual structure of a specific domain in the extant literature [38,39]. Bibliometric analysis helps to decipher and map the cumulative scientific knowledge and evolutionary nuances of well-established fields by making sense of large volumes of unstructured data in rigorous ways [29]. Therefore, bibliometric studies foster innovation and help to build a strong basis for advancing a research area in meaningful ways. Bibliometric analysis enables and empowers researchers to (1) have a one-stop overview, (2) investigate knowledge gaps, (3) design innovative ideas for investigation, and (4) position their intended contributions to the field.

Notwithstanding its merits, the deployment of bibliometric analysis does not fully realize its potential. This occurs when bibliometric studies rely on a limited set of data and techniques and provide only a piecemeal understanding of the field under study (e.g., performance analysis without science mapping: [40]). Although some studies are available regarding the bibliometric analysis of bacterial products in the textile industry [19,21,41], these studies lack some details on the most relevant issues to the field, such as (i) the most frequently explored research topics, (ii) publications with the most citations, and (iii) researchers with the most contributions to the body of knowledge in the field, among others. Given that there has not been such discipline-specific bibliometric research, we sought to analyze the literature on antibacterial treatments in the textile industry between 1998 and 2022 since, according to the search on the Web of Science, the first article appeared in 1998. Thus, we studied from the first article to the date of export of the data and conducted a bibliometric analysis of the literature in the mentioned timeframe to answer the following questions:

1. What are the major types of publications?
2. What are the most often explored topics?
3. What are the most highly cited publications and authors?
4. What are the most highly cited journals?
5. Which country/region has the most cited publications?

This bibliometric analysis demonstrates the importance of antibacterial textiles and highlights leading papers, authors, institutes, countries, and so on. We believe that this research can be helpful for researchers to understand and realize the importance of the domain, also providing a basic understanding of the level of work related to antibacterial textiles. Moreover, it will be helpful for industrials and professionals to endorse technology development and aid funding agencies in deciding on the novelty and innovations of antibacterial textiles.

2. Data Collection and Research Methodology

2.1. Data Collection

Web of Science (WoS), as one of the most comprehensive databases, contains articles from high-quality scientific journals. For bibliometric analysis, selecting appropriate keywords is crucial because they impact the research findings directly [42,43]; thus, we did not use the Baidu index [44,45], which is considered a reliable source in China. We extracted the raw data from the Web of Science Core Collection. Since the Web of Science does not allow the export of all the data at once, we chose 500 files at a time. A total of eight files were extracted; seven of them consisted of 500 documents, while the eighth consisted of 125 documents. These files were merged into a single file by using the Python program. The single file of all the documents was then saved to the Python directory with the txt extension. This single file was used for further data analysis.

2.2. Bibliometrix: An R-Tool

The number of academic publications is increasing rapidly, and it is becoming increasingly unfeasible to remain up to date with everything being published. Moreover, the emphasis on empirical contributions has resulted in voluminous and fragmented research streams. This hampers the ability to accumulate knowledge and actively collect evidence through earlier research papers. Therefore, literature reviews are increasingly assuming a crucial role in synthesizing past research findings to effectively use the existing knowledge base, advance a line of research, and provide evidence-based insight into the practice of exercising and sustaining professional judgment and expertise [46]. We examined the scientific productivity of articles, productive authors, citable documents, the most relevant institutions, cited countries, co-occurrence of keywords, thematic mapping, co-citations, and collaboration of authors and countries. We used Biblioshiny to conduct the performance and science mapping analyses [46].

2.3. Literature Search and Analysis

A comprehensive search of the primary scientific literature was conducted on 10 June 2022 on the Web of Science. The Web of Science is considered the most reliable and authentic database among the research community. The search was further modified to include other terms, such as "antibacterial textiles", "antimicrobial textiles", "antibacterial textile", "antimicrobial textile", "anti-bacterial textiles", "anti-microbial textiles", "anti-bacterial textiles", and "anti-microbial textile", to extract specific scientific literature. The following filters were manually activated for the search: (i) relevance, (ii) title and abstract, (iii) year of publication, which ranged between 1998 and 2022, (vi) researchers, (v) research categories, (vi) research type, (vii) source titles, and (viii) journal list. The source data's statistical details were downloaded as plain text files and processed with the bibliometric analysis tool Bibliometrix [47,48].

3. Results

3.1. Data Analysis

The documents are from 1998 to 2022 from a total of 878 sources. A brief description is provided in Table 1. A total of 3625 documents were found, of which 2898 were articles. A total of 11,192 authors contributed to the topic during the period under study. The document's average age was 5.14 years, and the average number of citations per document was 22.8, totaling 110,295 references. The author's keywords amounted to 7390, while the KeyWords Plus reached 4808. There were only 149 documents attributed to a single author. We found that one book chapter was published on the topic, along with three corrections and twenty meeting abstracts. There were 308 review papers, 2 book review chapters, and 240 proceeding papers. We found three corrections, one news item, and one letter (refer to Table 1).

Table 1. Main information about the data.

Description	Results
Main Information About Data	
Timespan	1998–2022
Sources (journals, books, etc.)	877
Documents	3625
Annual growth rate %	26.27
Document average age	5.14
Average citations per doc	22.8
References	110,295
Document Contents	
KeyWords Plus (ID)	4808
Author's keywords (DE)	7390
AUTHORS	
Authors	11,192
Authors of single-author docs	118
Authors Collaboration	
Single-author docs	149
Co-authors per doc	4.87
International co-authorships %	23.17
Document Types	
Article	2898
Article; early access	1
Article; book chapter	1
Article; early access	89
Article; proceedings paper	43
Correction	3
Editorial material	1
Letter	1

Table 1. *Cont.*

Description	Results
Meeting abstract	20
News item	1
Proceedings paper	240
Review	308
Review; book chapter	2
Review; early access	17

3.2. Annual Publications

We used the Bibliometrix sub-tool Biblioshiny and extracted the basic information on the publications. We can see a substantial increase in the number of publications each year, with an annual growth rate of 26.27%. Only one document was published in 1998. The article describes the synthesis, stability, surface activity, and antimicrobial properties of a new family of cationic surfactants (the long chain arginylalkylamide dihydrochloride salts) derived from the condensation of the amino acid arginine, and a long-chain alkylamine was described. Based on Google Scholar metrics, this article has been cited by 40 researchers as of the end of August 2022. The addition of documents to the literature was sparse until 2007. From 1999 to 2002, not a single document was published. In 2003, one more document was added to the literature. In this publication, the antibacterial activities of a newly formed compound were measured with *Escherichia coli* 8099 as a Gram-negative strain and *Staphylococcus aureus* ATCC 6538 as a Gram-positive strain. According to the journal's official metrics, this article has been cited by 39 researchers as of August 2022. No document was added for the next three years, i.e., 2004–2006.

There was a considerable boost in publications in 2007. A total of 53 documents were added to the literature in 2007 and 79 documents in 2008. There has been a continuous increase in the number of documents published each year. In 2021, a total of 534 documents were published, and by mid-2022, a total of 228 documents had been added to the literature, as shown in Figure 1.

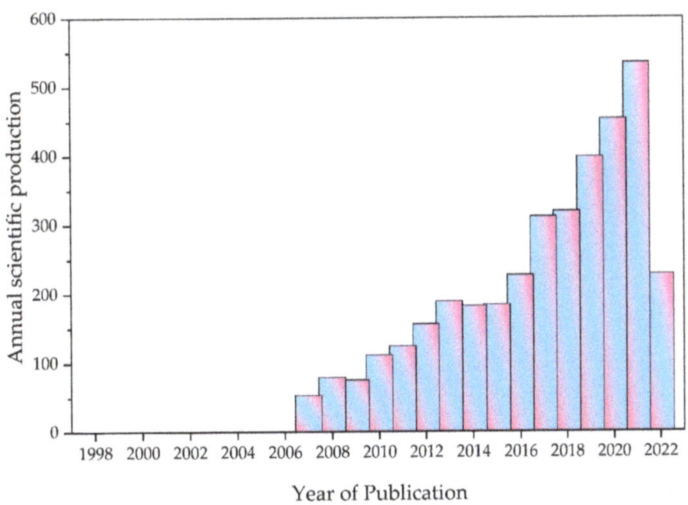

Figure 1. Annual scientific production during each year. (Bar chart created by Origin Pro 2022b, Northampton, MA, USA).

3.3. Annual Citations

As more and more research has been conducted over time, the citations of the articles has increased. In the beginning, there were few articles, and there was a decline in the literature. In 1998, there was only one citation, and no citations were available from 1999 to 2002. In 2003, a total of 1.8 citations per year were available. For the next three years, there was not a single citation available; however, in 2007, we can see exponential growth in the number of citations, as shown in Figure 2. Until 2021, we can see that there has been a diverse range of citations each year, increasing in some years and decreasing in others. However, we can see a considerable rise in citations during the calendar year 2022.

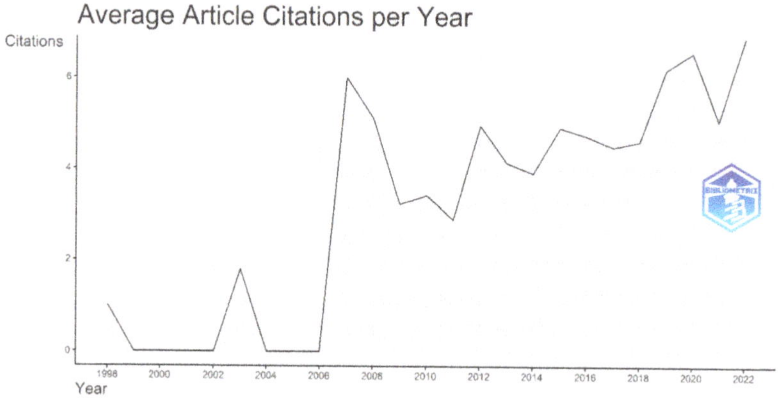

Figure 2. Average citations per year. The x-axis indicates years; the y-axis shows the average citations. In 1998, there was a single citation, and there was an exponential rise in citations in 2007.

3.4. Most Relevant Sources and Corresponding Authors' Countries

We performed an analysis to identify the most relevant sources. We found that the National Research Center, Egypt (NATL RES CTR), is at the top, with a total of 260 documents. The Islamic Azad University, Iran (ISLAMIC AZAD UNIV), is second, with a total of 177 documents; the third was Donghua University, China (DONGHUA UNIV), with a total of 131 documents (Figure 3).

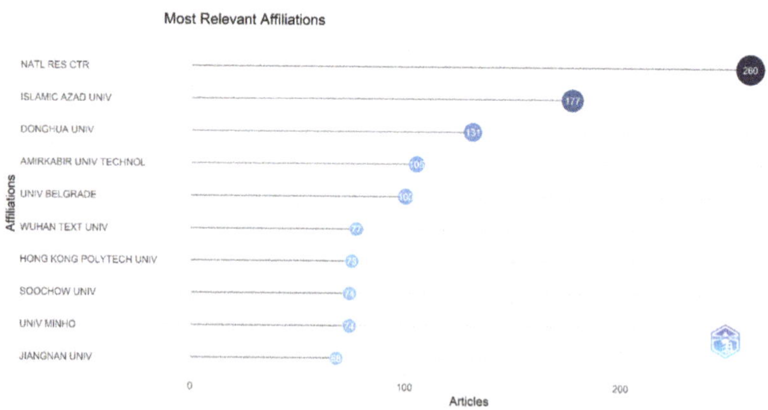

Figure 3. The most relevant affiliated institutes. The x-axis indicates the number of documents; the y-axis shows the institute name.

Similarly, the corresponding authors' countries were evaluated. As shown in Figure 4, China is the leading country regarding the corresponding authors, with a total of 144 multiple-country publication (MCP) documents and 578 single-country publication (SCP) documents, followed by India, with an MCP of 78 documents and an SCP of 409 documents.

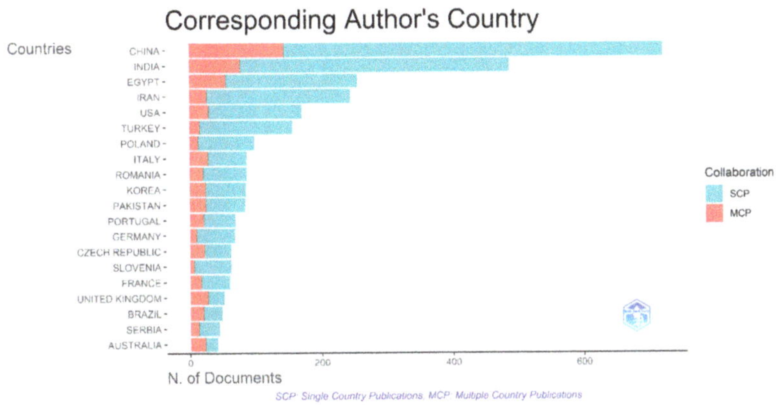

Figure 4. Corresponding authors' countries. The x-axis indicates the number of published documents; the y-axis shows the country names. China leads with a total of 144 MCP documents and 578 SCP documents.

3.5. Most Relevant Author and Author's Local Impact

The data were evaluated to identify the most relevant authors. Montazer M., with 71 documents, is the leading author, followed by Ibrahim N. A. and Simoncic B., with 34 documents (Figure 5). We also analyzed the data to determine the authors' production over time. The leading author, Montazer M., published two articles in 2007, one article in 2009, two articles during the calendar year 2010, five articles in 2011, four articles in 2012, seven articles in 2013, five articles each in 2014 and 2015, seven articles in 2016, nine articles in 2017, eight articles in 2018, six articles in 2019, two articles in 2020, six articles in 2021, and only one article in 2022. The highest number of articles he published in a calendar year is nine. Similarly, the authors' impact was evaluated. Montazer M., with a total of 3294 citations, is the leading author, followed by Mohammad F., with 1663 citations (Figure 6).

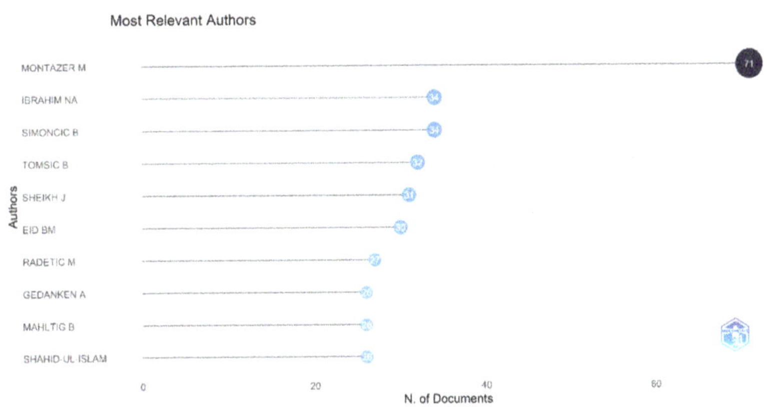

Figure 5. The most relevant authors. The x-axis indicates the number of documents; the y-axis indicates the authors' names. Montazer M. leads with a total of 71 documents.

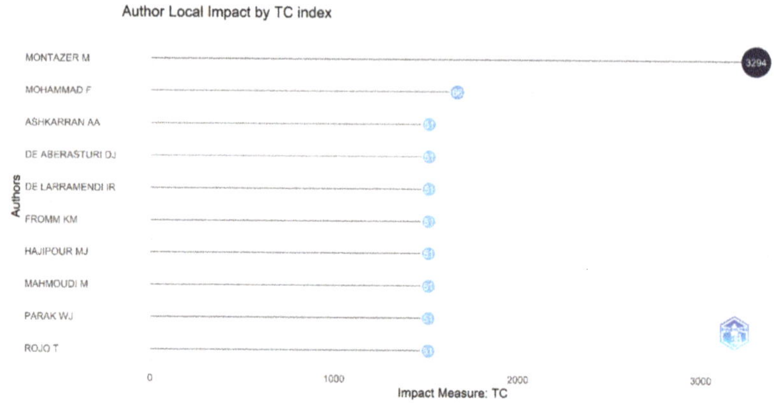

Figure 6. The most relevant authors concerning the total number of citations. The *x*-axis shows the number of citations; the *y*-axis indicates the authors' names.

3.6. *Most Frequent Words*

The most frequent words used in the literature from 1998 to date were "textiles", which led with a total count of 597, followed by "nanoparticles", with a total count of 457, and "antibacterial", with a total count of 445 (Figure 7).

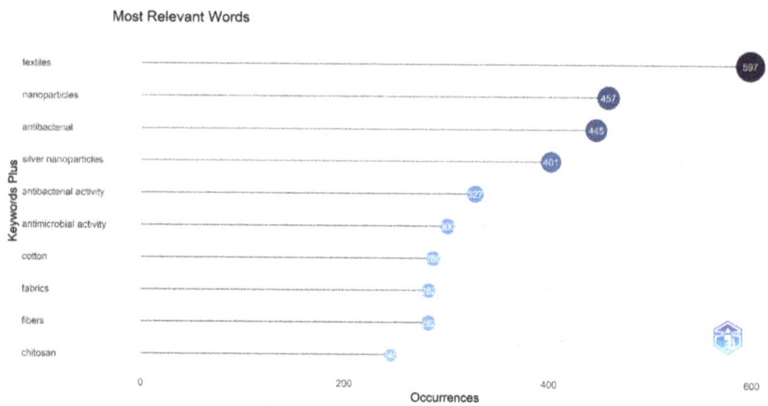

Figure 7. The most relevant words used in the literature. The *x*-axis shows the number of occurrences; the *y*-axis indicates the key terms.

3.7. *Word Cloud of KeyWords Plus*

The co-occurrence network was generated using the KeyWords Plus. The most relevant words were "textiles" with a frequency of 597, "cotton" with a frequency of 286, and "fabrics" and "fibers" with a frequency of 282 each, as shown in Figure 8.

Figure 8. Word cloud of KeyWords Plus. Textiles, with a frequency of 597, is the leading term, followed by cotton.

3.8. Most Cited Countries

We analyzed the data to identify the leading countries that have produced the most publications. We found that China was at the top, with a total of 12,985 citations having 1678 scientific productions, followed by India, with a total of 10,038 citations and 1080 scientific productions. Iran stands in the third position with a total of 8883 citations and 544 documents, as shown in Figure 9 and Table 2.

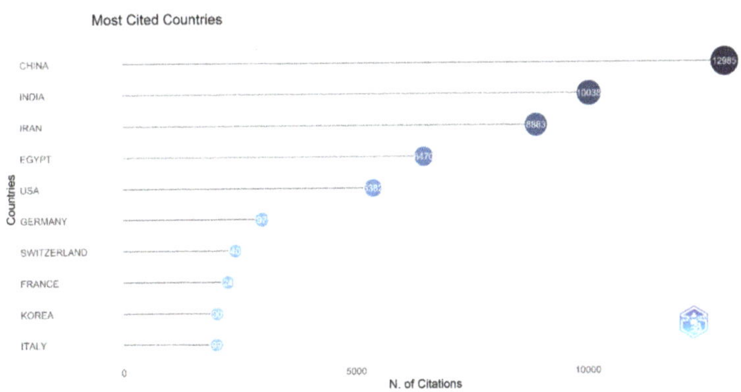

Figure 9. Most cited countries. The *x*-axis indicates the total number of citations; the *y*-axis shows the names of the countries. Only the top 10 countries were selected.

Table 2. Country frequency in terms of scientific production.

Country	Frequency
China	1678
India	1080
Egypt	610
Iran	544
USA	531
Turkey	362
Pakistan	310
Italy	263
Romania	253
France	245

3.9. Most Globally Cited Documents

We found that "Antibacterial properties of nanoparticles", a review article by Hajipour M. J. et al., 2012, is the most cited document in this field, published in the journal *Trends in Biotechnology* [49]. This article is a review of the literature about antimicrobial activity in multiple fields. The authors also mention the importance of nanoparticles in antimicrobial activities. So far, this article has been cited in 1511 documents (Table 3).

Table 3. The most cited documents worldwide.

Title	Authors	Journal	Total Citations	TC (Year)	Reference
Antibacterial properties of nanoparticles	Hajipour MJ et al.	*Trends Biotechnol.*, 2012	1511	137.36	[49]
Nanosilver: A nanoproduct in medical application	Chen X et al.	*Toxicol. Lett.*, 2008	1357	90.47	[50]
The Chemistry and Applications of Antimicrobial Polymers: A State-of-the-Art Review	Kenawy ER et al.	*Biomacromolecules*, 2007	1165	72.81	[51]
Curcumin: the Indian solid gold	Dastjerdi R et al.	*Colloid Surface B*, 2010	934	71.85	[52]

3.10. Most Relevant Authors and Authors' Production over Time

We found that Montazer M. is the most relevant author, with a total of 71 documents, followed by Simoncic B. and Ibrahim N. A., with a total of 34 documents for each author (Figure 10). Similarly, we analyzed the data to identify the authors' production over time. We found that Montazer M. published two articles in 2007 and received 10.94 citations. A significant boost was seen in his production in 2017, when he published nine articles in a calendar year (Figure 11).

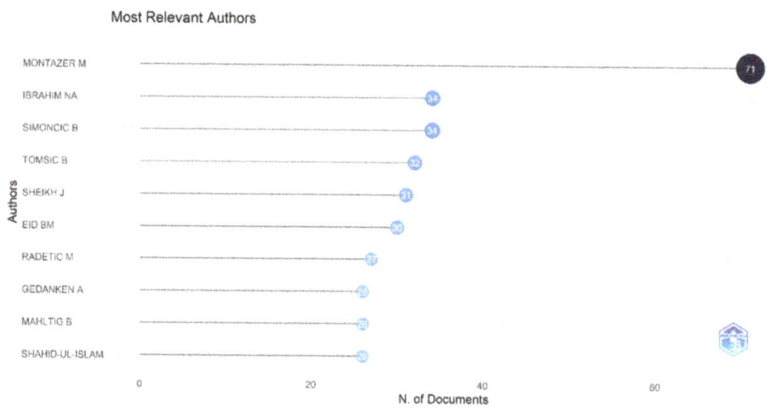

Figure 10. Most relevant authors. The x-axis indicates the number of documents; the y-axis indicates the authors' names.

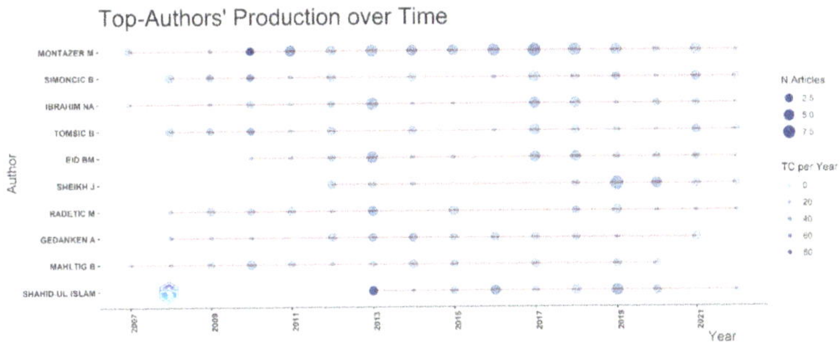

Figure 11. Authors' production over time. The x-axis indicates the year; the y-axis indicates the authors' names. The circle size shows the number of articles.

3.11. Most Relevant Sources

The most relevant sources were *Fibers and Polymers*, with a total of 156 documents, followed by *Cellulose* with a total of 137 documents and *The Journal of The Textile Institute* with a total of 104 documents, as shown in Figure 12. Similarly, the source of local impact was found. *Carbohydrate Polymers*, with an H-index of 43, is the most impactful source, followed by *Applied Surface Science*, as provided in Figure 13.

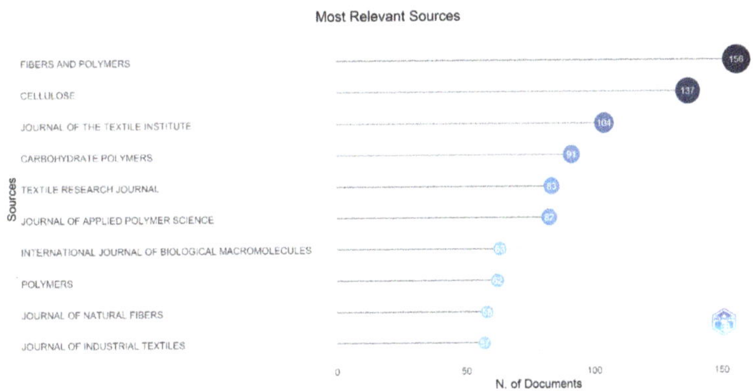

Figure 12. Most relevant sources. The x-axis shows the number of documents; the y-axis indicates the source name.

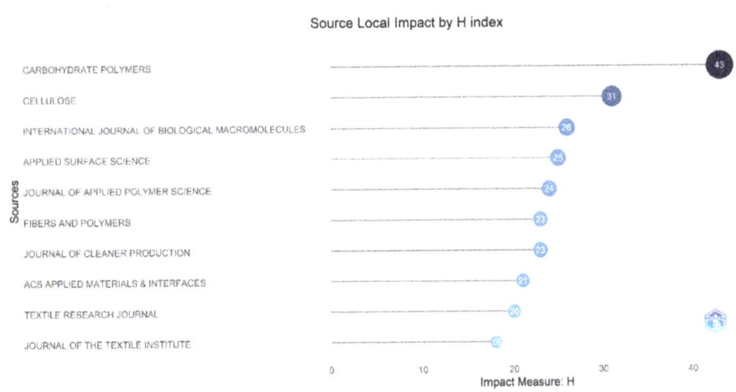

Figure 13. Source local impact based on H-index. The x-axis indicates the H-index measurement; the y-axis shows the source name.

3.12. Authors and Keywords

We identified the links between the authors, keywords, and sources. The most impactful author was Montazer M., and the most impactful source was *Cellulose*. The most frequent keyword was "cotton", followed by "antimicrobial", as shown in Figure 14.

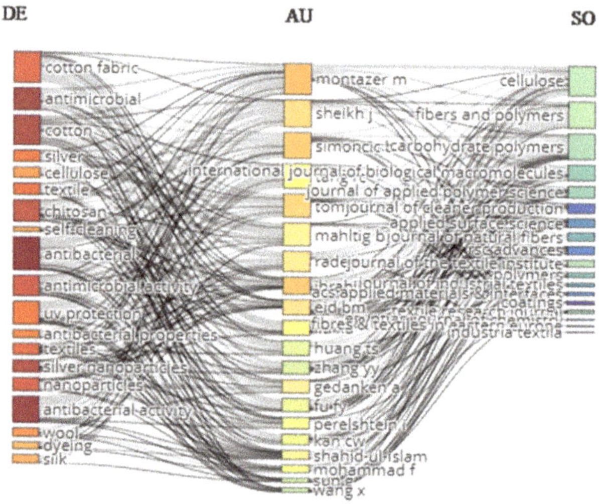

Figure 14. Links between authors, keywords, and sources.

3.13. Trending Topics

We analyzed the data to extract the trending topics from titles and abstracts. From the titles, "antibacterial" was the most trending topic, with a total frequency of 908. The second most trending topic was "antimicrobial", which appeared 690 times in the literature, as shown in Figure 15. The trending topics extracted from the abstracts were "antibacterial", with a frequency of 4299, followed by "fabrics", with a frequency of 3718, as shown in Figure 16. The other trending topics extracted from the abstracts were "properties", "antimicrobial", "activity", "surface", and so on.

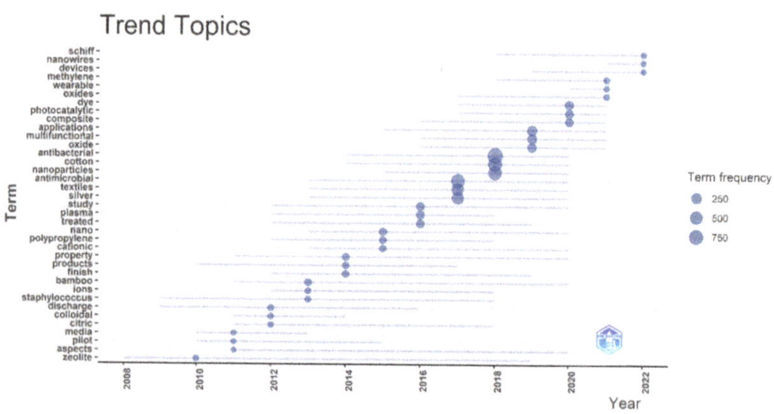

Figure 15. Trending topics extracted from titles. "Antibacterial", with a total frequency of 908, was the most trending topic, followed by "antimicrobial", with a total frequency of 690. The *x*-axis shows the years; the *y*-axis shows the topic. The circle size indicates the frequency.

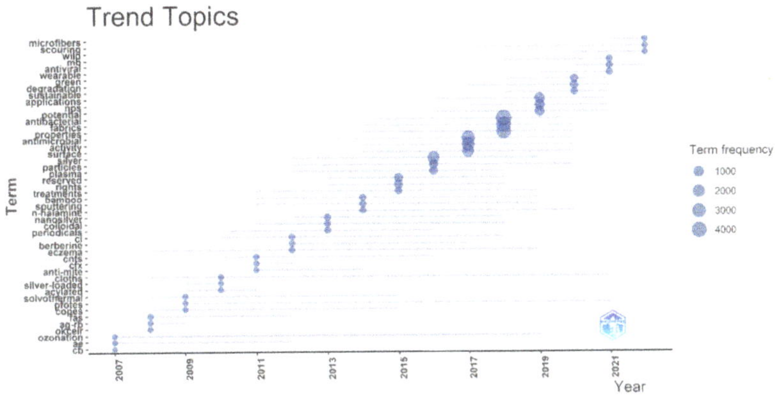

Figure 16. Trending topics extracted from abstract. The *x*-axis shows the years; the *y*-axis indicates the trending topics. The size of the circle indicates the frequency.

4. Discussion

Bibliometric analysis has been gaining immense recognition in research in multiple fields [19,21–23,31,53,54]. This research method can reveal the distribution patterns of articles published in databases within a given topic, field, institution, and country. Several databases can be used to retrieve the raw data for bibliometric analysis, such as Scopus, Web of Science, etc. The Science Citation Index Expanded from the Web of Science Core Collection of Clarivate Analytics (previously known as Thomson Reuters) is the most valuable and widely used data repository for analyzing scientific achievements across all research fields. Bibliometric analysis helps in deciphering and mapping the cumulative scientific knowledge and evolutionary nuances of well-established areas by making sense of large volumes of unstructured data in rigorous ways. In this study, we extracted the raw data from the Web of Science Core Collection. The documents, from a total of 878 sources, were published from 1998 to 2022.

A total of 3625 documents were found, and 2898 were articles. A total of 11,192 authors have contributed to this research field. We used the Bibliometrix sub-tool Biblioshiny and extracted the documents' basic information. We found a sizeable increase in the number of publications each year, with an annual growth rate of 26.27%. Almost every scientific field has significantly increased [55,56]. Our data date back to 1998, and only one document was published that year. From 1999 to 2002, not a single document was published. In 2003, one document was added to the literature. Not a single document was published for the next three years. However, there was a boost in 2007, when a total of 53 documents were added to the literature, followed by 79 documents in 2008. There has been a continuous increase in the documents published each year. In 2021, a total of 534 documents were published, and by mid-2022, a total of 228 documents had been added to the literature. As more and more research has been performed with time, the citations of the articles are boosted.

In the beginning of the period under study, there were few articles, and there was a decline in the literature. In 1998, there was only one citation; no citation was available from 1999 to 2002. In 2007, we observed exponential growth in the number of citations, as presented in Figure 3. There was a steep rise in citations during the calendar year 2022. the National Research Center, Egypt (NATL RES CTR), has published 260 documents, the leading institute among all. China is the top country in terms of the corresponding authors, with a total number of 278 multiple-country publication (MCP) documents and 1118 single-country publication (SCP) documents, followed by India with an MCP of 150 documents and an SCP of 780 documents. Montazer M. was the most impactful author, with a total of 6588 citations, followed by Mohammad M., with a total of 3326 citations. The most frequently used word in the literature from 1998 to date is "textiles", leading with a total count of 1160, followed by "nanoparticles" with a total count of 888 and antibacterial with a total count of 848. Overall, nanotechnology, including nanofibers, has been useful for anti-bacterial textiles [57,58]. The high number of publications shows that the policies and funding agencies in certain countries are interested in exploring this particular research area. Thus, reporting a country's frequency of scientific production is significant. We found that China was at the top, with a total of 25,970 citations having 3356 scientific productions, followed by India, with a total of 20,076 citations and 2160 scientific productions. The most relevant sources were *Fibers and Polymers*, with a total of 312 documents, followed by *Cellulose* with 274 and *The Journal of The Textile Institute* with 208.

5. Summary

5.1. Conclusions

This study summarizes the importance of antibacterial textiles and suggests that antibacterial textiles have been an important research topic over the past two decades. Our thematic analysis showed that introducing nanoparticles to textile materials is the most common approach for making antibacterial textiles. Bibliometric research is helpful for deciphering and mapping the cumulative scientific knowledge and evolutionary nuances of well-established fields by making sense of large volumes of unstructured data in rigorous ways. With an annual growth rate of 26.27%, we have seen a considerable boost in the literature. We suggest that a future study use multiple databases to extract data for analysis. As for every study, this research has certain limitations, as we focused only on data from WoS; thus, there might be some variation in the results if the data are taken from other sources. This study does not contain any information about reviewers or top editors in this field—since open publishing is being appreciated worldwide, we intend to analyze the information on reviewers and editors in the future.

5.2. Implications

With the rapid growth of publications worldwide in every field of science and technology, it has become difficult for researchers to read all papers and predict research trends. Bibliometric analysis is an effective way to precisely cover a wide field to predict research trends. For instance, in this study, we learned that silver nanoparticles have been widely explored for antimicrobial applications and used mostly on cotton fiber. Moreover, chitosan treatments are also widely used after the nanocoating finishing of textile materials. Thus, this study would be useful for new researchers looking to choose an appropriate approach. Moreover, it will help industrialists endorse the technology being studied the most. In addition, this work can be useful for funding agencies to gauge the influence and research gaps at a short glance, as this study summarizes numerous research works in one place.

5.3. Limitations and Future Research Directions

As with every scientific work, this study also has some limitations; for example, this study only covers the articles published on the Web of Science and thus overlooks many valuable works in other databases. Moreover, this study does not review the works of various authors, as often presented in review papers—it only analyzes research trends in the given field. Additionally, we did not analyze who reviewed the papers, such as who was the most influential reviewer in the given field. Furthermore, this study does not discuss whether the citations were self-citations and citations pushed by friends and workmates. Moreover, we did not discuss the usage count of the articles. A detailed study covering all these areas might be worth conducting to analyze the research direction more effectively. In addition, different tools or software might yield different results, and a comparison could be made. Finally, a similar field could be investigated by changing some keywords related to this research direction.

Author Contributions: Conceptualization, H.H. and H.M.; methodology, H.H.; software, H.H.; validation, H.H.; formal analysis, H.H.; investigation, H.H.; resources, H.H.; data curation, H.H.;

writing—original draft preparation, H.H.; writing—review and editing, H.M. and T.G.; visualization, H.H. and T.G.; supervision, T.G.; project administration, T.G. and H.M.; funding acquisition, H.M. All authors have read and agreed to the published version of the manuscript.

Funding: This work was supported by the Research Fund for International Scientists (RFIS-52150410416) and National Natural Science Foundation of China (no. 61673007).

Institutional Review Board Statement: Not applicable.

Informed Consent Statement: Not applicable.

Data Availability Statement: Data can be provided by corresponding author on request.

Conflicts of Interest: The authors declare no conflict of interest.

References

1. Wang, W.-Y.; Chiou, J.-C.; Yip, J.; Yung, K.-F.; Kan, C.-W. Development of Durable Antibacterial Textile Fabrics for Potential Application in Healthcare Environment. *Coatings* **2020**, *10*, 520. [CrossRef]
2. Kumar, M.N.R. A review of chitin and chitosan applications. *React. Funct. Polym.* **2000**, *46*, 1–27. [CrossRef]
3. Gupta, D.; Haile, A. Multifunctional properties of cotton fabric treated with chitosan and carboxymethyl chitosan. *Carbohydr. Polym.* **2007**, *69*, 164–171. [CrossRef]
4. Lan, S.; Zhang, J.; Li, J.; Guo, Y.; Sheng, X.; Dong, A. An N-Halamine/Graphene Oxide-Functionalized Electrospun Polymer Membrane That Inactivates Bacteria on Contact and by Releasing Active Chlorine. *Polymers* **2021**, *13*, 2784. [CrossRef]
5. Yuan, G.; Cranston, R. Recent Advances in Antimicrobial Treatments of Textiles. *Text. Res. J.* **2008**, *78*, 60–72. [CrossRef]
6. Granados, A.; Pleixats, R.; Vallribera, A. Recent Advances on Antimicrobial and Anti-Inflammatory Cotton Fabrics Containing Nanostructures. *Molecules* **2021**, *26*, 3008. [CrossRef]
7. Xue, C.-H.; Chen, J.; Yin, W.; Jia, S.-T.; Ma, J.-Z. Superhydrophobic conductive textiles with antibacterial property by coating fibers with silver nanoparticles. *Appl. Surf. Sci.* **2012**, *258*, 2468–2472. [CrossRef]
8. Broadhead, R.; Craeye, L.; Callewaert, C. The Future of Functional Clothing for an Improved Skin and Textile Microbiome Relationship. *Microorganisms* **2021**, *9*, 1192. [CrossRef]
9. Correia, J.; Rainert, K.T.; Oliveira, F.R.; de Cássia Siqueira Curto Valle, R.; Valle, J.A.B. Cationization of cotton fiber: An integrated view of cationic agents, processes variables, properties, market and future prospects. *Cellulose* **2020**, *27*, 8527–8550. [CrossRef]
10. Kramer, A.; Guggenbichler, P.; Heldt, P.; Jünger, M.; Ladwig, A.; Thierbach, H.; Weber, U.; Daeschlein, G. Hygienic relevance and risk assessment of antimicrobial-impregnated textiles. *Biofunct. Text. Ski.* **2006**, *33*, 78–109.
11. Bajpai, V.; Bajpai, S.; Jha, M.K.; Dey, A.; Ghosh, S. Microbial adherence on textile materials: A review. *J. Environ. Res. Dev.* **2011**, *5*, 666–672.
12. Lugoloobi, I.; Shahriari Khalaji, M.; Memon, H. Advanced Biological Applications of Modified Cotton. In *Cotton Science and Processing Technology: Gene, Ginning, Garment and Green Recycling*; Wang, H., Memon, H., Eds.; Springer: Singapore, 2020; pp. 473–500.
13. Lugoloobi, I.; Tebyetekerwa, M.; Memon, H.; Sun, C. Advanced Chemical Applications of Modified Cotton. In *Cotton Science and Processing Technology: Gene, Ginning, Garment and Green Recycling*; Wang, H., Memon, H., Eds.; Springer: Singapore, 2020; pp. 501–527.
14. Memon, H.; Wang, H.; Yasin, S.; Halepoto, A. Influence of Incorporating Silver Nanoparticles in Protease Treatment on Fiber Friction, Antistatic, and Antibacterial Properties of Wool Fibers. *J. Chem.* **2018**, *2018*, 4845687. [CrossRef]
15. Liu, Y.; Hussain, M.; Memon, H.; Yasin, S. Solar irradiation and Nageia nagi extract assisted rapid synthesis of silver nanoparticles and their antibacterial activity. *Dig. J. Nanomater. Biostruct.* **2015**, *10*, 1019–1024.
16. Salazar-Alemán, D.A.; Turner, R.J. Metal Based Antimicrobials: Uses and Challenges. In *Microbial Metabolism of Metals and Metalloids*; Hurst, C.J., Ed.; Springer International Publishing: Cham, Switzerland, 2022; pp. 77–106.
17. Simoncic, B.; Tomsic, B. Structures of novel antimicrobial agents for textiles—A review. *Text. Res. J.* **2010**, *80*, 1721–1737. [CrossRef]
18. Chen, S.; Chen, S.; Jiang, S.; Xiong, M.; Luo, J.; Tang, J.; Ge, Z. Environmentally Friendly Antibacterial Cotton Textiles Finished with Siloxane Sulfopropylbetaine. *ACS Appl. Mater. Interfaces* **2011**, *3*, 1154–1162. [CrossRef]
19. Yafetto, L. Application of solid-state fermentation by microbial biotechnology for bioprocessing of agro-industrial wastes from 1970 to 2020: A review and bibliometric analysis. *Heliyon* **2022**, *8*, e09173. [CrossRef]
20. Ho, Y.-S.; Fahad Halim, A.F.M.; Islam, M.T. The Trend of Bacterial Nanocellulose Research Published in the Science Citation Index Expanded From 2005 to 2020: A Bibliometric Analysis. *Front. Bioeng. Biotechnol.* **2022**, *9*, 795341. [CrossRef]
21. Han, M.-C.; Cai, S.-Z.; Wang, J.; He, H.-W. Single-Side Superhydrophobicity in Si3N4-Doped and SiO2-Treated Polypropylene Nonwoven Webs with Antibacterial Activity. *Polymers* **2022**, *14*, 2952. [CrossRef]
22. Borregan-Alvarado, J.; Alvarez-Meaza, I.; Cilleruelo-Carrasco, E.; Garechana-Anacabe, G. A Bibliometric Analysis in Industry 4.0 and Advanced Manufacturing: What about the Sustainable Supply Chain? *Sustainability* **2020**, *12*, 7840. [CrossRef]
23. Provin, A.P.; dos Reis, V.O.; Hilesheim, S.E.; Bianchet, R.T.; de Aguiar Dutra, A.R.; Cubas, A.L.V. Use of bacterial cellulose in the textile industry and the wettability challenge—A review. *Cellulose* **2021**, *28*, 8255–8274. [CrossRef]

24. Kuilang, Y.; Qian, X. Research on the innovation frontier of global intelligent textile technology based on patentometrics. *J. Silk* **2021**, *58*, 48–55. [CrossRef]
25. Noor, S.; Guo, Y.; Shah, S.H.H.; Halepoto, H. Bibliometric Analysis of Twitter Knowledge Management Publications Related to Health Promotion. In Proceedings of the 13th International Conference, Hangzhou, China, 28–30 August 2020; pp. 341–354.
26. Syed Hamd Hassan, S.; Saleha, N.; Atif Saleem, B.; Habiba, H. Twitter Research Synthesis for Health Promotion: A Bibliometric Analysis. *Iran. J. Public Health* **2021**, *50*, 2283–2291. [CrossRef]
27. Xiang, F.; Xiaopeng, W.; Xiaoxiao, Q.; Laili, W. Bibliometric analysis of literatures on textile and clothing footprint based on CiteSpace. *Adv. Text. Technol.* **2022**, *30*, 9–17. [CrossRef]
28. Mei, P.; Lizhu, G.; Zilin, K.; Lan, Z. Research review and prospects of VOSviewer-based textile plant dyeing. *J. Silk* **2021**, *58*, 53–59. [CrossRef]
29. Halepoto, H.; Gong, T.; Noor, S.; Memon, H. Bibliometric Analysis of Artificial Intelligence in Textiles. *Materials* **2022**, *15*, 2910. [CrossRef]
30. Xizhen, L.; Zhiqin, S. Research on the knowledge map and visualization of fashion design field in China based on CiteSpace. *J. Silk* **2020**, *57*, 25–34. [CrossRef]
31. Abuhassna, H.; Awae, F.; Bayoumi, K.; Alzitawi, D.U.; Alsharif, A.H.; Yahaya, N. Understanding Online Learning Readiness among University Students: A Bibliometric Analysis. *Int. J. Interact. Mob. Technol.* **2022**, *16*, 81–94. [CrossRef]
32. Alsharif, A.H.; Md Salleh, N.Z.; Baharun, R.; Rami Hashem, E.A. Neuromarketing research in the last five years: A bibliometric analysis. *Cogent Bus. Manag.* **2021**, *8*, 1978620. [CrossRef]
33. Ali, J.; Jusoh, A.; Idris, N.; Abbas, A.F.; Alsharif, A.H. Nine Years of Mobile Healthcare Research: A Bibliometric Analysis. *Int. J. Online Biomed. Eng.* **2021**, *17*, 144–159. [CrossRef]
34. Omotayo, T.; Moghayedi, A.; Awuzie, B.; Ajayi, S. Infrastructure elements for smart campuses: A bibliometric analysis. *Sustainability* **2021**, *13*, 7960. [CrossRef]
35. Wilk, V.; Cripps, H.; Capatina, A.; Micu, A.; Micu, A.-E. The state of# digitalentrepreneurship: A big data Leximancer analysis of social media activity. *Int. Entrep. Manag. J.* **2021**, *17*, 1899–1916.
36. Khan, S.; Rana, S.; Goel, A. Presence of digital sources in international marketing: A review of literature using Leximancer. *Int. J. Technol. Mark.* **2022**, *16*, 246–274. [CrossRef]
37. Van Eck, N.J.; Waltman, L. Software survey: VOSviewer, a computer program for bibliometric mapping. *Scientometrics* **2010**, *84*, 523–538. [CrossRef]
38. Donthu, N.; Kumar, S.; Pandey, N.; Lim, W.M. Research constituents, intellectual structure, and collaboration patterns in Journal of International Marketing: An analytical retrospective. *J. Int. Mark.* **2021**, *29*, 1–25. [CrossRef]
39. Verma, S.; Gustafsson, A. Investigating the emerging COVID-19 research trends in the field of business and management: A bibliometric analysis approach. *J. Bus. Res.* **2020**, *118*, 253–261. [CrossRef]
40. Brown, T.; Park, A.; Pitt, L. A 60-Year Bibliographic Review Of the Journal of Advertising Research Perspectives on Trends in Authorship, Influences, and Research Impact. *J. Advert. Res.* **2020**, *60*, 353–360. [CrossRef]
41. Fernandes, M.; Souto, A.P.; Dourado, F.; Gama, M. Application of Bacterial Cellulose in the Textile and Shoe Industry: Development of Biocomposites. *Polysaccharides* **2021**, *2*, 566–581. [CrossRef]
42. AlRyalat, S.A.S.; Malkawi, L.W.; Momani, S.M. Comparing bibliometric analysis using PubMed, Scopus, and Web of Science databases. *JoVE J. Vis. Exp.* **2019**, *152*, e58494. [CrossRef]
43. Khudhair, H.Y.; Jusoh, D.; Bin, A.; Abbas, A.F.; Mardani, A.; Nor, K.M. A review and bibliometric analysis of service quality and customer satisfaction by using Scopus database. *Int. J. Manag.* **2020**, *11*, 459–470.
44. Chen, X.; Yunhe, H.U. Analysis of public concern on traditional dyeing and printing techniques based on Baidu index. *J. Silk* **2020**, *57*, 41108.
45. Shansen, W.; Jianfang, L. The impact of COVID-19 on the attention to sustainable clothing consumption:Analysis of Baidu indexes on old clothes recycling, old clothes renovation and old clothes donation. *J. Silk* **2021**, *58*, 40–46. [CrossRef]
46. Aria, M.; Misuraca, M.; Spano, M. Mapping the Evolution of Social Research and Data Science on 30 Years of Social Indicators Research. *Soc. Indic. Res.* **2020**, *149*, 803–831. [CrossRef]
47. Aria, M.; Cuccurullo, C. bibliometrix: An R-tool for comprehensive science mapping analysis. *J. Informetr.* **2017**, *11*, 959–975. [CrossRef]
48. Koo, M. Systemic Lupus Erythematosus Research: A Bibliometric Analysis over a 50-Year Period. *Int. J. Environ. Res. Public Health* **2021**, *18*, 7095. [CrossRef]
49. Hajipour, M.J.; Fromm, K.M.; Ashkarran, A.A.; Jimenez de Aberasturi, D.; de Larramendi, I.R.; Rojo, T.; Serpooshan, V.; Parak, W.J.; Mahmoudi, M. Antibacterial properties of nanoparticles. *Trends Biotechnol.* **2012**, *30*, 499–511. [CrossRef]
50. Chen, X.; Schluesener, H.J. Nanosilver: A nanoproduct in medical application. *Toxicol. Lett.* **2008**, *176*, 1–12. [CrossRef]
51. Kenawy, E.-R.; Worley, S.D.; Broughton, R. The Chemistry and Applications of Antimicrobial Polymers: A State-of-the-Art Review. *Biomacromolecules* **2007**, *8*, 1359–1384. [CrossRef]
52. Dastjerdi, R.; Montazer, M. A review on the application of inorganic nano-structured materials in the modification of textiles: Focus on anti-microbial properties. *Colloids Surf. B Biointerfaces* **2010**, *79*, 5–18. [CrossRef]
53. Khan, M.A.; Pattnaik, D.; Ashraf, R.; Ali, I.; Kumar, S.; Donthu, N. Value of special issues in the journal of business research: A bibliometric analysis. *J. Bus. Res.* **2021**, *125*, 295–313. [CrossRef]

54. Alsharif, A.H.; Salleh, N.Z.M.; Baharun, R.; Abuhassna, H.; Hashem, E.A.R. Tendencias globales de investigación en neuromarketing: 2015–2020. *Rev. Comun.* **2022**, *21*, 15–32.
55. Yu, Y.; Li, Y.; Zhang, Z.; Gu, Z.; Zhong, H.; Zha, Q.; Yang, L.; Zhu, C.; Chen, E. A bibliometric analysis using VOSviewer of publications on COVID-19. *Ann. Transl. Med.* **2020**, *8*, 816. [CrossRef] [PubMed]
56. Cascella, M.; Monaco, F.; Nocerino, D.; Chinè, E.; Carpenedo, R.; Picerno, P.; Migliaccio, L.; Armignacco, A.; Franceschini, G.; Coluccia, S.; et al. Bibliometric Network Analysis on Rapid-Onset Opioids for Breakthrough Cancer Pain Treatment. *J. Pain Symptom Manag.* **2022**, *63*, 1041–1050. [CrossRef] [PubMed]
57. Fouda, M.M.G.; Abdel-Halim, E.S.; Al-Deyab, S.S. Antibacterial modification of cotton using nanotechnology. *Carbohydr. Polym.* **2013**, *92*, 943–954. [CrossRef]
58. Qiu, Q.; Chen, S.; Li, Y.; Yang, Y.; Zhang, H.; Quan, Z.; Qin, X.; Wang, R.; Yu, J. Functional nanofibers embedded into textiles for durable antibacterial properties. *Chem. Eng. J.* **2020**, *384*, 123241. [CrossRef]

sustainability

MDPI

Article

Sustainable Textiles in the Past "Wisdom of the Past: Inherited Weaving Techniques Are the Pillars of Sustainability in the Handloom Textile Sector of Sri Lanka"

Hafeezullah Memon [1,*], Gayathri Madubhani Ranathunga [2,*], Virajini Medagedara Karunaratne [2], Samudrika Wijayapala [3] and Nilhan Niles [3]

1 College of Textile Science and Engineering, International Institute of Silk, Zhejiang Sci-Tech University, Hangzhou 310018, China
2 Fashion Design & Product Development Degree Course, Department of Textile & Apparel Engineering, Faculty of Engineering, University of Moratuwa, Moratuwa 10400, Sri Lanka; virajinik@uom.lk
3 Department of Textile & Apparel Engineering, Faculty of Engineering, University of Moratuwa, Moratuwa 10400, Sri Lanka; samu@uom.lk (S.W.); niles@uom.lk (N.N.)
* Correspondence: hm@zstu.edu.cn (H.M.); gayathrir@uom.lk (G.M.R.)

Abstract: The paper aims to identify characteristics of the driving force of the domestic handloom textile industry of Sri Lanka. A qualitative data analysis methodology was chosen for the research. Data were gathered from observational studies and semi-structured interviews. The observational study was carried out in museum collections, scholarly written books, and research journal articles. Semi-structured interviews were carried out with weavers, designers, technical instructors, and administrative officers. Data were analyzed according to observational study, coding, and concept development. Identified characteristics are interlinked with inherited craft knowledge, Buddhist culture, and the networks of people who live and work in a particular industrial society. The industry is developed through hands-on experience and prolonged engagement, where human resources are the key factor. The research scrutinized seven concepts related to the sustainable survival of the weaving structures. Learning is the direct enforcer. The infrastructure is supplied by the Textile Department. The research suggests recognition of a consistent infrastructure supply chain as national policy. The paper recommends to policymakers with the factors found to connect culture with an active action plan.

Keywords: inherited knowledge system; human resource; learning; training; hands-on experience; prolonged engagement; culture and tradition; handloom industry of Sri Lanka

Citation: Memon, H.; Ranathunga, G.M.; Karunaratne, V.M.; Wijayapala, S.; Niles, N. Sustainable Textiles in the Past "Wisdom of the Past: Inherited Weaving Techniques Are the Pillars of Sustainability in the Handloom Textile Sector of Sri Lanka". *Sustainability* **2022**, *14*, 9439. https://doi.org/10.3390/su14159439

Academic Editor: John Carman

Received: 29 June 2022
Accepted: 25 July 2022
Published: 1 August 2022

1. Introduction

The objective of the paper is to identify characteristics of the driving force of the domestic handloom textile industry [1] (p. 3) of Sri Lanka. The objective is examined by addressing two questions which explore the relationship between inherited wearing structures of the domestic textile industry and their contribution to the cultural sustainability of the country. The research questions are as follows,

How does the sustainability of the domestic textile industry relate to its inherited knowledge culture?

What is the relationship between the market demand and the technical application of design structures of their productions?

This research paper focuses on inherited handloom textile knowledge of Sri Lanka. The significance of the research is concluded in several aspects. The basic definition and concept of cultural sustainability and its behavioral pattern are discussed in the context of Sri Lanka. Research findings are significant to the stakeholders of the handloom industry, research, and policy making. The socio-cultural structure is analyzed. The Sri Lankan handloom textile industry then and now has been analyzed to understand the continuous narration of the industry since ancient times. We record nine weaving structures that are used in Sri Lankan handloom textiles today with the history of development, weaving structure, technical drawing, and color. Characteristics which are identical to the Sri Lankan local context are forwarded. It is emphasized that inherited knowledge is being adapted to concurrent trends that are allowed for sustainable survival. Further, distinguishing characteristics which show an intimately connected link between the inherited craft, the craftsman, the product, and the market have been emphasized. It is viewed as a process of a steady evolution in which human ability grows in terms of establishing new structures, resolving issues, adjusting to ongoing social and cultural change, and working consciously and imaginatively to realize new objectives.

First, the introduction discusses key phases, inspirations, and structural changes of the domestic handloom industry of Sri Lanka from ancient times to today. Secondly, the introduction discusses the relationships between sustainability and tradition and development. A definition and explanation of sustainability is summarized by the views of scholars. It discusses the relationship of culture with some selected creative industries of the world.

1.1. The Domestic Handloom Textile Industry of Sri Lanka

The domestic handloom textile industry of Sri Lanka is economically viable. This identified success is suggested to be sustainable in this research. There are many contributory factors to this success. Of these factors, a large contribution comes from knowledge of inherited weaving structures. This knowledge has been inherited from generation to generation since ancient times and is now driven by a well-structured government body. The inherited knowledge gives identity, cultural diversity, and emotional experiences which is the target selling point in textile marketing of the consumer society today. The industry caters to two main market levels: mass productions for the larger group of consumers and designer productions for a niche consumer market. The second-level niche market is defined by its own unique needs, the exclusiveness of designs, preferences, or identity that makes it different from the market at large, but the cultural identity is higher there. The official website of the Textile Department of Sri Lanka notes its linkage with the heritage which dates to the time of Kuweni, the Princess of the Yaksha clan, who did textile spinning in Sri Lanka in the 5th century BC and the continuing historical relationship with the inherited textile production [2] (p. 123).

Sri Lanka possesses a successful domestic textile history which has continued since ancient times and later it became the most successful handloom industry that earns profits. Traditional knowledge of design and technology has been documented in textbooks that give resource intensity to future usage. Coomaraswamy [3] (pp. 232–236) in *Medieval Sinhalese Art* discusses hand-woven textiles and their inheritance, and families who had the patronage of the King during the period of the Kandyan Kingdom of Sri Lanka (14th Century AD—1815), the material used, color, product categories, domestic looms, and tools and equipment are recorded with illustrations. Is it because scholars of the early 20th century saw the direction of Sri Lankan art and craft headed and discussed a productive future? Sri Lankan traditional art and crafts of today are associated with Kandyan art and craft which was developed during the Kandyan reign (14th Century AD to 1815) of Sri Lanka. Kandyan art has a living culture. Ananda Coomaraswamy records participant observations about Kandyan art and craft in the book on *Medieval Sinhalese Art*. His attempt was like how William Morris developed new consumer art from Victorian art in Britain [4–6]. Coomaraswamy, who was much influenced by William Morris's work, envisioned a similar resurgence of traditional arts in Ceylon [7] (p. 2).

According to Ananda Coomaraswamy [3] (p. vi), the rise of commercialism is to blame for the arts' downfall because the system of production has separated art from labor.

When nations grow old
The arts grow cold
And commerce settles on every tree
[3] (p. vi)

His observations paved the way for the sustainable survival and development of art in the future. He defines the differences between traditional art and consumer art as follows: "Today commercial production forbids a union of art with labor" [3] (p. vii). This is the main characteristic of the art and craft production of traditional society. Therefore, tradition needs to join with modern technology to forward ideal innovations to the future as suggested by Huntington [8]. "It is essential that mechanical production should in the future be not abandoned but controlled in the real interests of humanity" [3] (p. vii). For a reunion of art with labor, that machinery should be controlled. "Traditional society art was developed by hand-work on it. To stop competition between machine and hand work in today's consumer society, we need to intelligently define and delimit the proper sphere by restoring the entire control of production to the actual workers, as in the old guild system. In history, the king or the overlord of the traditional community was the patron. Appointing as associations servants, its craftsmen, and its artists, just as it now appoints its judges, its preachers, its professors". Coomaraswamy [3] (p. viii) says that art from prophecy to amusement strikes at the root of any possible revival of true art, which has to do, not with imitation, but with imagination. To reunite imagination with technique is as necessary as to reassociate art with labor. Today, the hand-woven textile industry of Sri Lanka is driven by these principles.

1.1.1. Tradition of Textiles in the Kandyan Kingdom of Sri Lanka (14th Century AD—1815)

Design, color, material, and tools and techniques characterize the textile tradition of the past. Woven structures have been handed down from inherited families whose families served the King during the Kandyan Kingdom. During the Kandyan Kingdom, all lands belonged to the King [9] (p. 134). The civilians worked for the King and lived on the King's land. The caste and land tenure system known as Rajakariya, which assigned professions according to caste, organized the ancient Sri Lankan society. The King granted land (Nindagam) to the service castes in favor of giving them a variety of roles. Each caste had the responsibility to provide the King with a specific type of labor [10] (p. 132). Dumbara and Thalgune, two villages in Sri Lanka's hill region known for their handloom textile weaving, have carried on the hereditary legacy of indigenous textiles since the days of the Kandyan era.

The weaving of homespun cotton continues in the Kandyan region and its surroundings. Coomaraswamy [11] (p. 203) says "Sri Lankan work is essential cotton". Cotton yarn was often used for weaving but occasionally silk and linen were also used [3] (p. 238). At Sri Lankan museums, a variety of items are on display. These include bedspreads, handkerchiefs, tuppotti (men's body cloth), hela (women's saree), diya-kacci (king's loincloth), and kerchiefs (mottappiiliya—a cloth for head covering), shawls, pillowcases, belts, mats, and carpets.

Color composition is created through the structure used. Blue and red stripes on white cloth are common. The cloth was colored by dyeing. Colors used were red and blue on white. Many designs and motifs are associated with the Kandyan Kingdom and have continued in the current society.

In the case of weaving two classes, the first was limited in amount, for the court and aristocracy and Indian in style, the second more strictly indigenous seen. It is shown that more Indian weaving techniques penetrated the local tradition. It is clear from the Indian term such as *dik-jala* (horizontal jala) which is used in weaving structures today. *Jala* means net-like pattern which may have been derived from India. Woven textile productions have been revived since the early 20th century. It is seen that hand-woven textile production has been incorporated into a successful domestic industry which is incorporated into the local economy. Hand-woven textile productions have been revived since the early 20th century. It has been observed that hand-woven textile production has tried to be incorporated into the local economy as a successful domestic industry. Several attempts are recorded. Mrs. Coomaraswamy taught a class of Kandyan girls about inherited embroidery techniques in 1905 and 1906. Prizes were offered through the Ceylon Society of Arts and annual exhibitions were held in 1905 and 1906. Coomaraswamy [3] (p. 242) concluded that "the best evidence of the value of such work, was the interest awakened in old traditions and design".

1.1.2. Sri Lankan Handloom Textile Industry Then and Now

The handloom industry has developed in leaps and bounds over the years. By the early 20th century, a proper industry had not been developed. Coomaraswamy [3] (p. 242) describes the current market demand for handloom textiles in 1905. If it were not for the demand for bed linens produced by the Kandy Art Association, the industry would struggle to thrive. Later, traditional textile crafts emerged as the top provider of consumer demand.

The handloom industry in Sri Lanka is concentrated especially in the Western, Eastern, Northwestern, and Southern Provinces. Handloom is a household or cottage industry in Sri Lanka, where experienced hands manufacture high-value items. Ukkubanda Dolapihille was sent to Japan in 1906 by Anagarika Dharmapala, who paid for the trip out of his riches, to receive training in textile industry technologies. As a result, the Hewawitharana Training School in Rajagiriya, Sri Lanka, was created on 4 December 1912 as the country's first-ever textile industry education provider. According to records, weaving inspectors were appointed during the British administration in 1920, and by 1945 the country had 198 display centers, 81 weaving schools, and 55 handloom workshops [12] (p. 70). There are 14 textile industry training schools established to sustain the human resources of the industry today. There are two design schools in operation under the Department of Textile Industry aimed at producing trained personnel for the handloom textile industry. The first offers a six-month Basic Training Course and the second a Final Certificate Course in Textile Industry (NVQ Level 04) over 13 months. Male and female pupils who complete the Final Certificate Course in Textile Industry (NVQ Level 04) are recruited to a course (12 months) which is called the Advanced Textile Industry Training Schools of Design. At present, the textile industry training and research activities are carried out here making use of the natural colors in the Rajagiriya Natural Color Mixing Unit. Further, the Research, Training, Designs, and Services Institute was established in Katubedda, Moratuwa. This institute comprises divisions such as the Training School of Design in Katubedda, Museum, Designs Division and Lab. The handloom industry in the nation reached its zenith in the 1960s and early 1970s due to support and encouragement from the ruling party at the time.

In 1965, the Cabinet authorized a plan to establish 1000 handloom centers throughout the country, with the primary goal of employing 25,000 people. The budgeted amount was 18.2 million rupees [12] (p. 80). Each center was supposed to have six looms and teach 25 students how to weave. The weavers of each center were to form co-operative society at the end of the six-month training session. The initiative began in 1966 to promote "Swadeshi Textiles". Volunteer organizations responded magnificently, with 420 building sites for these centers said to have been provided by individuals. The number of handlooms in the country was predicted to increase to 70,000 after this initiative was completed. Around 70,000 installed looms were recorded in 1967 [13] (p. 22). According to the Industrial Product Regulation Act of 1965 and restrictions on foreign exchange, the country's fundamental textile and garment needs provided a protected domestic market for handloom producers.

The institutions engaged in handloom and power loom activities were responsible for meeting the apparel needs of Sri Lankans until the Department of Textile Industries was founded on 7 July 1977. Before the implementation of free economic policies in 1977, the World Bank performed a survey that found the highest number of installed looms to be 115,000 [13] (p. 2) However, due to different issues, only around 30% of these looms were operational. The Handloom Industry was concentrated in specific sections of the island during this time, primarily in the Western, Northwestern, Central, Southern, and Eastern Provinces. Cabinet approval of a five-year development plan for 790 handloom villages came in 1990. This plan would add 16,000 more looms to the inventory and employ close to 24,000 people [13] (p. 5). The objectives were then changed to include 40,000 new looms, 60,000 new jobs, and a 60-million meter increase in the annual production of hand-loomed textiles. However, only 193 villages had been built by the end of June 1991, and only about 3000 weavers were employed. The Western Province is the largest contributor, followed by the Eastern and Central Provinces. Approximately 962 private producers are operating, with small, medium, and large-scale units. The export business is dominated by four main corporations. A total of 771 production center were owned by eight provincial councils. Around 15,000 people are employed in the industry, with a large proportion of them being women [12] (p. 80).

Sri Lanka produces ready-made clothing, soft toys, hand-woven carpets, tapestries, upholstery fabrics and curtaining, dress textiles, bed linen, table linen, and stationery items such as writing pads, books, notebooks, and albums with handcrafted covers.

Small and medium-scale producers dominate the handloom sector. There are also some large-scale producers. The following are the three categories in which the industry can be classified export-focused private businesses, indirect exporters, subcontractors, and individual suppliers; provincial councils that oversee the government's industrial facilities; and cooperative–cooperative societies.

The next political regime began in 2005 and lasted until 2014; political imperatives and programs for the development of crafts are considered as being intertwined. The Minister of Industry and Commerce established a Task Force in 2012 to create "a vibrant, globally competitive, and sustainable Handloom Industry with an internationally renowned brand identity" by the president's Mahinda Chintana or Vision for a New Sri Lanka, a policy framework. The Task Force was charged with examining industry issues and developing a business model for the industry's long-term sustainable development. The Task Force concentrated its research on four major areas: marketing, training, design, technology, and entrepreneurship development. They analyzed the whole sector and produced the above cause-and-effect graph showing seven components that have an immediate effect on the resilience of Sri Lanka's handloom sector.

The product, market, profitability, environment, and the 3Ms, or man, machine, and material, are all important considerations. The Task Force, determined the seven factors such as product, market, profitability, environment, manpower, machinery and material have a direct impact on the sustainability of the Sri Lankan handloom sector [13] (p. 11).

Furthermore, the Task Force encourages and facilitates the marketing of handloom products to ensure the industry's long-term viability and maximize benefits to all stakeholders as the Ministry of Industry & Commerce, Export Development Board and (EDB) producers [13] (p. 14).

The research Task Force made two recommendations for achieving the goals of Sri Lanka's handloom textile industry.

1. An export promotion handbook has been produced by EDB and the Department of Commerce. Establish a brand identity, including a name and logo, active local marketing targeted at the tourism industry, and high-end sales through upscale stores and department stores.
2. Involve well-known supermarket companies in production and distribution.

The total value of handloom exports in 2018 was 0.96 million USD, while it was 0.935 million USD in 2019 (Jan to Nov). The top markets for Sri Lankan handloom textiles are as follows: European Union (Austria, Italy, Belgium, Switzerland, Germany, the United Kingdom, Sweden, and France), United States, Australia, Japan, United Arab Emirates, Norway, and the Maldives. The following organizations collaborate with government agencies to address issues relating to the handloom industry to support it: Sri Lanka Institute of Textile and Apparel, University of Moratuwa (UOM), National Enterprise

Development Authority (NEDA), Department of Textiles [14]; Ministry of Industry and Commerce (p. 5).

1.2. Basic Definition and Concept of Sustainability

The basic definition of sustainability is the enduring well-being of human societies and communities [15] (p. 88). The concept is widely discussed in a consumer society in many disciplines such as science, agriculture, engineering, medicine, art, and culture extensively in such wide areas today [16] (p. 215), [17] (p. 3), [18] (pp. 1–2), [19] (p. 877). Therefore, sustainability has become an important topic in every aspect of life, and it has conjoined with the process of development. The history of sustainable development has also evolved as a concept, originating from the science of economics [20] (p. 2).

According to the definition of development, it is "an evolutionary process in which human potential improves in terms of creating new structures, coping with issues, adapting to continual change, and working consciously and creatively to reach new goals" [21] (p. 92)". It is a socio-economic situation in a country where the demands of the populace are met through the sensible and sustainable utilization of natural systems and resources. Although at times it was mostly focused on economic activity that led to environmental deterioration and the depletion of natural resources, that definition has evolved in recognition of the close connections between society and the environment [8].

Intergenerational equity is a sustainability trait. Intellectual aspects of a society that have been passed down from generation to generation are processed by cultural capital [22] (p. 15).

This results in three interconnected realms or spheres of sustainability that explain the connections between the environmental, economic, and social facets of sustainable development [20] (p. 8), [23] (p. 681). It was observed that as a progressive movement, human potential is important in development. Further, human capital is key and without it political or economic development is impossible. Culture stands for the norms, values, and identity of a society. Culture stands for the best solutions and practices achieved through coping, adapting, and creating such processes. Therefore, it is observed that culture makes a pervasive influence in all societies. It is because culture itself is an indicator of development.

1.2.1. Culture Suggests an Intellectual Property Economy in Sustainable Development

According to Tipps [24] (p. 208), the traditional cultures' development is hampered by the rules, beliefs, and values that surround them. Developmental tradition assesses as changing processes [24] (p. 212). Native social and cultural aspects are a significant source of modernity. Considering the structural mechanisms underlying the interaction between societies is advised by Tipps [24] (p. 212). This concept provides the connection with the objective of the paper which is to identify characteristics of the driving force of the domestic economic institute. Further, additional key factors that explain the nature of their political and economic autonomy include the effects of colonial dominance, international political relations, trade, and cross-national flow.

The culture of contemporary civilizations, which is characterized by the accumulation of capital and industrialization, which are compatible with development, must thus be imitated by traditional communities for them to advance [25] (p. 49). In general, this approach aims to raise traditional societies' living standards through economic development and the introduction of modern technologies [8].

This idea urges us to conjoin traditional knowledge with current technology even in 1976. The concept of sustainable development has expanded. Some theorists stressed on the cultural pillar in sustainable development [26] (p. 33), [27] (p. 5), [28] (p. 5).

The cultural industry, which is another economic area with growth potential, includes the intellectual property economy. Many people agree that the cultural industries and the arts sector serve as platforms for artistic expression and as indicators of regional and societal identity [26] (p. 41). The cultural industries, often known as mono-culture economies, enable a more competitive development platform [26] (p. 46). The field of sustainable development has been acknowledged as one that protects the environment and uses fewer natural resources [27] (p. 5). The intellectual and emotional components of society, such as values, traditions, and cultural standards, are what some academics refer to as the "culture dimension" of sustainable development [28] (p. 11). Culture is emphasized as the economy of intellectual property in this discussion.

1.2.2. Sustainability Objectives Are Based on Economy, Community, Social Values, and Culture

A critical component of the conversation on sustainability is centered on social and cultural values [27] (p. 5). Scholars argue that sustainable development is incomplete without culture. Keywords of sustainability such as slow lifestyle, recycling, up-cycling, eco-friendly production, and consumption had been merely rooted in life then. In 1999, Sweden adopted the concept that cultural heritage is a driving force of a sustainable society. The Swedish government believes that a connection between natural resources and cultural heritage is essential to the achievement of a long-lasting, high-quality environment. Tourism largely depends on the well-being of inherited art and culture of the country [29] (p. 7). Culture gives the art, heritage, and cultural identity for sustainable development [26] (p. 36). Addressing the "social and economic opportunities and requirements to mainstream investments in cultural heritage and the living arts" is the major goal of the relationship between culture and sustainable development [26] (p. 35).

Cultural marketing in Ubud, Bali [30] (p. 4209) Dakshina-Chitra in Southern India [31] (p. 694), and Chinese lacquer art, China [32] (p. 423) achieved the highest tourist attractions.

In Tokyo, Japan, a special initiative that began in 2002 attempts to inspire kids to try their hand at various crafts during lessons offered in 60 schools and taught by acknowledged professional craft MSEs in silk weaving and ceramics [33] (p. 16). Thailand's creative industries have been encouraged to use local knowledge and skills to produce handicrafts under the "One Tambon One Product" (OTOP) idea. In 2004, 29,000 productive communities signed up for the initiative, a significant rise from 16,000 in 2003. By 2004, their sales had reached 874 million USD, and by the end of 2005, they were projected to reach 1 billion USD [33] (p. 12). Traditional art and crafts have been based on distinguished communities since ancient times in Sri Lanka. Reed works in Pasyala, pottery in Molagoda, Brassware in Pilimathalawa, handloom textile in Thalagune, and mat weaving in Dumbara are famous tourist destinations which resemble traditional art and craft values. Skills and knowledge are considered as inheritance and the wisdom of the ancestors. The knowledge is strictly given by father to son. They serve at the annual procession at the temple of the tooth relic in Central Province without any monetary gain. Nurse [26] (p. 35) in his book "Culture" identifies four disputed definitions of culture that point to the adoption of attitudinal changes to achieve sustainable growth.

1. A mental condition that has developed, as in "a person of culture" or "a cultured person".
2. The development processes, such as "culture interests" and "cultural activities."
3. The tools used in these processes are referred to as "the arts" and "humane intellectual. works" in culture.
4. Finally, as "a comprehensive way of life," "a signifying system," and a means of expressing, experiencing, and exploring a social order.

According to Mazzocchi [15] (p. 77), "indigenous knowledge may supply, evaluating the principles that govern indigenous relationships with nature, such reciprocity and caretaking." Through indigenous, perceptive principles, and multiple cultural viewpoints, different cultures and their knowledge systems contribute to sustainability.

Learning from indigenous ideas such as customs and ancient norms and values may aid modern cultures in making better decisions and policies [15] (p. 89).

"We show respect for a plant by accepting its offering with respect" [15] (p. 81). Many indigenous civilizations from around the world share concepts like interconnectedness, reciprocity, and caretaking. Ideal sustainability calls for a giving, reciprocal, and caring attitude as well as a sense of interconnectedness and dependency on the natural environment.

1.2.3. Sustainable Textiles in the Past

Textiles are the world's largest waste today (dyeing, finishing, decomposed material). Therefore, sustainability in clothing and textile has become a well-known area for seeking sustainable ways of production systems. Scholars focus on the concept of an ideology of use rather than materialistic answers. It implies that usage ideology determines the durability, which is helped by materials, design, and construction [34] (p. 236).

Scholars are invited to find better solutions from traditional practices from inherited knowledge [35,36] (p. 80). Scholars try to identify the transcendent qualities of traditional practices. According to Fletcher [34] (p. 230), "the ability of a traditional design process to reach into the life world of the user and influence behavior appears to be weak, and in the context of fashion clothes, weakened further by fashion's social nature, which sees it influenced by human exchanges and actions as well as material objects, rather than just physical objects. There are long-lasting clothes, but they are more influenced by an ideology of usage than by the physical durability or the quality of the interaction between the user and the object".

Traditional knowledge is interwoven with the environmental, cultural, social, and economic four pillars of sustainability, those who are interested in fostering the development of sustainability. In fact, according to the World Intellectual Property Organization, traditional knowledge can aid in the sustainable growth of a community [37] (p. 84). According to Article 25 of the United Nations Charter, "Indigenous people have the right to maintain, control, defend, and develop their traditional knowledge, including to protect and develop their intellectual property over such traditional knowledge [37] (p. 84)".

Traditional literary, artistic, or scientific works, performances, innovations, scientific discoveries, designs, marks, names, and unreleased information would all fall under this category. Environmental, biological, cultural, social, and economic aspects are all linked by traditional wisdom. Traditional knowledge has economic worth, which cannot be separated from other values.

Matale, Thalagune village, and Dumbara designs at Dumbara are famous for unique traditional handloom textiles. They live as textile communities reluctant to share or publish skills, techniques, and designs that they protect as ancient wisdom.

These members also serve in the textile industry with the employees who newly gain education from the textile school. It is observed that they have an identical problem-solving system which is unique to their living surroundings. Their sustainable practices, culture, and tradition are bound with customs, beliefs, and norms, therefore they differ from other countries. Therefore, it is seen that the way of developing a sustainable lifestyle differs due to cultural factors. Handloom weaving and related techniques are supportive factors in their living culture.

Documentation enables traditional knowledge to be ascertained for future references by our next generation. The holders of such traditional knowledge are growing old, and many die taking their knowledge with them to the grave [38] (p. 24). Therefore, documentation of traditional knowledge has its advantages.

The inherent characteristics of the Meghalayan artists in the northeastern part of India indicated a great love for textile art through ancestry, enthusiasm, inventiveness, and pride. External considerations made clear the importance of the local administration, the economy, and the ease of living in maintaining the traditional textile arts in this area [39].

Bali is home to a wide variety of traditional textiles whose knowledge and folklore allude to the abilities and philosophies that emerge in society as a result of the interplay between the diverse cultures, religions, social environments, and natural elements [40]. Analysis showed that despite the ingenuity and invention of the Balinese handwoven culture business, which improved people's desire for preserving their tradition and strengthened their feeling of identity, the cultural worth of handwoven textiles was still present. Further investigation based on the cultural value found that the sustainability of Balinese handwoven textiles is largely affected by local knowledge.

Nationalist groups were gaining ground in colonial nations all over the world in the 19th century. The resurgence of local languages, indigenous faiths, and cultural expressions was a key component of these nationalist movements. Irish traditional crafts were revived in part owing to William Morris (1834–1896) of the Arts and Crafts Movement. Local wisdom plays a pivotal role in sustainable development all around the world.

It is observed from the literature review that sustainability has been a lifestyle in traditional society. Sustainability and culture are inextricably linked together. The evolution of traditional cultures is hampered by the norms, beliefs, and values that bind them. Therefore, traditional societies must imitate to advance. The achievement of sustainability goals depends on local economies, social and cultural values, and a change in attitude.

Indigenous knowledge is based on caretaking which is crucial to indigenous relationships with the environment. According to academics, indigenous concepts including ancient norms, beliefs, and rituals may aid modern cultures in developing better policies and decisions.

The literature review suggests the concept of an ideology of use rather than materialistic answers in analyzing sustainable textiles as a creative process.

2. Materials and Methods

The objective of the paper is to identify characteristics of the driving force of the domestic handloom textile industry of Sri Lanka. In UNCTAD 2011 textiles are part of the valuation of crafts and their contribution to the sustainable development goals focus on south Asia. It focuses on policies for sustainable tourism, strengthening human and institutional capacity, and optimizing the developmental potentials of services [41] (p. 53).

To better understand the issue and improve the reliability of the results, this study combined qualitative research approaches. The following discussion covers data collecting and analysis methods.

2.1. Data Gathering Techniques

In this study, the data were gathered using two different methods: observational studies of museum collections, scholarly written books, and research journal articles and data were gathered from semi-structured interviews. The two methods are briefly discussed below.

2.2. Observational Studies: Museum and Handloom Teaching Observation

Observational visits were made to the Textile Museum in Katubedda attached to the Handloom Teaching School in Moratuwa, in Colombo District during March and April 2022, which showcased the history of handloom textiles in Sri Lanka for nearly 45 years, exhibited an array of woven structures utilized in the handloom industry, Sri Lanka. Observation is useful for understanding people's daily routines and for giving the researcher a glimpse into the interviewee's perspective [42] (p. 39). The visits were made to observe the product categories, materials, different handloom textiles with finishing techniques, woven structures, National Award-winning handloom products, manufacturing machinery and equipment, and portfolios of weaving structures that were utilized in the recent past of Sri Lankan textile history. Additionally, relevant objects and woven structures were visually recorded. The findings were written down on a sheet. The teaching of handloom textile weaving in the school is also observed on the same premises. Studies of the two-year curriculum of handloom weaving and weaving structured observations carried out. Different looms and machinery were utilized for weaving. All the looms were photographed, and descriptions were noted down on a sheet. Woven structures and productions were recorded.

Collections of ancient textiles are on display at the National Museum of Colombo and the Museum of Kandy. Woven structures, design and motifs, colors, and product categories were examined. Textiles belong to the Kandyan Kingdom of Sri Lanka.

2.3. Semi-Structured Interviews

Semi-structured interviews are a common qualitative research method [43] (p. 194) to get the answers to two inquiries that look at the connections between inherited woven structures of the handloom textile industry and their contribution to the cultural sustainability of the industry. In a semi-structured interview, questions are prepared that are consistent and methodical and are guided by defined themes to elicit more in-depth responses to help steer the conversation toward the subjects and concerns that the interviewers want to learn more about, and to incorporate several major themes that will be covered during the interview [44] (p. 245).

- How does the sustainability of the domestic textile industry relate to its inherited knowledge culture?
- What is the relationship between the market demand and the technical application of design structures of their productions?

The interviews ended with an open-ended question asking for additional comments on a variety of handloom woven structures. For conducting interviews, interviewees were selected from the industry employees who were considered ideal since they were awarded National Textile Awards for many years for handloom products of traditional woven patterns with a considerable bent to crafting modern patterns on handloom products. Further, the study was conducted on exploring the sustainability related to its inherited knowledge culture and the connection between the market demand for both the productions' technical implementation to the sustainability of the industry.

The purpose of the interviews was to identify the inherited knowledge and its continuation throughout history, we explore an array of questions to the employees of the industry. Qualitative researchers often rely on the interview as the basic method of data gathering [45] (p. 646). What are the weaving structures that practices in the industry? What are the tools and techniques, material, color, design and motifs, and products related to each structure? Who are the industry employees? What are thier linkages to inherited craft families or traditional education obtained from the industrial school? How do they adapt to the consumer demands of today? What are the infrastructure plans implemented by the administration to earn profit from the market? What are the market levels that they cater to? How to design knowledge and the skill vary according to the market level Such key points were questioned. The interviewees responded in their way. Although the information obtained from qualitative interviews cannot be generalized, its exploratory nature enables the gathering of rich information that can address issues about which little is now known [46] (pp. 391–392). According to Seidman "the desire to comprehend other people's experiences and the significance they place on them lies at the heart of interviewing [47] (p. 3). The ability for interviewees to respond in their own words, according to how they think and use language, is important. If the researchers seek to comprehend how the respondents view the social milieu under study, it tends to be extremely helpful [44] (p. 243). Semi-structured interviews, for instance, can be used to better understand how people interpret and construct meanings about their roles and work environments [44] (p. 246). We recorded interviews by writing clear descriptions of their experiences and thought processes.

The information gleaned from interviews provides another analytical choice for the "documentary method of interpretation (DMI)" [48] (p. 6)". Interviewees were selected according to institutional hierarchy and the expert knowledge of each weaving technique. Additionally, the actual interviewing procedure is strictly controlled to ensure that researchers are impartial and consistent throughout all interviews [49] (p. 4). The craft artisans, such as Weavers and Designers, officers such as Provincial Directors, Technical Officers, and Instructors employed in the handloom textile industry, are known for the history and development of the handloom textile clusters in Sri Lanka. Accordingly, 11 interviewees, both male and female, were selected. It was assumed that all participants could accurately identify the woven structures and known evolution of the handloom textile industry in Sri

Lanka. Interviews were held in April and May 2022 with four Weaving Instructors, three Instructors, one Professor of Textile, one senior academic of Textile Engineering, one design in charge, weavers, and one officer in the Textile Department.

Details of the interviewees and the purpose of the interviews are given in Table 1.

Table 1. Interview summary.

Code	Interviewee Details	Purpose of Interview
Interviewee 1	Weaver	Self-experiences in handloom weaving To understand the woven material behavior and different structures The knowledge gained from Training Sessions
Interviewee 2	Weaver	Self-experiences in handloom weaving Behaviorand final look of the different yarns in textile products The knowledge gained from Training Sessions Method of designing Identification of demanding products and designs
Interviewee 3	Provincial Director	Availability of material for the production Production strategy and production target Annual turn over Marketing and selling the products Organizing the Annual Textile Competition Organize buyer -seller forum Capacity enhancement of MOU with State Universities Forecasting new trends Strategies used to motivate weavers
Interviewee 4	Weaver	Traditional weaving practices that hailed from inheritance Current market demand for traditional woven products Traditional folklore about weaving Types of Dumbara woven structures
Interviewee 5	Professor of Textile Engineering	Innovations—Banana fiberhandloom products
Interviewee 6	Senior academic with a doctoral in Textile Engineering	Availability of raw material The efficiency of the Supply Chain in the handloom industry Utilizing advanced technology in the handloom industry comparison of the infrastructure available in the past and present
Interviewee 7	Instructor, Textile School	Educational qualification for a professional weaver Quality of the human resources for the handloom industry
Interviewee 8	Retired Teacher, Textile School	An ability to learn woven structures, produce a technical drawing, costing the end-product, and mechanical knowledge of textile production
Interviewee 9	Weaving Instructor	Identification of capability to blend two different woven structures Accuracy of using proportionate woven designs The ability of self-learning capacity Experience gained from overseas training and inspiration from Indian designs
Interviewee 10	Designer In Charge	The institutional procedure of designing and producing handloom textile products Evaluating customer feedback Costing the end-product Considering customer satisfaction and demand
Interviewee 11	Weaving Instructor	The capability of handling weaving shraft designs (native word—*polu rata*) Own experimentation in handloom weaving Achieved awards for handloom The knowledge gained from Training Sessions Utilizing new knowledge for the Annual Textile Competition The self-satisfaction of the job

2.4. Data Analysis

Data were analyzed according to steps of observational study, coding, and concept development. First, handloom textile-related items were observed and recorded. Then gathered items were categorized according to weaving structures, color, design and motifs, tools and techniques used. Observational study of research items helped to identify the links between research items and the factors that affect items. Memo-ing was carried out while observing. Identified links revealed themes or codes. Codes themselves show links with the socio-cultural context of Sri Lanka. Coded data were analyzed with socio-cultural factors. Derived coded data explore concepts. The chart below shows how data were analyzed and concepts scrutinized.

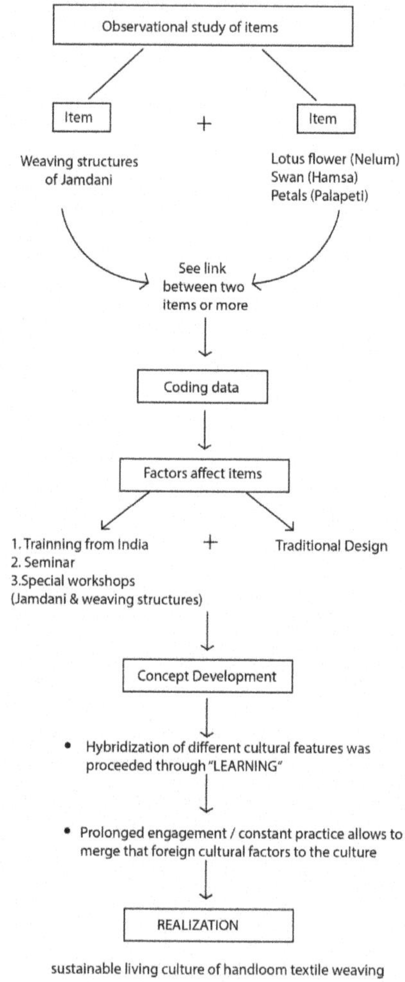

3. Results

3.1. *How Does the Sustainability of the Domestic Textile Industry Relate to Its Inherited Knowledge Culture?*

Data were categorized into different themes which showed similar characteristics to the two research questions of the research.

3.1.1. Weaving Culture Shows a Strong Connection with Buddhist Customs, Rituals, Norms, and Values

Historical roots and development show that inherited weaving structures have been handed down from generation to generation. This section sets out the technical drawings of each basic weaving structure which are currently in use. Dumbara Rata, Mal Baama, Plain weave, and Shaft designs (*Polu* rata) show direct links with native culture, folklore, and tradition. Weaving culture shows a strong connection with the Buddhist ideology such as customs, rituals, norms, and values. Intricate design patterns and skills are not given by the artisans for recording as they considered it as their hereditary wisdom.

- 'Dumbara Rata' weaving technique

Dumbara Rata weaving shows folklore-based design and motifs, color and weaving structure. Rata means designs. Dumbara Rata weaving structures were originally woven in black, ash, and white. According to folk stories, the pattern has a history which goes back to the dates of Kuweni (5th century BC) (a warrior queen of the country). It is said that when Prince Vijaya came from north India, he met her while she was spinning cotton threads. According to the legendary story, she owned a cotton farm in Dumbara in the central hills of the country. When she was expelled by Prince Vijaya later on, she cursed and then burnt the field of her cotton farm. The ashes rose high in the sky and blanketed the land with patterns. Dumbara Rata weaving patterns have taken inspiration from the ancient pattern and continued since then. Beliefs, customs, and norms became a lifestyle, and it is the intellectual property of the handloom weavers today.

During the Kandyan Kingdom, a village known as Dumbara was famous for mat weaving [3] (p. 243,334), (plateXXXIV). It was observed that now they applied mat weaving designs to textiles. Weaving culture has changed over time according to market demand.

> *"My great ancestors served for the last king of Kandy, Sri Wickrama Rajasinghe. He provided Dumbara mats for the Temple of the Tooth relic annually. That is our responsibility and service to the great Buddha. We live on lands donated by the King. Now we make only wall hangings with many colors. Wall hangings of Dumbara Rata are demanding among both locals and foreigners. We all know folk stories about the history of Dumbara Rata. Our grandmother and relatives told us. I will carry on weaving further".*
>
> *(Interview 4, Weaver, 2022)*

It is seen that weaving is their lifestyle. Artisans guard inherited craft knowledge. They have identified the value of ancient craft. It is observed that traditional customs, norms, values, and Buddhist culture are the direct factors of the sustainable survival of the weaving culture.

Figure 1 showcases a method of patterning where some weft threads perform the function of creating the design motifs while other weft threads weave the ground fabric. Traditionally the warp and the weft of the ground fabric were both of unbleached cotton yarn. The design picks were colored red, and dark blue, green, and black.

- 'Mal Baama' weaving technique.

This is the most common and extensively demanding style in handloom consumers of Sri Lanka since the Kandyan era. Ananda Coomaraswamy records many historical handloom textiles with this technique [3] (plates XXVIII and XXIX). Many ancient textile samples and products are on display at the National Museum of Colombo and the Museum of Kandy. Most of the designs resemble Kandyan-era flora and fauna designs and motifs. Artisans collections, museum collections, consumer's collections and items that are in the outlets for selling show similar consistency in design and motifs. The total appearance of the art looks like a decoration. Figure graphics are artistically and symbolically developed based on natural figure forms. The most common floral design is the lotus. Sri Lankan culture is based on Buddhism. The lotus is a sacred flower which is interpreted in many abstract forms in every item of the handloom fabrics. Swan is the famous animal motif in handloom textiles. It is believed that swans are competent in distilling milk from water. It is an influential animal that can identify good merit from bad attitudes. Moreover, geometric designs are the most common design element. A traditional form of creepers called 'palapethi' is common in this weaving technique.

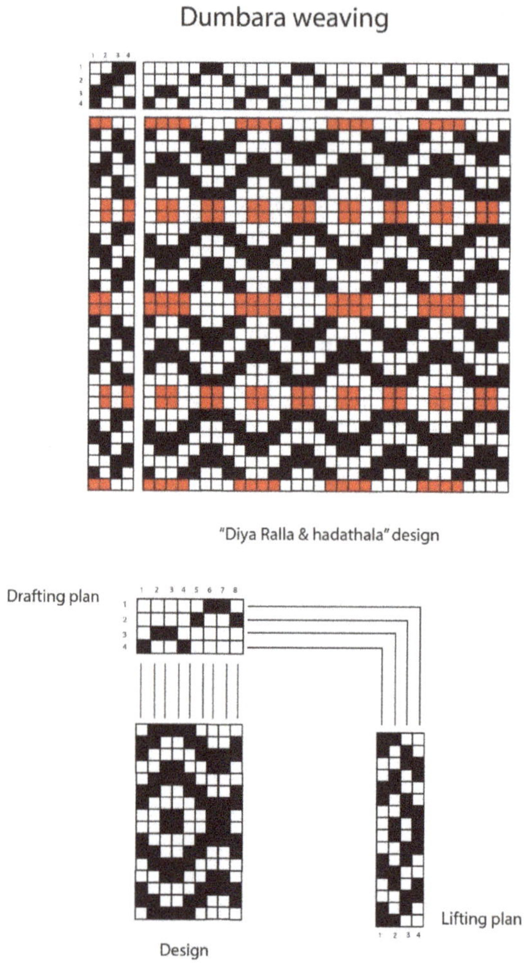

Figure 1. Technical drawing and the drafting plan of Dumbara weaving structure (drawing by KPN Bhaggya).

Mal Baama is shown in Figure 2. Fabrics with intriguing textures can be produced by selecting a weave pattern that varies the proportion of under and over threads. Materials with visual patterns can be created by employing colored threads and, in a similar manner, selecting which threads should appear across the design at particular times. Weft thread should be joined with three shafts and weaved by the pattern. By choosing a weave pattern that mixes the ratio of under and over threads, fabrics with interesting textures can be created. Materials with visual patterns can be fashioned by using different colored threads and choosing which threads appear at the top of the design (over the threads) at certain points in a similar way. Three shafts should be combined and weft thread should be weaved according to the design.

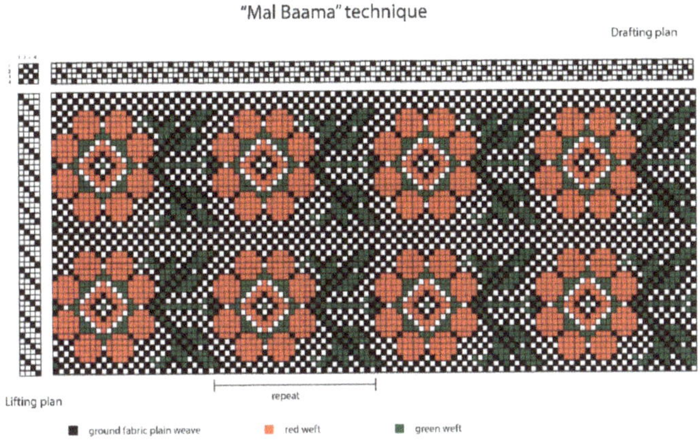

Figure 2. Technical drawing of 'Mal Baama' technique (drawing by KPN Bhaggya).

- Plain weave technique

Figure 3 is known as plain weave which is the oldest and most commonly used weaving structure. Plain weave is woven with each weft thread passing over and beneath each warp thread. Clothes of plain weave were white or limited colors during the Kandyan Kingdom of the 18th Century. The famous Kandyan blue cloth which is traditionally called *'kalu kangan'* was also made of plain weave [3] (p. 233). Except for green (seen very rarely indeed), only the three colors, red, white, and blue are found even in the most ornamental woven cloths. The materials were also made into betel bags. Today, this weaving technique is applied to produce color combination sarees which is the most demanded category among customers. A spectrum of colored sarees is produced today. It was observed that plain weaving has experimented with color application. Natural dye is another area which has been given more attention by the industry. Sustainability presented by the use of natural dyes has been an age-old tradition of Sri Lanka. A dye lab has been established in Katubedda Textile Weaving School for research and innovation in natural dye. Natural dye application is another category in the National Handloom Competition. Recent research by the authors of this article presented natural dye substances, dye recipes, and natural mordants which show the traditional practice of handloom textiles in Sri Lanka [50] (pp. 58–83).

- Shaft designs (*Polu rata*) weaving technique.

Figure 4 shows shaft designs (*Polu rata*). rata means designs.This weaving technique uses different heald shafts and foot treadles to weave the fabric design [51] (p. 128). According to the drafting thread pattern and lifting foot treadles pattern, the design is woven in the fabric. Shaft designs *(Polu* rata) are often done in association with four or six shafts *(polu)* to create colorful delicate designs. Six shafts create more intricate designs than four. These designs are known as *mal baama* (creating designs with shafts).

3.1.2. Cross-Fertilization Features to Enrich the Collection of Design and Motifs and Woven Structures from Time to Time

Different techniques and tools have been introduced from time to time. Shaft designs (*Polu* rata), Dumbara Rata (Indigenous weaving designs), Mal Baama, and Plain weave were seen in the retrieved artifacts inherited from the Kandyan era of Sri Lanka. Some products are on display in the curatorial department at the National Museum of Colombo. Most of the original objects are recorded in photographs and sketches. Coomaraswamy [3] (p. 235) says that "some design techniques called *mal petta, para mala* and *dethi mala* are mentioned.

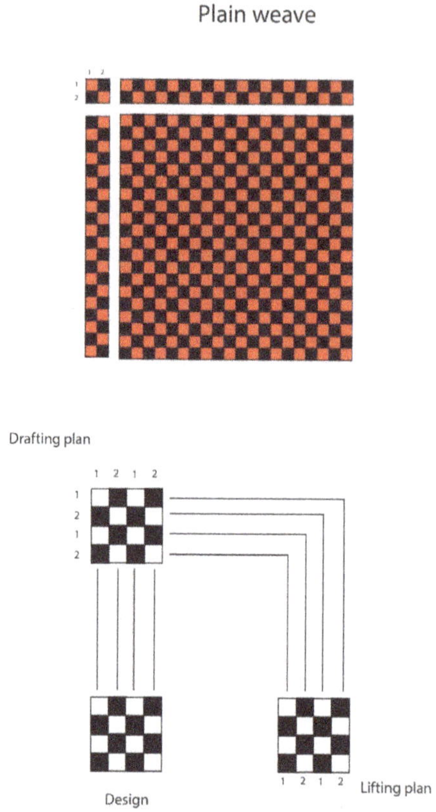

Figure 3. Technical drawing and the drafting plan of the plain weave structure (drawing by KPN Bhaggya).

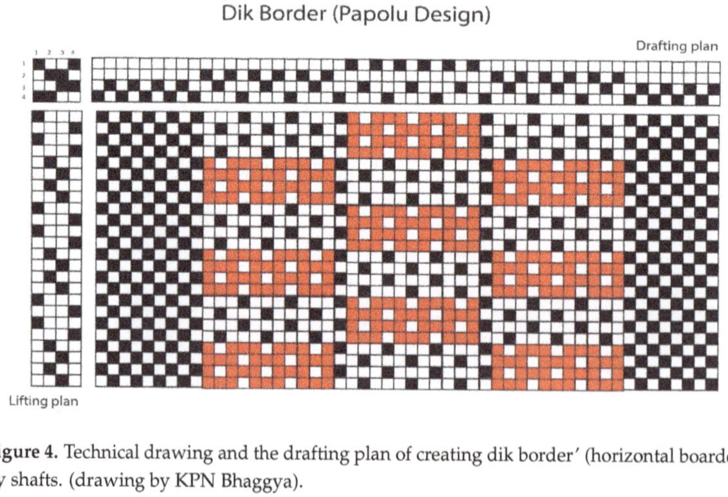

Figure 4. Technical drawing and the drafting plan of creating dik border′ (horizontal boarder) design by shafts. (drawing by KPN Bhaggya).

Mala gives the meaning of the flower. The names of the many patterns found on old clothes are now forgotten. Old weaving is considered as a combination of weaving and tapestry and colorswere limited to red and blue".

New weaving techniques were introduced in the 1970s. Interviewers claim that Dobby, Jacquard, Jamdani, Dik border structure, (Dik border is also known as Dik Jala), and Warp printing were introduced by India.

- Dobby weaving technique

Figure 5 is called Dobby weaving technique and consists of small figurative designs (floral or geometric) woven repeatedly into the cloth, created with a dobby attachment on the loom and a combination of two or more basic weaves. A plastic tape with punched holes that controls the raising and lowering of warp yarns controls the weaving pattern. It has a maximum of 32 harnesses. Dobby is commonly used for borders of sarees.

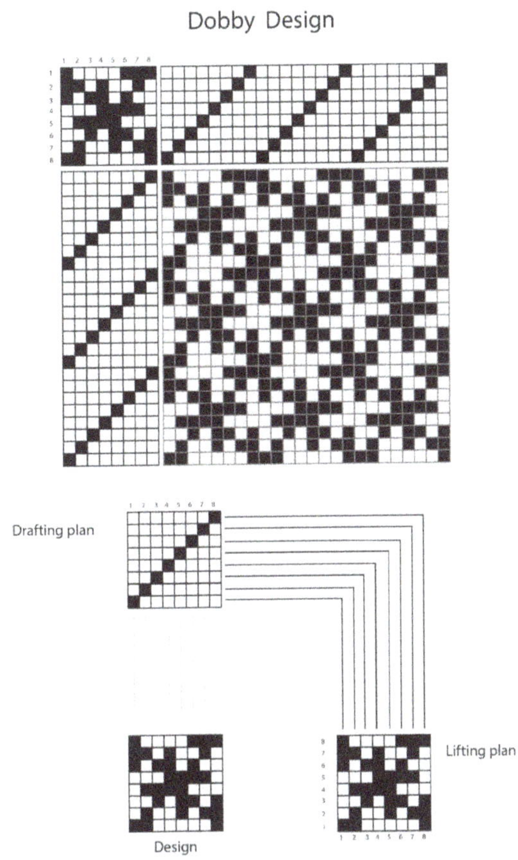

Figure 5. Technical drawing and the drafting plan of dobby weaving technique (drawing by KPN Bhaggya).

- Jacquard weaving technique

The Jacquard weaving technique, which is depicted in Figure 6, is used to create all-over figured fabrics such as brocades, tapestries, and damasks. It is woven on a loom with a Jacquard attachment to control individual warps. Two or more basic weaves are combined to create the Jacquard weave, with different weaves used for the design and background. The "punch card" control system is what makes Jacquard remarkable. The extra weft or extra warp in handloom Jacquard design is the design thread that creates the embossed impression on the fabric. One weft yarn is used to weave the ground cloth, and an additional weft is introduced at intervals to form a decorative pattern on the surface of the cloth in an extra weft structure [52] (p. 2864). To accentuate the extra weft, the ground fabric is normally woven with a simple structure such as a plain or twill weave, but various structures can be employed. A set of warp yarns is used for the ground cloth in an extra warp structure, and another set of warp yarn is put up on separate shafts to allow it to be woven independently from the ground cloth [53] (p. 128). Jacquard is commonly used for borders of sarees.

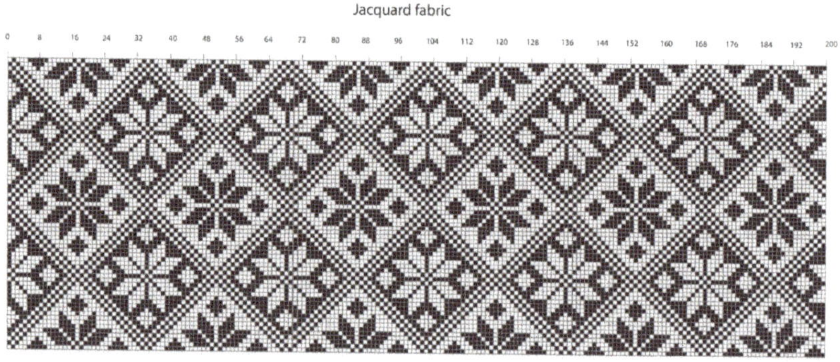

Figure 6. Technical drawing and the drafting plan of Jacquard weaving technique (drawing by KPN Bhaggya).

- Jamdani weaving technique.

Jamdani is a weft weaving technique showcased in Figure 7 and is the practice of delicate "Parsi Gara" on a loom. By manually combining a denser thread with the delicate warp threads of Jamdani, designs are inlaid into the cloth. This method is famous in Bengal, India. Jamdani is a weaving method that employs additional or supplemental weft to produce a motif. Interviewers claim that this weaving technique was introduced in India. Now, this technique is used with Sri Lankan traditional designs for handloom sarees. It was observed that the hybrid formation of indigenous design with foreign influence became Sri Lankan. On the other hand, traditional designs meet new technology. These two factors affect their traditional way of life as handloom weavers. These two factors directly affect the cultural sustainability of the handloom textile industry.

- Dik Jala (Horizontal jala) weaving technique

Figure 8 shows the Dik Jala design structure. The Jala technique in weaving involves raising the Warp threads via the Jala. The weaving technique is used to weave the borders of the sarees. Four heald shafts and four -foot treadles are used to weave the "dik jala" design. The key benefit of this method is that it is simple to create a corner design element at the intersection of a border and cross-border design. interviewees claim that this weaving technique was introduced in India. Now, this technique is used with Sri Lankan traditional designs for handloom sarees.

> *"I have made another machine part for Dik Jala. This is my innovation. It worked. Now I weave neat, beautiful dik jala sarees which are demanding among customers".(Interview 1, weaving Instructor, 2022)*

It was observed that the prolonged engagement of the same skill teaches them solutions. They have their solutions.

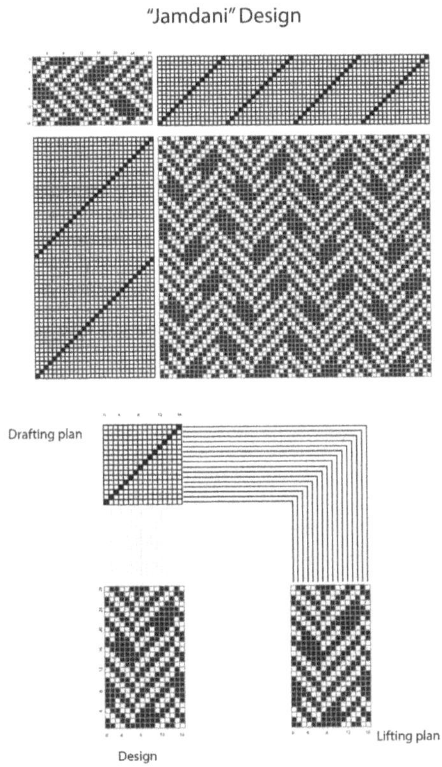

Figure 7. Technical drawing and the drafting plan of 'Jamdani' structure (drawing by KPN Bhaggya).

- Warp printing technique.

Warp printing is a method of placing the design on the warp threads shown in Figure 9. The fabric must be a plain woven fabric. The warp threads are painted with fabric dyes according to design motifs. After drying the warp threads, the weaving process can proceed. Interviewers claim that this weaving technique was introduced in India. Now, this technique is used with Sri Lankan traditional designs for handloom sarees.

> *"I learnt from my father. My grandmother is very skilled in handloom textile weaving. Then I joined the industry. I had 37 years of working experience in the handloom textile industry. Jamdani, Dig Jala, Haras jala, catch card and Warp printing such weaving techniques were introduced in 1970 from India. During 1990 we had a training program in South India. They did not train us; they just wove in front of us. We spent lots of time observing how they weave beautiful designs. As we are skillful at weaving, we could learn those techniques by ourselves. After 1970 textile products were given variety with these newly introduced techniques". (Interview 8, Retired Teacher, Government Textile school, 2022)*

It is seen that education has been continuous during 37 years of experience. Cross-fertilization features entered the weaving culture through education.

3.1.3. Experiential Learning in Weaving Structures with New Yarns

The weaving structure is visualized through yarn, color, and design application. Rayon yarns were introduced in the early 21st century, as were fancy yarns (Interview 1, 2, 3, 4, 8, and 11, 2022). Smooth, rough, glossy, and cozy textile qualities were introduced.

Fancy yarns and yarns made from rayon fibers are used in ceremonial sarees such as wedding sarees. Whole rayon sarees are woven. Rayon and cotton are mixed in designing sarees. Weavers and designers do experiments using fancy yarns in different weaving structures. They are well-experienced, educated, and trained artisans. Some weavers have experimented with combinations of fancy yarns with weaving structures.

> *"The Provincial Director gave linen fiber. I wove a sarong and a shirt. But I feel that the material is hard in texture. I think if I wash it with soap it will soften. I will try natural softener as well".(Interview 1, weaver, 2022)*

> *"I had 12 years of experience, we had only cotton yarn until 2003. Then we were introduced to rayon and fancy yarns. It is a glossy fiber. Now I am using fancy yarn for wedding sarees".(Interview 2, weaver,2022)*

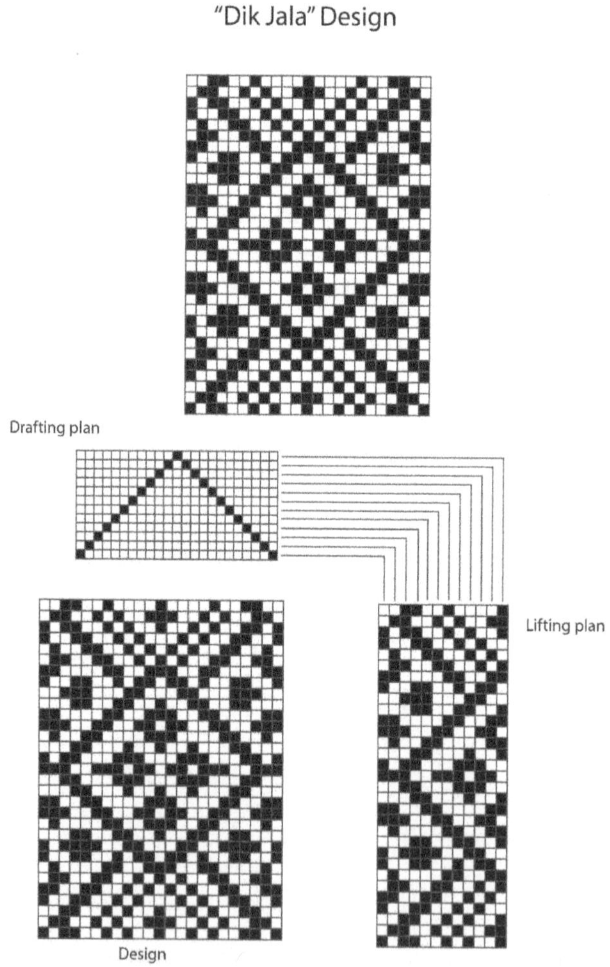

Figure 8. Technical drawing and the drafting plan of 'Dik Jala' weaving technique (drawing by KPN Bhaggya).

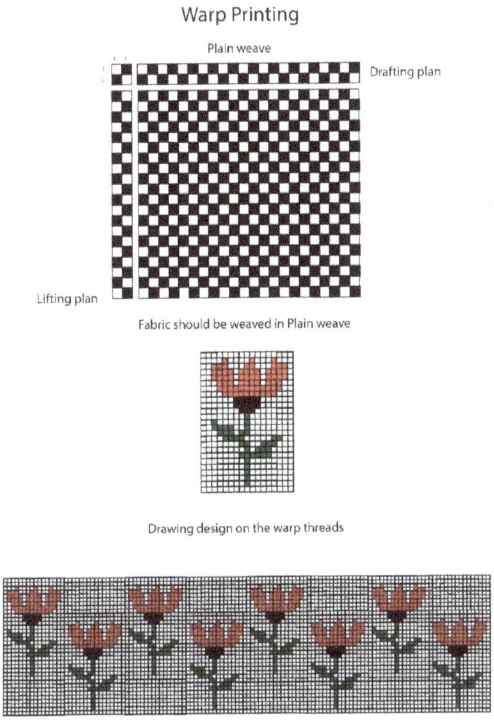

Figure 9. Technical drawing and the drafting plan of 'Dik Jala' structure (drawing by KPN Bhaggya).

- Mixing two structuresmake an innovation: learning through experiments

Employees who had systematic education and were trained at the government textile weaving school tend to experiment with new weaving structures. They do trial and error experiments, change the normal order of the traditional weaving process to create new techniques or mix two techniques to create a new one. New creations were accepted by the design committee and allowed permission for taking orders. These combined design sarees attract customers and receive orders (Interview 1, weaver, 2022). The employee trained the new employees as well.

> *"I intentionally did. I wove the border of the saree from twill to strengthen the border. I changed the normal order of the weaving of Shaft designs (Polu rata) technique. I combined Shaft designs (Polu rata) and the twill techniques. I did the border from Dik Jala and designs were added to the center of the saree from Jamdani".(Interview 1, weaver, 2022)*

Weavers learn and their skill has been enhanced from their prolonged engagement in the same work.

> *"I have 22 years of experience. First, I wove designs with plain weave. First, I created designs from only 2 shafts. Now I have experienced doing 2 to 6 or 8 shafts. Because I have been doing this for 22 years".(Interview 11, Weaving instructor, 2022)*

> *"I have 22 years of experience. I learn from trial-and-error experiments. Hakaback and plain weaving structures don't go together. Huckaback is good for towels and pillowcases. Shaft designs (Polu rata) go well with plain weaving techniques. Twill weaving technique gives delicate soft fabric which is suitable for clothing material".(Interview 9, Weaving instructor, 2022)*

- Mixing two yarns: An innovative yarn

In 2020, a new fiber created from a mixture of banana and cotton was experimented and it was announced by the Ministry of Industry and Commerce that the attempt had succeeded (Interview 5, Professor in Textile, 2022). Research scholars state that raw material, dye supply chain, tools, techniques, and advanced technology infrastructure should ensure supply consistency to fulfil the demand and supply of innovation. (Interview 6, Senior Lecturer, 2022). This fact is considerable even in the handloom industry today. All yarns and dye are imported from India today (Interview 3, Provincial Director, 2022). In the past, the infrastructure was established in the local context. Sri Lankan woven textiles are essentially cotton in ancient times. It is said that cotton was extensively grown in Sri Lanka since ancient times [3] (p. 235).

3.1.4. New Technology for Traditional Tasks of Weaving

Traditional tasks associated with weaving are modified with new technology. Spinning is a time-consuming task. In 2019, with the collaboration of the University of Ruhuna, the Southern Province of Textile Department developed an effective machine for spinning. It reduced the cost and saved time and energy (Interview 3, Provincial Director, 2022). Semi-automated winding machines were developed with the collaboration of the University of Ruhuna. Three machines were given to three centers in three districts: Poddala center in Galle, Vellamadama center in Matara, and Dammulla center in Hambantota.

3.1.5. Inherited Traditional Knowledge Is Protected by Imposing Rules

European designs and motifs are not allowed for the Annual National Textile Competition. Rules and regulations of the competition are documented and handed over to panel of judges before the commencement of the judging. Competitors are given awareness through Instructors and Designers of each Provincial Council Department before the preparation of the products for the annual competition. Weavers and designers are given traditional Sri Lankan designs from the book of Ananda Coomaraswamy's *Medieval Sinhalese Art* at their annual training, seminars, and the weaving textile school. Therefore, as the norm European designs are hardly practiced in handloom textiles even for the local market. Due to prolonged engagement of the system, all employees are pertinent to the survival of the structure of the existence of the industry.

> *"I have participated in several workshops where I had a good opportunity to observe traditional designs of birds, flowers, animals, creepers, and geometric designs. During the workshop, I drew a technical drawing of traditional swan (hamsa) design, lotus flower and geometric designs. because those designs are demanding in the local market. It took time. Because there were so many details to be drawn accurately and proportionately. Then in the evening, I experimented on a table loom to practice my technical drawing of (swan) design. I learned a lot. Now I am practicing these detailed designs for sarees".(Interview 9, Weaving instructor, 2022)*

> *"I absorbed many advanced techniques by observing how Indian weavers did, sometimes I copied Indian designs and used them in weaving. But I am careful not to take too many Indian-looking designs".(Interview 9, Weaving instructor, 2022)*

> *"In 2000 pethampili (a weaving technique) designs were at a basic level. Those days we didn't do weft pethampili, we did only pethampili for warp. Now pethampili is done for both warp and weft lines. I do lots of experiments with pethampili. I experimented with pethampili with natural dyes and tie-dyes. I won a Golden Award at the Annual National Handloom Competition for a saree of pethampili that was made of both warp and weft lines".(Interview 11, Weaving Instructor, 2022)*

3.1.6. Both Traditional and Contemporary Knowledge Is Institutionally Structured

The artisans who have education from the textile schools are recruited in each district's department. A range of subjects weaving, designing, coloring, raw material, machines, marketing, and administration particular to the handloom textile industry are taught at these schools.

> *"I studied the basics of textile weaving. Maths related to textiles. There, recognition of yarn, yarn counting, yarn numbering techniques, costing, technical drawings, around 40 weaving techniques and mechanical drawings are taught".(Interview 8, Retired Teacher, Government Textile School, 2022)*

Employees who complete and get through the examination after two years are recruited to the Textile Department. Therefore, the industry comprises educated, trained, and skillful employees (Interview 7, Instructor of the Government Textile school, 2022). New professions are created according to thedemand such as dye master, designer, merchandisers, and development officers. According to the administrative regulations, Designers, Instructors, and Marketing Officers are given comprehensive training programs throughout the year from time to time. Apart from that, according to the regulation all the employers of each category of the Textile Department must undergo twelve hours of training per year. The authorities make sure to conduct these training schedules each year to further the development of the industry. The progress of the training programs is evaluated in a five-year strategic plan (Interview 3, Provincial Director, 2022). It was observed that experience and training help them to identify more advanced techniques and designs.

> *"I had training in India. I absorbed many advanced techniques by observing how Indian weavers did, sometimes I copied Indian designs and used them in weaving".(Interview 9, Weaving instructor, 2022)*

3.2. What Is the Relationship between the Market Demand and the Technical Application of Their Productions?

3.2.1. Production Priority Is Given to Demanding Textile Designs and Products at the Market

The product range has increased due to customer demand throughout history. During the Kandyan Kingdom, items such as mats, quilts or sheets, carpets, pillowcases, aprons or bathing drawers for men, kerchiefs or shawls, or towels were produced [3] (p. 232). Today, the product catalogue of the Annual Textile Competition indicates 19 product categories. The weavers and designers are being educated about world handcrafts by the internet. They tend to create new designs for the annual textile competition taking inspiration from global demanding products, shapes, techniques, colors, and materials (Interview 9, Weaving instructor, 2022).

Designing is often dependent on the availability of material and Color. Designs are copied from the textbook given by the weaving textile school of the Department. The design committee also recommends designs. There are resources which provide designs to weavers such as textbooks, design schools, and design committees.

The Textile Department has noticed that during the last 5–6 years (2018–2022), demand factors have risen dramatically for design sarees (Interview 2, weaver, 2022). There was a shortage of importing thick threads (2/30) which were used for designing. Therefore, they have stopped fewer demanding products which were designed as pillowcases and created sarees only. This situation has increased during the COVID-19 pandemic period and due to the economic crisis of the country.

3.2.2. Annual, Monthly Events Forecast New Customer Demands

The Annual National Exhibition forecasts new customer demands. The award-winning sarees are reproduced on customer demand. Therefore, instructions are given to keep a technical sketch of the saree. In 2021 the Mahamodara outlet of Galle District was able to sell 90% of award-winning design sarees. There was a dramatic demand for *pethampili* sarees during 2019/2020. The demand is also high for color combination sarees and bedroom sets (Interview 3, Provincial Director, 2022).

The annual competition is set in two steps. Products are selected from the provincial level by a particular panel of judges. Then, selected products are sent to the national level and judged by the national team of judges. A range of products is presented for the competition. After the competition at the provincial exhibition and at the national exhibition many productions are sold at a high rate. There are a considerable number of weavers who are competent in intricate woven structures. It was observed that weavers tend to create intricate designs by using complex structures only for the Annual National Textile Competition known as *"Ransalu"*.

> *"I utilize the knowledge acquired from training sessions to present an excellent product for the Annual Competition. I use simple weaving techniques to produce sarees for today's normal orders".(Interview 11, Weaving Instructor, 2022)*

However, they do not often practice them. There is a fixed rate paid for each product at their weaving centers. The weaving centers assigned a target number of products for each weaver for each month. One weaving center is given 750 production targets for a month and depending on the above adjustments it is enabled to fulfill the target (Interview 2, weaver, 2022). The most demanding products are made due to the availability of material and color. Each weaver who exceeded the target number is paid an incentive. Incentives are paid for several tasks, including dyeing and weaving. The weaver is paid 5% for 1 m (Interview 3, Provincial Director, 2022). The target consumer is the government official who often wears an official dress *saree, ohori*, or a shirt for duty. The Textile Department organizes a Trade Fair once a month in government offices, and schools, and allows consumers to buy products on an easy payment system which is very famous among the government officers in Sri Lanka. This easy payment system removed all the barriers of cash in the hand buying system. The total amount for the products will be deducted from their monthly salaries at a minimum rate (Interview 3, Provincial Director, 2022). This system is convenient to the consumer and ultimately, they tend to buy local handloom products. Therefore, it is always exceeding the expected profit of the Handloom Textile Industry.

> *"We have enough work, and we are well paid for the target work".(Interview 11, Weaving Instructor, 2022)*

In this system, it was observed that some common product categories with moderate designs are the target selling point of the mass market. Weavers are reluctant to sacrifice their overpaid incentives to carry out experiments with intricate woven structures which take time and a lot of energy and have not been paid an extra amount. As a result of the lack of weaving in such intricate and complex woven structures designs, there are hardly any such products available in the current mass market.

Customer demand is observed by the feedback given by the customers to the in-house weaving technical officer at the outlets. There is a monthly committee in each province to evaluate the progress of their sales composed of Provincial directors, Instructors, and Designers. The Chief Designer of the committee and the technical experts evaluate the customer feedback and give necessary guidelines to fulfill the demand. At this meeting, the committee is focused on the cost per meter against the required design. They tried to give the best solution to the customer demand and satisfaction. (Interview 10, Designer In charge, 2022).

The buyer-seller forum is important in identifying the textile demand (a buyer-seller program that permits pre-scheduled interactions between exhibitors and buyers from domestic traders) (Interview 3, Provincial Director, 2022). This forum is set up by the Provincial Council after the Provincial competition is over, inviting outside business parties to buy their existing products. This forum allows businesses to connect directly with key buyers to meet and do business.

4. Discussion

There is a direct correlation between sustainable survival of the handloom weaving structures with Buddhism, inheritance, knowledge, customs, folklore, norms, values, traditional education, modern education, traditional institutions, and today's government institutional structure. The direct factors are discussed according to two questions of the research.

4.1. How does the Sustainability of the Domestic Textile Industry Relate to Its Inherited Knowledge Culture?

4.1.1. Traditional Customs, Norms, and Values Are Based on Buddhism

Most common designs and motifs resemble the philosophy of Buddhism, and they are represented in cultural practices of Buddhist religion and lifestyle as discussed in the results points in Sections 3.1.1 and 3.1.2. Sri Lankan art and craft were based on Buddhist culture. The most common designs in the handloom textile products are the lotus flower and swan. The lotus signifies purity in Buddhist culture. The swan is a symbol of the good merits and bad. These textile motifs have been continued since ancient times to the present day emphasizing the cultural link between the past and present. "The values seem to be path dependent" [23] (p. 49). Therefore, the belief systems are so robust and enduring, it is impossible that the traditional value system will lose its power [23] (p. 49).

4.1.2. Inherited Craft Knowledge Is Preserved by Artisans

The artisans are reluctant to expose their skills, tools, and knowledge for public usage as pointed out in Sections 3.1.1 and 3.1.2. The inherited textile knowledge is considered a treasure and the ancient wisdom of their ancestors. On the other hand, they use their maximum capacity to compete in the annual competition to produce the most identical traditional handloom product. The artisans identify the demanding value of inherited knowledge. They take care of the knowledge; the knowledge takes care of them. This attitudinal caretaking is seen in *Dumbara Rata* weaving technique (Indigenous weaving technique). There is a living culture in the concept of art and living. Designs, colors, and weaving structures are recorded in folklore which gives Sri Lankan cultural identity.

4.1.3. Cross-Fertilization Features to Enrich the Collection of Design and Motifs and Woven Structures from Time to Time

Traditional design and motifs, tools, techniques, materials, product categories, color, and market have been influenced and increased by foreign trends from time to time. Foreign training, seminars, and workshops helped to increase the cross-fertilization features of the industry as pointed out in Section 3.1.3. This shows how inherited knowledge developed through hybrid culture and fertilized with different cultural values. It was observed that traditional craft knowledge has been modified by modern technology as pointed out in Section 3.1.4. Traditional handloom textiles were influenced by multifaceted machines such as Dobby, Jacquard, Jamdani, Dik border, and Warp printing, the researchers identify that traditional Sri Lankan designs are woven in Indian introduced handloom machine called 'Jamdani' signifies cross-cultural features of a living culture. "Tradition changes; it is not stagnant" [23] (p. 49).

4.1.4. The Knowledge of the Weaving Structures Survived through the Constant Practice of Artisans

Artisans are Systematically Involved in Constant Learning as pointed out in Sections 3.1.1–3.1.5.

The practice is demanded by a sound market. Both traditional and contemporary knowledge is institutionally established. In the traditional society, youngsters learn from the elders of the family. Today, it is identified that systematized training and pedagogical structure lead the way in the industry. As in Japan, and Thailand, identified learning and training are proactive tools in protecting the heritage [31] (p. 16). The Department arranges annual workshops, seminars, and foreign training, and implements a research center for each province to encourage employees to increase their creativity. Human resources are craft artisans who are the employees in the industry. Technology will only be a facilitator. The human touch of a product is unique. Identity and authenticity are the most demanding selling points in both high-end and mass markets. "Products must transform into crafts of use and cultivate and amplify the skills, habits of mind, and abilities of users to generate more inventiveness and longevity [32] (p. 236)".

4.1.5. Traditional Technology was Modified with Concurrent Technology

It was shown that traditional technology was modified with concurrent technology during the course of time as Pointed out in Sections 3.1.3, 3.1.4 and 3.1.6. That helped the survival of the traditional knowledge, innovations, and adaptation to consumer society. Techniques were combined to create an innovation. New material introduction and increase in experiments are key turning points in adapting the traditional system to modern society. Inherited knowledge successfully walked from the traditional society to consumer society. This fact suggests that tradition needs to cooperate with concurrent technology.

4.2. What Is the Relationship between the Market Demand and the Technical Application of Design Structures of Their Productions?

4.2.1. Production Priority Is Given to Demanding Textile Designs and Products at the Market

Textile designs are the demanding factor as pointed out in Section 3.2.1. It is seen that product range has been increased due to gradual consumer demand.

4.2.2. Annual Competition and Exhibition, Monthly Trade Fair, Buyer–Seller Forum, and Design Committee Meeting Constantly Identify the Most Demanding Customer Selling Point in Handloom Textile Productions

National exhibitions and the competition added value to the industry. Methodically and systematically selected artisans are awarded annually. Imposing rules and regulations on the handloom industry employees of the traditional design culture are protected. It was observed that artisans identify the event where their art and craft knowledge and skills are evaluated. Their target customer is penetrated through lower level to high-end designer level from annual competitions, trade fairs, and exhibitions. Traditional weaving techniques are presented in a variety of products, at their peak of traditional art and craft value. Therefore, it is seen that the handloom textile market is stabilized in the local cultural context.

5. Conclusions

The discussion section scrutinized seven concepts related to the sustainable survival of the weaving structures' handloom sector in Sri Lanka. Learning is the direct enforcer. These concepts are recommended as the survival-driven forces of the industry. The weaving structures, design and motifs, and color of the handloom industry were based on Buddhist culture. There is an existing living culture. The living culture of handloom textile has been continuing through continuous learning of many weaving structures since ancient times. The learning enabled making an identity and demand in weaving structures. Artisans are persuaded to learn weaving techniques, design and motifs, and color from their education system. It is seen that there has been a consistency in forwarding design applications since ancient times. The same traditional design and motifs have been used since ancient times. In the 1970s, a variety of weaving techniques were introduced. This collection of weaving techniques provided ample opportunities to produce various products with traditional designs and motifs. The product range has been increased. Inherited knowledge was passed down from ancestors to the new generation from learning and practicing. The new technology of weaving structures is fused with traditional knowledge. It was noticed that newly learnt knowledge of weaving structures became a tradition through practice. Then, the next factor shows that training, seminars, and workshops with such a variety of learning styles enable a bridge between traditional knowledge and new technology. Innovation on material tools developed through a creative learning process which is being tried to address the millennium customer desires. Annual national exhibition and competition, monthly trade fair, buyer-seller forum, and design committee meetings where customer demand, availability of material, durability, and cost factors overlap and juxtaposed give insight into the most economically and culturally viable products.

This found identical authentic learning structure has been adapted in today's industry. The ancient textile industry was taken care of by the King. In ancient times, their art and craft were modified and evolved through education gained from inheritances. Artisans who were involved in learning art and craft were involved in the occupation as well. Traditional social organizations protected both art and artisans. Today, the same structure has been reformed and organized to protect the artisan and the industry. The artist is accommodated in a structured institute where the artisan acquires constant learning as training. The learning introduced them to new technology, foreign trends, tools, techniques, and materials such as infrastructure, which industry requires constantly. Learning gives insight and the ability to the artisan to adapt to today.

The industry is developed through hands-on experience, and prolonged engagement, where human resources are the key factor. The infrastructure is supplied by the Textile Department. The research suggests recognition of a consistent infrastructure supply chain as national policy. The paper recommends policy makers with the factors found to connect culture with an active action plan.

Author Contributions: H.M. and G.M.R. conceptualization and methodology; G.M.R. and V.M.K. collected and analyzed the data; G.M.R., V.M.K. and N.N. writing—original draft preparation; S.W. and N.N. writing—review and editing; H.M. funding acquisition. All authors have read and agreed to the published version of the manuscript.

Funding: The authors would like to acknowledge the Research Fund for International Scientists (RFIS-52150410416), National Natural Science Foundation of China.

Institutional Review Board Statement: Not applicable.

Informed Consent Statement: Not applicable.

Data Availability Statement: Data can be provided by corresponding author, i.e., GM Ranathunga, on request.

Acknowledgments: KPN Bhaggya, Teaching Assistant, Center for Fashion, Textile and Clothing Technologies, School of Technology, Sri Lanka Technological Campus, Sri Lanka produced technical drawings. All the directors and relevant officers who take care of designing and decision making on production and sales of Textile Departments of Provincial Councils in Sri Lanka, Interviewers (designers, weavers, instructors, university and professor) and other instructors of other Handloom Weaving Schools of Sri Lanka.

Conflicts of Interest: The authors declare no conflict of interest.

References

1. Yan, X.; Chen, L.; Hafeezullah, M. Introduction. In *Textile and Fashion Education Internationalization. Textile Science and Clothing Technology*; Yan, X., Chen, L., Memon, H., Eds.; Springer: Singapore, 2022. [CrossRef]
2. Wijayapala, U.G.S.; Alwis, A.A.P.; Ranathunga, G.M.; Karunaratne, P.V.M. Evolution of Sri Lankan Textile Education from Ancient Times to the 21st Century. In *Textile and Fashion Education Internationalization. Textile Science and Clothing Technology*; Yan, X., Chen, L., Memon, H., Eds.; Springer: Singapore, 2022. [CrossRef]
3. Coomaraswamy, A. *Medieval Sinhalese Art*; National Museum: Colombo, Sri Lanka, 1959.
4. Harvey, C.; Press, J.; Maclean, M. William Morris, cultural leadership, and the dynamics of taste. *Bus. Hist. Rev.* **2011**, *85*, 245–271. [CrossRef]
5. MacCarthy, F. *William Morris: A Life for Our Time*; Faber & Faber: London, UK, 2015.
6. Thompson, E.P. *William Morris: Romantic to Revolutionary*; PM Press: Oakland, CA, USA, 2011.
7. Romain, J. Ananda Coomaraswamy and the formation of the Sri Lanka collection at the Los Angeles County Museum of Art. *J. Hist. Collect.* **2016**, *28*, 465–478. [CrossRef]
8. Huntington, S. The Change to Change: Modernization, development and politics. In *Comparative Politics*; New York Free Press: New York, NY, USA, 1976; Volume 30–31, p. 45.
9. Davy, J. *An Account of the Interior of Ceylon and of Its Inhabitants*; Longman: London, UK, 1821.
10. D'Oyly, J. A Sketch of the Constitution of the Kandyan Kingdom. Turner, L.J.B., Ed.; Tisara Publishers: Dehiwala, Sri Lanka, 1975; Volume 24.
11. Coomaraswamy, A. *Art and Crafts of India and Ceylon*; Today and Tomorrow's Printers and Publishers: Delhi, India, 1913.
12. Pararajasingham, E. *Textile History of Sri Lanka*; Paranan Associate Private Limited: Colombo, Sri Lanka, 2006.

13. *Report of the Task Force for a Sustainable Development of The Handloom Industry in Sri Lanka*; Textile Industry Development Division Ministry of Industry & Commerce: Colombo, Sri Lanka, 2012; Available online: https://fdocuments.net/document/report-of-the-task-force-appointed-for-a-sustainable-loompdfreport-of-the-task.html?page=1 (accessed on 20 April 2022).
14. *Industry Capability Report: Sri Lankan Handloom Sector*; Export Development Board: Colombo, Sri Lanka, 2013; Available online: http://www.srilankabusiness.com/pdf/incapreporthandloom.pdf (accessed on 20 April 2022).
15. Mazzocchi, F. A deeper meaning of sustainability: Insights from indigenous knowledge. *Anthr. Rev.* **2020**, *7*, 77–93. [CrossRef]
16. Redclift, M. Sustainable development (1987–2005): An oxymoron comes of age. *Sustain. Dev.* **2005**, *13*, 212–227. [CrossRef]
17. Derlukiewicz, N.; Mempel-Śnieżyk, A.; Mankowska, D.; Dyjakon, A.; Minta, S.; Pilawka, T. How do clusters foster sustainable development? An analysis of EU policies. *Sustainability* **2020**, *12*, 1297. [CrossRef]
18. Stancu, C.M.; Grønhøj, A.; Lähteenmäki, L. Meanings and motives for consumers' sustainable actions in the food and clothing domains. *Sustainability* **2020**, *12*, 10400. [CrossRef]
19. De Bakker, E.; Dagevos, H. Reducing meat consumption in today's consumer society: Questioning the citizen-consumer gap. *J. Agric. Environ. Ethics* **2012**, *25*, 877–894. [CrossRef]
20. Mensah, J. Sustainable development: Meaning, history, principles, pillars, and implications for human action: Literature review. *Cogent Soc. Sci.* **2019**, *5*, 1653531. [CrossRef]
21. Du Pisani, J.A. Sustainable development–historical roots of the concept. *Environ. Sci.* **2006**, *3*, 83–96. [CrossRef]
22. Throsby, D. Sustainability and culture some theoretical issues. *Int. J. Cult. Policy* **1997**, *4*, 7–19. [CrossRef]
23. Purvis, B.; Mao, Y.; Robinson, D. Three pillars of sustainability: In search of conceptual origins. *Sustain. Sci.* **2019**, *14*, 681–695. [CrossRef]
24. Tipps, D. Modernization Theory and the Comparative Study of National Societies: A Critical Perspective. *Comp. Stud. Soc. Hist.* **1973**, *15*, 199–226. [CrossRef]
25. Inglehart, R.; Baker, W.E. Modernization, cultural change, and the persistence of traditional values. *Am. Sociol. Rev.* **2000**, *65*, 19–51. [CrossRef]
26. Nurse, K. Culture as the fourth pillar of sustainable development. *Small States Econ. Rev. Basic Stat.* **2006**, *11*, 28–40.
27. Powter, A.; Ross, S. Sustainability for Heritage Properties. *APT Bull.* **2005**, *36*, 5–11. Available online: https://www.jstor.org/stable/40003157 (accessed on 20 April 2022).
28. Shi, L.; Han, L.; Yang, F.; Gao, L. The evolution of sustainable development theory: Types, goals, and research prospects. *Sustainability* **2019**, *11*, 7158. [CrossRef]
29. Butler, R.W. Sustainable tourism: A state-of-the-art review. *Tour. Geogr.* **1999**, *1*, 7–25. [CrossRef]
30. Pradana, G.Y.K.; Arcana, K.T.P. Balinese traditional homestay in a sustainable tourism entering millennial era. *J. Xi'an Univ. Archit. Technol.* **2020**, *12*, 4208–4217.
31. Hancock, M. Subjects of heritage in urban southern India. *Environ. Plan. D Soc. Space* **2002**, *20*, 693–717. [CrossRef]
32. Li, Y. Research on the Development of Creative Tourist Industry in Yantai Taking the Development of Creative Tourist Craft Design Industry as an Example. In *2015 2nd International Conference on Education, Language, Art and Intercultural Communication (ICELAIC-15)*; Atlantis Press: Zhengzhou, China, 2015; pp. 423–426.
33. *Report on Creative Industries and Micro & Small Scale Enterprise Development A Contribution to Poverty Alleviation*; United Nations Industrial Development Organization (UNIDO): Vienna, Austria, 2005; Available online: https://www.unido.org/sites/default/files/2009-03/69264_creative_industries_0.pdf (accessed on 9 July 2022).
34. Fletcher, K. Durability, fashion, sustainability: The processes and practices of use. *Fash. Pract.* **2012**, *4*, 221–238. [CrossRef]
35. Fletcher, K. Other fashion systems. In *Routledge Handbook of Sustainability and Fashion*; Routledge: London, UK, 2014; pp. 33–42.
36. Correa, M.E. Fashion, Design and Sustainability. New Horizons in the Ways of Conceiving Production Processes. In *Sustainable Fashion and Textiles in Latin America*; Springer: Singapore, 2021; pp. 219–237. [CrossRef]
37. Francis, J.N.; Beninger, C.; Beninger, S. Traditional Knowledge as a Genetic Resource: Protections for Sustainable Development in Africa (Accès aux ressources génétiques et partage des avantages résultant de leur exploitation, RADE, Revue Africaine de Droit de, l' Environnement). *Afr. J. Environ. Law* **2014**, *79*, 657.
38. Zuallcobley, R.W.; Nik, M.Z.A. Safeguarding Indigenous Knowledge and Cultural Tradition while Promoting Sustainable Tourism. *Asian J. Arts Cult. Tour.* **2020**, *1*, 22–25.
39. Ramkumar, B.; Dias, R.M. Sustaining traditional textile art among the Indigenous Nongtluh women of north-eastern India: An interpretative phenomenological analysis. *Fash. Style Pop. Cult.* **2022**, *9*, 1–2. [CrossRef]
40. Parameswara, A.; Saskara, I.A.N.; Utama, I.M.S.; Setyari, N.P.W. Exploring Cultural Value and its Sustainability of Balinese Handwoven Textiles. *TEXTILE* **2022**, 1–24. [CrossRef]
41. *UNCTAD Annual Report*; United Nations Conference on Trade and development: Geneva, Switzerland, 2011; Available online: https://unctad.org/system/files/official-document/dom2012d1_en.pdf (accessed on 8 July 2022).
42. Regan, C.L.; Kincade, D.H.; Sheldon, G. Applicability of the engineering design process theory in the apparel design process. *Cloth. Text. Res. J.* **1998**, *16*, 36–46. [CrossRef]
43. Alvesson, M. Beyond neopositivist, romantics and localists: A reflective approach to interviews in organizational research. *Acad. Manag. Rev.* **2003**, *28*, 13–33. [CrossRef]
44. Qu, S.Q.; Dumay, J. The qualitative research interview. *Qual. Res. Account. Manag.* **2011**, *8*, 238–264. [CrossRef]

45. Fontana, A.; Frey, J.H. The interview: From structured questions to negotiated text. In *Handbook of Qualitative Research*; Denzin, N.K., Lincoln, Y.S., Eds.; Sage: Thousand Oaks, CA, USA, 2000; pp. 645–672.
46. Nathan, S.; Newman, C.; Lancaster, K. Qualitative interviewing. In *Pranee Liamputtong*; Handbook of Research Methods in Health Social Sciences; Springer: Singapore, 2019; pp. 391–410.
47. Seidman, I.E. *Interviewing as Qualitative Research: A Guide for Researchers in Education and the Social Sciences*; Teachers College Press: New York, NY, USA, 1991.
48. Philipps, A.; Mrowczynski, R. Getting more out of interviews. Understanding Interviewees' Accounts in Relation to Their Frames of Orientation. *Qual. Res.* **2021**, *21*, 59–75. [CrossRef]
49. Knox, S.; Burkard, A.W. Qualitative research interviews. *Psychother. Res.* **2009**, 566–575. [CrossRef]
50. Wijayapala, U.S.; Ranathunga, G.M.; Karunaratne, P.V.M. *The Buddhist robe: The Path Pointing to Natural Dyes and Possible Establishment of a Natural Dyeing Industry in the Apparel Sector of Sri Lanka with Special Reference on the Buddhist Robe*; S. Godage & Brothers (Pvt.) Limited: Colombo, Sri Lanka, 2019.
51. Lu, S.; Zhou, J. The practice of digital innovative design of traditional ornamental brocade. *J. Silk* **2021**, *58*, 127–133. [CrossRef]
52. Yan, T. Weft-knitting double-sided jacquard fabric and production practice of its one-bath multi-color dyeing process. *Adv. Text. Technol.* **2020**, 2862–2865. [CrossRef]
53. Wu, M.; Zhang, A.; Zhang, Y. Automatic layout and CUDA parallel implementation of warp knitted jacquard shoe upper. *J. Silk* **2021**, *58*, 126–132. [CrossRef]

sustainability

MDPI

Article

Factors That Influence Consumers' Sustainable Apparel Purchase Intention: The Moderating Effect of Generational Cohorts

Pei-Hsin Lin * and Wun-Hwa Chen

Department of Business Administration, National Taiwan University, Taipei 106, Taiwan; andychen@ntu.edu.tw
* Correspondence: jilllin329@gmail.com

Abstract: The circular economy is one of the crucial issues in fashion because the fashion industry is a major global polluter. Many consumers are adopting a more sustainable lifestyle and it shows in their buying preferences and behaviors. This study aims to predict sustainable fashion apparel consumption using an extended version of the belief–attitude–intention framework, by investigated the moderating effect of generational cohorts. Particularly, the study emphasizes the rental apparel, second-hand apparel, and recycled apparel markets. Survey data were collected from 135 Generation X consumers, 134 Generation Y consumers, and 139 Generation Z consumers in Taiwan. Structural equation modeling and the bootstrapping method were applied to test the hypothesized relationships. The findings determined environmental consciousness, perceived value, and perceived risk as key predictors of consumers' sustainable apparel purchase intentions. The findings also showed that the generational cohort negatively moderated the relationship between environmental consciousness and sustainable apparel purchase intentions. Therefore, fully understanding consumers' purchase intentions regarding sustainable apparel is an indispensable topic for both academia and industry in a circular environment. Moreover, the fashion industry should concentrate more on promoting sustainability and ecologically friendly apparel products as well as developing multi-generational marketing strategies.

Keywords: fashion industry; sustainable apparel; environmental consciousness; perceived value; perceived risk; purchase intention; generational cohort

Citation: Lin, P.-H.; Chen, W.-H. Factors That Influence Consumers' Sustainable Apparel Purchase Intention: The Moderating Effect of Generational Cohorts. *Sustainability* **2022**, *14*, 8950. https://doi.org/10.3390/su14148950

Academic Editors: Hafeezullah Memon, Xiaoke Jin, Tian Wei and Chengyan Zhu

Received: 14 June 2022
Accepted: 18 July 2022
Published: 21 July 2022

1. Introduction

The fashion industry is widely believed to be the second largest polluting industry on the planet, and the environmental damage is increasing as the industry grows [1,2]. According to Niinimäki et al. [3], garment production is responsible for around 20% of industrial wastewater pollution from dyeing and finishing textiles, 8–10% of carbon dioxide emissions, and more than 92 million tonnes of waste produced per year. Meanwhile, fast fashion has caused a considerable rise in the quantity of garments produced and thrown away [4]. Recently, these problems throughout the textile and fashion supply chains have appealed to consumers' growing concerns and demands for sustainable products made according to ecological and social principles, and thus a growing number of firms are committing to minimizing their detrimental impacts on ecosystems and societies [5–7]. However, consumers' understanding of sustainable apparel are often vague, and their buying decisions between sustainable and non-sustainable apparel often depend on aesthetic, functional, and financial benefits, resulting in low involvement of consumers' sustainable apparel consumption [8–10]. For that reason, it is critical to clearly identify the important factors affecting sustainable apparel purchase intentions.

In the field of sustainable fashion consumption, renting and buying second-hand or recycled clothing are considered as sustainable options, and there is an increasing body of literature that recognizes the relationship between consumers' environmental consciousness, perceived values, perceived risks, and purchase intentions [11–13]. In addition, recent evidence confirms that consumers of different generations have different sustainable consumer behaviors due to their different living backgrounds and environments [14–17]. So far, however, much uncertainty still exists about the relationship between environmental consciousness and purchase intention in the sustainable apparel market, and it is still not known how different generations moderate sustainable apparel purchase intentions. Therefore, this study attempts to show how an individual's environmental consciousness affects their sustainable apparel purchase intention through perceived values and perceived risks, and to recognize the moderating roles of the generations. In this study, sustainable fashion apparel covers rental clothing, second-hand clothing, and recycled clothing, and the targeted sample group consists of Generation X, Generation Y, and Generation Z. The study presumes that consumers who have high environmental consciousness may present a stronger behavioral intention to purchase sustainable fashion apparel, and a moderating effect of generational cohort may exist in the consumers' sustainable apparel purchase intention.

This study makes an important contribution to the circular fashion industry and provides valuable insights for fashion brands or retailers to understand the different generations' sustainable apparel consumption preferences and behaviors regarding different sustainable apparel products. The rest of the study is organized as follows: Section 2 presents the literature review on circular fashion, environmental consciousness, perceived value, perceived risk, and generational cohort theory; Section 3 introduces the research methodology and sample profile; Section 4 displays the hypothesis testing results for the three types of sustainable apparel, and in particular, the moderating effects of generational cohorts are presented; Section 5 examines the empirical findings in detail; Section 6 summarizes the main findings and limitations of the current study, with future research recommendations.

2. Theoretical Framework

2.1. Circular Fashion

The circular economy aims to change the linear material flow, which directs many manufacturing operations, into closed-loop models that emphasize that resource and residual waste are shared, repaired, reused, or recycled [18,19]. A circular model for fashion generates green products and services for customers, while contributing to a prosperous fashion industry and environmental regeneration [20,21]. Given the consumer's growing awareness of sustainable consumption, there has been an increasing amount of literature on alternate models of clothing consumption [22–25]. Kim et al. [12] classified sustainable fashion products into three types: second-hand clothing, upcycled clothing, and recycled clothing. Machado et al. [26] demonstrated that the second-hand clothing business model is a way of decreasing resource use where waste in the form of used goods is reused, and product lifespans are extended through transferring ownership. Additionally, Clube and Tennant [27] and Shrivastava et al. [28] showed that online rental of used clothing is an emerging business model that bolsters circular fashion practices linked to environmental and economic sustainability. As a result, the circular fashion industry aims to promote the re-use and recycling of clothing. In the current study, we investigated the three different types of sustainable apparel consumption: rental clothing, second-hand clothing, and recycled clothing.

2.2. Environmental Consciousness

Environmental consciousness is considered as a complex concept built upon cognitive, attitudinal, and behavioral components as well as environmental knowledge [29,30]. Accordingly, environmental consciousness is an element of the belief system that contributes to specific mental influences linked to one's tendency to participate in the eco-friendly behavior regime [31]. To date, many recent studies have highlighted that environmental consciousness has become a vital component of the consumer decision-making process in the sustainable consumption context [32–35]. For example, Kautish and Sharma [36] and Zhang et al. [37] validated the relationships among environmental consciousness, perceived values, and behavioral intentions for green products. Wang et al. [38] revealed that perceived value regarding quality and price was a mediator of environmental consciousness and green purchase behaviors. Similarly, Hasan et al. [39] confirmed the positive relationship between environmental concerns, consumer attitudes, and a willingness to purchase organic cotton clothing. Souza et al. [40] posited that environmental consciousness, perceived values, and risk factors are key predictors of environmental purchase behavior. Szabo and Webster [41] indicated that environmental beliefs have positive effects on perceived value and negative effects on perceived risk. Therefore, these subsequent hypotheses are proposed:

H1: *Environmental consciousness is positively associated with perceived value.*

H2: *Environmental consciousness is negatively associated with perceived risk.*

2.3. Perceived Value and Perceived Risk

Perceived value is a customer's personal assessment of a product's utility [42]. According to the consumption values theory, consumers' product choice behaviors are established on five perceived values: functional, emotional, social, epistemic, and conditional [43]. In the textile and apparel domain, a multidimensional perspective of perceived value has been widely adopted to investigate consumers' sustainable clothing purchase behaviors. Bielawska and Grebosz-Krawczyk [44] analyzed data from 496 Polish consumers and concluded that emotional, conditional, and environmental values had significantly positive influences on purchase behavior regarding eco-friendly clothing. Baek and Oh [45] studied consumers' adoption intentions regarding fashion apparel rental services and revealed that functional, economic, and emotional values boost attitudes and intentions. Chun et al. [46] suggested that social value, emotional value, functional value, economic value, and eco value all have positive influences on behavioral intentions regarding recycled fashion products. Chi et al. [47] identified the perceived green value of recycled polyester-made athleisure apparel products as a multidimensional construct, including functional value, social value, emotional value, conditional value, and epistemic value. Considering this evidence, it seems that perceived value positively affects behavioral intention regarding sustainable apparel products or services. Therefore, the following hypothesis is proposed:

H3: *Perceived value is positively associated with sustainable apparel purchase intention.*

In previous studies on consumer decision-making behaviors, researchers conceptualized perceived risk as customers' beliefs about the uncertainty and potential negative outcomes of purchasing a product or service [48–50]. To date, several studies have highlighted various types of risk that are associated with consumers' purchase intentions, such as time, financial, social, psychological, physical, and performance risk [51–55]. In the case of online fashion rental services, Lee et al. [13] demonstrated that financial risk, performance risk, and social risk have a negative effect on usage intention. Similarly, Yoo et al. [56] proposed that key obstacles to consumer purchase intentions regarding upcycled apparel were social, financial, and performance risks. Moreover, hygiene concerns are suggested to be a strong barrier to the purchase intention of sustainable clothes [24,57–59]. Park and Choo [60] and Kim et al. [12] proved that sanitary risk negatively affects the intention to purchase second-hand, upcycled, and recycled clothing. Collectively, these studies suggest that perceived risk negatively influences behavioral intentions regarding sustainable apparel products or services. Therefore, the following hypothesis is proposed:

H4: *Perceived risk is negatively associated with sustainable apparel purchase intention.*

2.4. Generational Cohort Theory

Mannheim [61] proposed the generational cohort theory (GCT) and stated that a generational cohort is a set of individuals born within a particular span of time who share a similar age and stage of life. More specifically, a generation refers to a cohort of people who have comparable age and experience similar social, economic, political, and cultural events [62–64]. As a result, each generation would inherit a collective consciousness and develop unique values, belief systems, peer personalities, and behavioral tendencies [65]. According to Dimock [66], the generational cohort can be divided into five age groups: the Silent Generation (born 1928–1945), Baby Boomers (born 1946–1964), Generation X (born 1965–1980), Generation Y (also known as Millennials; born 1981–1996), and Generation Z (born 1997–2012). With members of Generation X and Y representing the largest percentage of today's workforce, and the oldest members of Generation Z are entering the workforce, the targeted sample of this study was Generations X, Y, and Z. In essence, Generation Xers are those born before the widespread adoption of digital technology; Generation Yers are also known as digital natives and technology is part of their everyday lives; Generation Zers have been exposed to the internet, to social networks, and to mobile systems.

Thus far, many previous studies in the field of marketing have demonstrated that an individual's decision-making processes [67,68], online shopping orientations [69–71], social media and technology usage [72,73], brand engagement and loyalty [74], and sustainable consumption behaviors [17,75] could be modified by that individual's generational identity. Much of the current literature on sustainable fashion apparel consumption pays particular attention to generational differences. Kim [76] pointed out that there are significant differences in the cognitive, affective, and behavioral responses to fashion luxury products between Millennials and Baby Boomers. Gazzola et al. [15] confirmed that the younger generations today are more aware of sustainability and the circular economy. Liang and Xu [77] found that Generation X demonstrated a high resistance to second-hand clothing products, while younger generations held stronger perceived values and purchase intentions regarding second-hand clothing than their older counterparts. Overall, the evidence reviewed here seems to suggest a moderating role for generational cohorts. Therefore, the subsequent hypotheses are proposed:

H5: *Generational cohorts moderate the direct effects of environmental consciousness on (a) perceived value, and (b) perceived risk.*

H6: *Generational cohorts moderate the direct effects of (a) perceived value, and (b) perceived risk on sustainable apparel purchase intention.*

2.5. Conceptual Model

Drawing on the belief–attitude–intention framework and the generational cohort theory, the purpose of this study is to explore the impact of environmental consciousness on perceived value, perceived risk, and sustainable apparel purchase intentions, and to investigate how generational differences moderate the association between environmental consciousness and sustainable apparel purchase intention. The conceptual model is displayed in Figure 1, which encompasses the above-mentioned five dimensions.

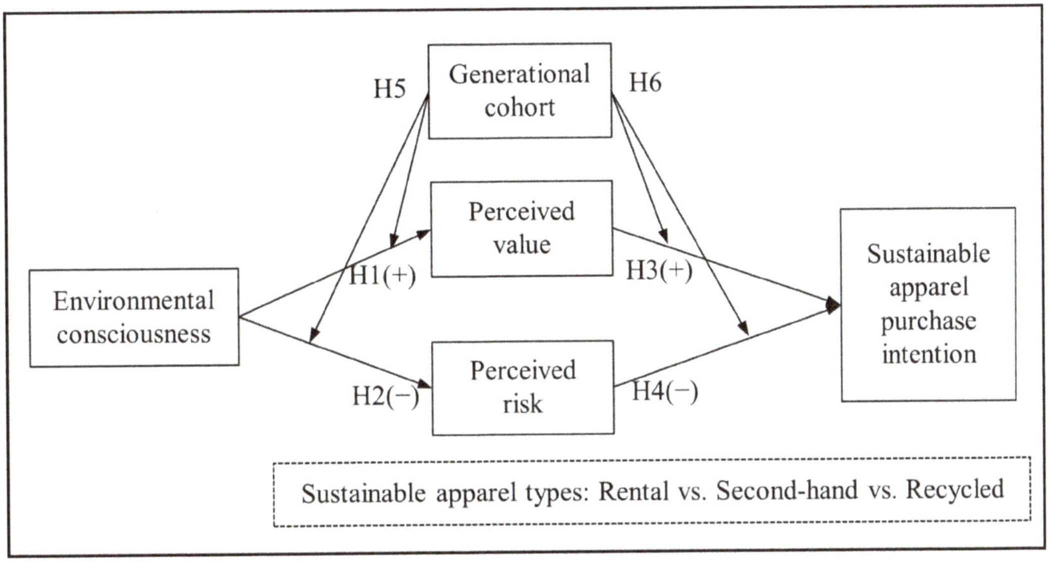

Figure 1. Conceptual model of this study.

3. Methods

3.1. Sample and Procedure

The current research investigated the determinants of consumers' sustainable apparel purchase intentions, and further examined the moderating role of generational cohorts. A survey approach was used to conduct this quantitative study, with the samples for the study drawn from Taiwan. Individuals aged below 56 were the targeted sample in this study, and the sample size of 200 or more for this study was determined because a minimum sample size of 200 is recommended for survey studies of behavior/cognition [78]. This study employed the SurveyCake online survey platform to collect the target samples. SurveyCake (https://www.surveycake.com/ (accessed on 3 September 2021)) is a market research platform that surveys internet users, and was used because it contains millions of users and offers representative sampling techniques where questionnaires can be sent to specific demographic groups to obtain accurate audience samples and target numbers. Participants who completed all scales for three types of sustainable clothing received TWD 100 as an incentive. A total of 408 valid responses were obtained between October and December 2021 and used for statistical analysis.

3.2. Measurement and Analytic Method

The survey started with a brief outline of the investigation's purpose and illustrations of three types of sustainable apparel: rental clothing, second-hand clothing, and recycled clothing. The questionnaire included two sections: (1) demographic: gender, age, marital status, monthly discretionary income, average monthly apparel expenditure and sustainable apparel purchase experience; (2) survey: items measuring environmental consciousness, perceived value, perceived risk, and sustainable apparel purchase intention. All multi-item instruments were measured utilizing 5-point Likert scales (1 = Strongly disagree, 3 = Neutral, 5 = Strongly agree) and were well-established measurements adapted from previous studies. Environmental consciousness was evaluated using three items from Wang et al. [38]. To access the perceived value in purchasing sustainable apparel products, five items related to functional, emotional, social, epistemic, and environmental value were measured referring to Kim et al.'s [12] and Bielawska and Grebosz-Krawczyk's [44] studies. As for perceived risk, four items measuring social, financial, performance, and sanitary risk were adopted from Yoo et al. [56] and Kim et al. [12]. Items measuring sustainable apparel purchase intention (3 items) were taken from Lee et al. [13] and Yoo et al. [56].

For data analysis, descriptive analysis, confirmatory factor analysis (CFA), structural equation modeling (SEM) analysis, and moderation analysis were used. SPSS software was utilized in the study to analyze demographic data and moderation effects. A confirmatory factor analysis (CFA) was used to evaluate how well the survey results matched the hypothesized measurement model. CFA enables researchers to systematically test specific prior hypotheses about the structure underlying survey results and to compare alternative measurement models with respect to explanatory power; this makes CFA a valuable tool for theory testing and building [79]. In addition, the hypotheses were tested by applying SEM, which is the most commonly used estimation method for robust analysis of data in the behavioral and social sciences [80]. CFA and SEM were performed adopting AMOS software.

3.3. Sample Profile

The sample profile shown in Table 1 was obtained via a descriptive analysis of the data. The sample consisted of 33.1% (n = 135) generation X, 32.8% (n = 134) generation Y, and 34.1% (n = 139) generation Z, indicating that the sample sizes for the three cohorts were equivalent. Of the 408 respondents, 245 (60%) were female and 163 (40%) were male. Most respondents in Generation X were married (67.4%) and the Generation Z respondents were predominantly (77.7%) single. As for their discretionary income, more than 60% of Generation X respondents had over TWD 20,000 a month, while the Generation Z respondents indicated a monthly discretionary income of less than TWD 20,000 (60.4%). In terms of average monthly apparel expenditure, between TWD 1001 and TWD 4000 was the most common, followed by less than TWD 1000. For the sustainable apparel purchase experience, most respondents had purchased second-hand and recycled clothing but had never used clothing rental services.

Table 1. Sample profile.

Variable		Generation X (Ages 41–56)	Generation Y (Ages 25–40)	Generation Z (Ages 9–24)
		Number (%)	Number (%)	Number (%)
Gender	Male	58 (43.0%)	54 (40.3%)	51 (36.7%)
	Female	77 (57.0%)	80 (59.7%)	88 (63.3%)
Marital status	Married	91 (67.4%)	57 (42.5%)	31 (22.3%)
	Single	44 (32.6%)	77 (57.5%)	108 (77.7%)
Monthly Discretionary Income	<TWD 10,000	13 (9.6%)	25 (18.7%)	42 (30.2%)
	10,001–20,000	37 (27.4%)	35 (26.1%)	42 (30.2%)
	20,001–30,000	35 (25.9%)	35 (26.1%)	38 (27.3%)
	30,000–40,000	29 (21.5%)	21 (15.7%)	14 (10.1%)
	>TWD 40,000	21 (15.6%)	18 (13.4%)	3 (2.2%)
Average monthly apparel expenditure	<TWD 1000	39 (28.9%)	33 (24.6%)	43 (30.9%)
	1001–4000	63 (46.7%)	71 (53.0%)	65 (46.8%)
	4001–7000	21 (15.6%)	16 (11.9%)	18 (12.9%)
	7001–10,000	9 (6.7%)	6 (4.5%)	12 (8.6%)
	>TWD 10,000	3 (2.2%)	8 (6.0%)	1 (0.7%)
Ever use clothing rental services?	Yes	60 (44.4%)	60 (44.8%)	57 (41.0%)
	No	75 (55.6%)	74 (55.2%)	82 (59.0%)
Ever buy second-hand clothing or resale used clothing?	Yes	84 (62.2%)	92 (68.7%)	83 (59.7%)
	No	51 (37.8%)	42 (31.3%)	56 (40.3%)
Ever buy recycled clothing?	Yes	89 (65.9%)	90 (67.2%)	76 (54.7%)
	No	46 (34.1%)	44 (32.8%)	63 (45.3%)

4. Results

4.1. Reliability and Validity

A confirmatory factor analysis was utilized to examine the reliability and validity of the measurement model. Table 2 exhibits mean and estimated factor loadings (FL) for each item and Cronbach's α, composite reliabilities (CR), and average variance extracted (AVE) for each dimension. All FL values ranged from 0.709 to 0.914, and the AVE for all constructs was between 0.512 and 0.815, surpassing the 0.5 cutoff value. The value for CR should exceed 0.7. The coefficient values of CR in this study were between 0.758 and 0.930. All estimated FL, AVE, and CR values met the relevant criteria, providing significant evidence of convergent validity [81]. In addition, Cronbach's α was used to evaluate the internal reliability of the measurements. The Cronbach's α values for all dimensions were in the range of 0.825–0.928, surpassing the suggested threshold of 0.7 [82]. Thus, these results reveal that the measurement of this study was reliable and valid. From the table, it can be seen that the highest AVE, CR and Cronbach's α values were for the sustainable apparel purchase intention dimension (AVE = 0.512, CR = 0.758, Cronbach's α = 0.825), while the lowest AVE, CR and Cronbach's α values were for the environmental consciousness dimension (AVE = 0.815, CR = 0.930, Cronbach's α = 0.928).

Table 2. Results of reliability and validity analyses.

Dimension	Items	Mean	FL	Statistics
Environmental consciousness	1. The balance of nature is very delicate and can be easily upset.	4.219	0.709	CR = 0.758, AVE = 0.512, Cronbach's α = 0.825
	2. I have switched products for ecological reasons.	4.127	0.711	
	3. When I have a choice between two equal products, I purchase the one less harmful to other people and the environment.	4.153	0.725	
Perceived value	1. This sustainable clothing is well made and worth the money.	3.784	0.874	CR = 0.917, AVE = 0.690, Cronbach's α = 0.889
	2. Purchasing this sustainable clothing makes me feel good.	3.699	0.909	
	3. Purchasing this sustainable clothing can give its owner social approval.	3.599	0.801	
	4. This sustainable clothing offers uniqueness.	3.756	0.815	
	5. This sustainable clothing helps save resources.	3.953	0.744	
Perceived risk	1. This sustainable clothing would not be durable.	3.221	0.840	CR = 0.890, AVE = 0.668, Cronbach's α = 0.889
	2. This sustainable clothing is unlikely to be hygienic.	3.600	0.788	
	3. I would not feel comfortable wearing this sustainable clothing in public.	3.421	0.783	
	4. I think it is not worthwhile to spend money on this sustainable clothing.	3.210	0.857	
Sustainable apparel purchase intention	1. I am willing to visit a store that sells this sustainable clothing.	3.662	0.883	CR = 0.930, AVE = 0.815, Cronbach's α = 0.928
	2. I am willing to visit the website of this sustainable clothing.	3.544	0.914	
	3. I am willing to recommend this sustainable clothing to others.	3.653	0.911	

Sustainable clothing corresponds to rental clothing, second-hand clothing, and recycled clothing. The mean is the average of the numbers of the three types of apparel. Significance level: $p < 0.05$; $p < 0.01$; $p < 0.001$.

4.2. Hypothesis Verification

Structural equation modeling (SEM) was carried out to examine the interrelationships between environmental consciousness, perceived value, perceived risk, and sustainable apparel purchase intention using the AMOS software. Table 3 shows the main research hypotheses for overall sustainable apparel; except for H2, all other research hypotheses were supported. In detail, environmental consciousness had a significant positive impact on perceived value ($\beta = 0.364$, $p < 0.001$), and perceived value had a significant positive impact on sustainable apparel purchase intention ($\beta = 0.919$, $p < 0.001$), whereas perceived risk had a significant negative impact on sustainable apparel purchase intention ($\beta = -0.122$, $p < 0.001$). No significant influence was found between environmental consciousness and perceived risk ($\beta = 0.014$, $p > 0.05$). In terms of the differences between apparel types, for rental and second-hand apparel, H1, H3, and H4 were supported, but H2 was not supported. As for recycled apparel, H1 and H3 were supported but H2 and H4 were not supported, indicating that perceived risk did not significantly mediate the relationship between environmental consciousness and recycled apparel purchase intention. Moreover, the results for model fitness yielded from AMOS were as follows: $\chi 2/df = 2.487$, RMSEA = 0.060, RMR = 0.034, GFI = 0.913, AGFI = 0.890, CFI = 0.932, IFI = 0.933, NFI = 0.903, RFI = 0.907, and NNFI = 0.923. All values met the threshold suggested by Hu and Bentler [83]. It can therefore be concluded that the proposed model reasonably explained the collected data.

Table 3. Hypothesis testing results for the three types of sustainable apparel.

Hypothesis	Path	Rental Apparel	Second-Hand Apparel	Recycled Apparel	Overall Sustainable Apparel
H1	EC→PV	Supported (+)	Supported (+)	Supported (+)	Supported (+)
H2	EC→PR	Not supported	Not supported	Not supported	Not supported
H3	PV→SAPI	Supported (+)	Supported (+)	Supported (+)	Supported (+)
H4	PR→SAPI	Supported (−)	Supported (−)	Not supported	Supported (−)

EC: Environmental consciousness; PV: Perceived value; PR: Perceived risk; SAPI: Sustainable apparel purchase intention; Overall sustainable apparel covers the three types of sustainable apparel. Significance level: $p < 0.05$; $p < 0.01$; $p < 0.001$.

4.3. Moderating Effects of Generational Cohorts

The bootstrapping method was performed using SPSS Process Macro to examine if generational cohorts moderated the relationship among environmental consciousness, perceived value, perceived risk, and sustainable apparel purchase intentions. Table 4 provides the hypothesis testing results for H5 and H6. Regarding overall sustainable apparel, a moderating effect of generational cohorts was found in the relationship between environmental consciousness and perceived value ($\beta = -0.140$, CI = −0.2683~−0.0120), indicating that the relationship between environmental consciousness and perceived value was significantly more positive for the younger generations compared to Generation Xers. A moderating effect of generational cohorts was found in the relationship between environmental consciousness and perceived risk ($\beta = -0.313$, CI = −0.4827~−0.1445), indicating that the younger generations would show stronger negative associations between environmental consciousness and perceived risk than would Generation Xers. Thus, H5 was fully supported. Additionally, results also show that generational cohorts moderate the influence of perceived risk on sustainable apparel purchase intention ($\beta = -0.161$, CI = −0.2889~−0.0325). That is, younger generations would show stronger negative associations between perceived risk and sustainable apparel purchase intention than would Generation Xers. However, the moderating effect of generational cohorts on the relationship between perceived value and sustainable apparel purchase intention was not significant ($\beta = 0.018$, CI = −0.2159~0.0002). H6 was therefore partially supported.

Table 4. Hypothesis testing results for the moderator.

Hypothesis	Path	Rental Apparel	Second-Hand Apparel	Recycled Apparel	Overall Sustainable Apparel
H5 (a)	EC→PV (EC × GC)	Supported (−)	Supported (−)	Supported (−)	Supported (−)
H5 (b)	EC→PR (EC × GC)	Supported (−)	Supported (−)	Supported (−)	Supported (−)
H6 (a)	PV→SAPI (PV × GC)	Not supported	Not supported	Not supported	Not supported
H6 (b)	PR→SAPI (PR × GC)	Supported (−)	Supported (−)	Not supported	Supported (−)

EC: Environmental consciousness; PV: Perceived value; PR: Perceived risk; SAPI: Sustainable apparel purchase intention; GC: Generational cohorts; Overall sustainable apparel covers the three types of sustainable apparel. Significance level: $p < 0.05$; $p < 0.01$; $p < 0.001$.

5. Discussion

This study set out to examine the relationship between environmental consciousness and sustainable apparel purchase intention and to determine whether generational differences moderate the relationship. We divided current sustainable apparel into three types (rental clothing, second-hand clothing, and recycled clothing) and targeted Generations X, Y, and Z. In terms of Hypotheses 1 and 3, we tested the relationship between environmental consciousness, perceived value, and sustainable apparel purchase intention, and found that environmental consciousness strongly affected sustainable apparel purchase intention through perceived value. These results are consistent with studies that examined the influence of environmental concerns on consumers' value perceptions and purchase intentions, specifically those conducted by Hasan et al. [39] and Baek and Oh [45], who found that environmental concerns and perceived values were directly related to consumer decision-making on purchasing eco-friendly clothing. A possible explanation for this might be that environmental consciousness is viewed as an important component of the consumer decision-making process in the context of sustainable consumption [34,35]. That is, the more environmentally conscious the consumers, the stronger their value perceptions and behavioral intentions become regarding sustainable apparel goods.

As for Hypotheses 2 and 4, we tested the relationship between environmental consciousness, perceived risk, and sustainable apparel purchase intention. Surprisingly, this study did not find a significant association between environmental consciousness and perceived risk. In other words, high environmental consciousness does not lead to a lesser risk perception of sustainable clothing. This inconsistency may be due to the growing level of environmental awareness, and the increasing number of fashion brands and retailers committed to providing services and products that combine environmental awareness with quality and performance [5–7]. In terms of the relationship between perceived risk and sustainable apparel purchase intention, the results support a negative direct effect of perceived risks relating to rental and second-hand clothing on purchase intention; however, no evidence of a remarkable effect of perceived risks relating to recycled clothing was detected. A possible explanation for this might be that rental and second-hand clothing may be perceived to be more unsanitary than products made of recycled materials [12,59]. These results reflect those of Yoo et al. [56] who also found that the risks perceived by buyers differed according to the product characteristics when purchasing sustainable fashion products.

For Hypotheses 5 and 6, we tested the moderating effects of generational cohorts in all paths and found these cohorts weakened the effect of environmental consciousness on perceived value, and perceived risk and the effects of perceived risk on sustainable apparel purchase intentions. The results revealed that Generation Zers would show stronger associations between belief, attitude, and intention compared to Generation Yers and Xers, presenting evidence of moderation by generational cohorts. These results are in line with those of previous studies [15,17,75,76]. For instance, Liang and Xu [77] confirmed that the younger generations perceived higher values and held higher purchase intentions regarding sustainable apparel products than their older counterparts. In summary, although not all the hypotheses were confirmed, the empirical results strongly support a positive effect of the environmental consciousness and perceived value on sustainable apparel purchase intention, and a negative effect of the perceived risk on sustainable apparel purchase intention. These results imply that the consumers' willingness to purchase sustainable clothing is significantly related to environmental consciousness and valence perceptions. Additionally, the findings of this study suggest that consumers' sustainable consumption behaviors could be modified by generational cohorts, proving that young consumers are driving the market for sustainable products and services.

6. Conclusions

This study was conducted to disentangle the factors determining the consumers' sustainable apparel purchase intentions including rental, second-hand, and recycled clothing, and to explore the moderating effect of generational cohorts. The most obvious finding to emerge from the study is that environmental considerations and perceptions are important factors in explaining consumer sustainable apparel purchase intention, and a moderating effect of generational cohorts exists in the consumers' sustainable apparel purchase intentions. The present study generates theoretical contributions by synthesizing five research dimensions in the sustainable apparel consumption domain. Specifically, this study advances our understanding of sustainable apparel consumption by exploring consumers' environmental consciousness, perceived value, perceived risk, and purchase intention for different sustainable apparel products. Moreover, this study is the first comprehensive investigation of the moderating role of generations in consumers' pro-environmental attitudes and behaviors regarding sustainable apparel products. Overall, the present study provides a deeper insight into the rapidly expanding field of circular fashion consumption and suggests that sustainable apparel purchase intention varies across market segments and generations.

This study has two main practical contributions. Firstly, it can be a guideline for retailers and marketers in the fashion industry to understand the factors to be considered in promoting sustainable apparel. Particularly, enhancing environmental and functional value may be an effective way of facilitating consumers' sustainable apparel buying intentions. Moreover, it is possible to increase buying intention of sustainable apparel products by reducing consumers' risk perception through promoting specific and genuine environmental friendliness in products from eco-friendly brands. Secondly, it provides a deeper insight into how consumers' sustainable shopping attitudes and behaviors can vary across the different generations. For younger generations, marketers should promote the idea of buying less, and produce fashion products that are made with minimal and environmentally friendly resources. Meanwhile, to make older generations more knowledgeable about sustainable apparel products, marketers and policy enforcers should spread information about sustainable issues through legal guidelines, media reports, and public relations in the manufacturing and marketing sectors of companies. However, there are certain limitations that restrict the generalizability of these findings. Firstly, the sample size and the target sample impeded the generalization of the results. A larger sample size is required to improve the generalizability of research findings. This study was conducted in Taiwan, and further research could be conducted in other countries to extend the understanding in sustainable fashion consumption from a global perspective. Secondly, this study relies on data collected using a questionnaire. Future studies should be undertaken to provide an in-depth view regarding the customers' perceptions and behaviors when purchasing sustainable apparel, by utilizing qualitative techniques such as open-ended survey questionnaires and interviews. Thirdly, this study does not consider the impact of demographic factors such as gender, monthly discretionary income, and marital status. Future work is needed to validate demographic differences in consumers' environmental awareness and sustainable behaviors. Finally, unifying the four well-known constructs may not fully elaborate the consumers' sustainable behaviors, so future studies should consider incorporating other psychographic and social factors (e.g., lifestyles, values, subjective norms) associated with behavior regarding green consumption.

Author Contributions: Conceptualization, P.-H.L.; Formal analysis, P.-H.L.; Investigation, P.-H.L.; Methodology, P.-H.L.; Data curation, P.-H.L.; Writing—original draft, P.-H.L.; Writing—review & editing, P.-H.L.; Supervision, W.-H.C. All authors have read and agreed to the published version of the manuscript.

Funding: This research received no external funding.

Institutional Review Board Statement: Not applicable.

Informed Consent Statement: Informed consent was obtained from all subjects involved in the study.

Data Availability Statement: The data presented in this study are available on request from the corresponding author. The data are not publicly available due to privacy reasons.

Conflicts of Interest: The authors declare no conflict of interest.

References

1. Jia, F.; Yin, S.; Chen, L.; Chen, X. The Circular Economy in the Textile and Apparel Industry: A Systematic Literature Review. *J. Clean. Prod.* **2020**, *259*, 120728. [CrossRef]
2. Muthukumarana, T.T.; Karunathilake, H.P.; Punchihewa, H.K.G.; Manthilake, M.M.I.D.; Hewage, K.N. Life Cycle Environmental Impacts of the Apparel Industry in Sri Lanka: Analysis of the Energy Sources. *J. Clean. Prod.* **2018**, *172*, 1346–1357. [CrossRef]
3. Niinimäki, K.; Peters, G.; Dahlbo, H.; Perry, P.; Rissanen, T.; Gwilt, A. Author Correction: The Environmental Price of Fast Fashion. *Nat. Rev. Earth Environ.* **2020**, *1*, 278. [CrossRef]
4. Buzzo, A.; Abreu, M.J. Fast fashion, fashion brands & sustainable consumption. In *Textile Science and Clothing Technology*; Muthu, S.S., Ed.; Springer: Singapore, 2019; pp. 1–17. [CrossRef]
5. Islam, M.M.; Perry, P.; Gill, S. Mapping Environmentally Sustainable Practices in Textiles, Apparel and Fashion Industries: A Systematic Literature Review. *J. Fash. Mark. Manag.* **2020**, *25*, 331–353. [CrossRef]
6. Li, J.; Leonas, K.K. Sustainability Topic Trends in the Textile and Apparel Industry: A Text Mining-based Magazine Article Analysis. *J. Fash. Mark. Manag. Int. J.* **2021**, *26*, 67–87. [CrossRef]
7. Peters, G.; Li, M.; Lenzen, M. The Need to Decelerate Fast Fashion in a Hot Climate-A Global Sustainability Perspective on the Garment Industry. *J. Clean. Prod.* **2021**, *295*, 126390. [CrossRef]
8. Chang, H.J.; Watchravesringkan, K.T. Who Are Sustainably Minded Apparel Shoppers? An Investigation to the Influencing Factors of Sustainable Apparel Consumption. *Int. J. Retail Distrib. Manag.* **2018**, *46*, 148–162. [CrossRef]
9. Henninger, C.E.; Alevizou, P.J.; Goworek, H.; Ryding, D. *Sustainability in Fashion: A Cradle to Upcycle Approach*; Springer: Berlin/Heidelberg, Germany, 2017. [CrossRef]
10. Rahman, O.; Koszewska, M. A Study of Consumer Choice Between Sustainable and Non-sustainable Apparel Cues in Poland. *J. Fash. Mark. Manag. Int. J.* **2020**, *24*, 213–234. [CrossRef]
11. Chaturvedi, P.; Kulshreshtha, K.; Tripathi, V. Investigating the Determinants of Behavioral Intentions of Generation Z for Recycled Clothing: An Evidence from a Developing Economy. *Young Consum.* **2020**, *21*, 403–417. [CrossRef]
12. Kim, I.; Jung, H.J.; Lee, Y. Consumers' Value and Risk Perceptions of Circular Fashion: Comparison between Secondhand, Upcycled, and Recycled Clothing. *Sustainability* **2021**, *13*, 1208. [CrossRef]
13. Lee, S.E.; Jung, H.J.; Lee, K.H. Motivating Collaborative Consumption in Fashion: Consumer Benefits, Perceived Risks, Service Trust, and Usage Intention of Online Fashion Rental Services. *Sustainability* **2021**, *13*, 1804. [CrossRef]
14. Dabija, D.C.; Bejan, B.M.; Dinu, V. How Sustainability Oriented is Generation Z in Retail? A Literature Review. *Transform. Bus. Econ.* **2019**, *18*, 140–155.
15. Gazzola, P.; Pavione, E.; Pezzetti, R.; Grechi, D. Trends in the Fashion Industry. The Perception of Sustainability and Circular Economy: A Gender/Generation Quantitative Approach. *Sustainability* **2020**, *12*, 2809. [CrossRef]
16. Ivanova, O.; Flores-Zamora, J.; Khelladi, I.; Ivanaj, S. The Generational Cohort Effect in the Context of Responsible Consumption. *Manag. Decis.* **2019**, *57*, 1162–1183. [CrossRef]
17. Kamenidou, I.E.; Stavrianea, A.; Bara, E.Z. Generational Differences toward Organic Food Behavior: Insights from Five Generational Cohorts. *Sustainability* **2020**, *12*, 2299. [CrossRef]
18. Brydges, T. Closing the Loop on Take, Make, Waste: Investigating Circular Economy Practices in the Swedish Fashion Industry. *J. Clean. Prod.* **2021**, *293*, 126245. [CrossRef]
19. Rathinamoorthy, R. Circular fashion. In *Circular Economy in Textiles and Apparel*; Muthu, S.S., Ed.; Woodhead Publishing: Sawston, UK, 2019; pp. 13–48.

20. Dahlbo, H.; Aalto, K.; Eskelinen, H.; Salmenperä, H. Increasing Textile Circulation—Consequences and Requirements. *Sustain. Prod. Consum.* **2017**, *9*, 44–57. [CrossRef]
21. Koszewska, M. Circular Economy—Challenges for the Textile and Clothing Industry. *Autex Res. J.* **2018**, *18*, 337–347. [CrossRef]
22. Colasante, A.; D'Adamo, I. The Circular Economy and Bioeconomy in the Fashion Sector: Emergence of a "Sustainability Bias". *J. Clean. Prod.* **2021**, *329*, 129774. [CrossRef]
23. Mukendi, A.; Henninger, C.E. Exploring the Spectrum of Fashion Rental. *J. Fash. Mark. Manag.* **2020**, *24*, 455–469. [CrossRef]
24. Silva, S.C.; Santos, A.; Duarte, P.; Vlačić, B. The Relevant Role of Social Embarrassment, Sustainability, Familiarity, and Perception of Hygiene on Secondhand Clothing Experience. *Int. J. Retail Distrib. Manag.* **2021**, *49*, 717–734. [CrossRef]
25. Vehmas, K.; Raudaskoski, A.; Heikkilä, P.; Harlin, A.; Mensonen, A. Consumer Attitudes and Communication in Circular Fashion. *J. Fash. Mark. Manag.* **2018**, *22*, 286–300. [CrossRef]
26. Machado, M.A.D.; de Almeida, S.O.; Bollick, L.C.; Bragagnolo, G. Second-hand Fashion Market: Consumer Role in Circular Economy. *J. Fash. Mark. Manag.* **2019**, *23*, 382–395. [CrossRef]
27. Clube, R.K.; Tennant, M. Exploring Garment Rental as a Sustainable Business Model in the Fashion Industry: Does Contamination Impact the Consumption Experience? *J. Consum. Behav.* **2020**, *19*, 359–370. [CrossRef]
28. Shrivastava, A.; Jain, G.; Kamble, S.S.; Belhadi, A. Sustainability through Online Renting Clothing: Circular Fashion Fueled by Instagram Micro-celebrities. *J. Clean. Prod.* **2021**, *278*, 123772. [CrossRef]
29. Kollmuss, A.; Agyeman, J. Mind the Gap: Why Do People Act Environmentally and What are the Barriers to Pro-environmental Behavior? *Environ. Educ. Res.* **2002**, *8*, 239–260. [CrossRef]
30. Schlegelmilch, B.B.; Bohlen, G.M.; Diamantopoulos, A. The Link between Green Purchasing Decisions and Measures of Environmental Consciousness. *Eur. J. Mark.* **1996**, *30*, 35–55. [CrossRef]
31. Zelezny, L.C.; Schultz, P.W. Promoting Environmentalism. *J. Soc. Issues.* **2000**, *56*, 365–372. [CrossRef]
32. Golob, U.; Kronegger, L. Environmental Consciousness of European Consumers: A Segmentation-based Study. *J. Clean. Prod.* **2019**, *221*, 1–9. [CrossRef]
33. Kautish, P.; Paul, J.; Sharma, R. The Moderating Influence of Environmental Consciousness and Recycling Intentions on Green Purchase Behavior. *J. Clean. Prod.* **2019**, *228*, 1425–1436. [CrossRef]
34. Kautish, P.; Sharma, R. Determinants of Pro-environmental Behavior and Environmentally Conscious Consumer Behavior: An Empirical Investigation from Emerging Market. *Bus. Strategy Dev.* **2020**, *3*, 112–127. [CrossRef]
35. Lin, S.T.; Niu, H.J. Green Consumption: Environmental Knowledge, Environmental Consciousness, Social Norms, and Purchasing Behavior. *Bus. Strategy Environ.* **2018**, *27*, 1679–1688. [CrossRef]
36. Kautish, P.; Sharma, R. Study on Relationships among Terminal and Instrumental Values, Environmental Consciousness and Behavioral Intentions for Green Products. *J. Indian Bus. Res.* **2018**, *13*, 1–29. [CrossRef]
37. Zhang, Y.; Xiao, C.; Zhou, G. Willingness to Pay a Price Premium for Energy-saving Appliances: Role of Perceived Value and Energy Efficiency Labeling. *J. Clean. Prod.* **2020**, *242*, 118555. [CrossRef]
38. Wang, J.; Pham, T.L.; Dang, V.T. Environmental Consciousness and Organic Food Purchase Intention: A Moderated Mediation Model of Perceived Food Quality and Price Sensitivity. *Int. J. Environ. Res. Public Health* **2020**, *17*, 850. [CrossRef]
39. Hasan, M.M.; Cai, L.; Ji, X.; Ocran, F.M. Eco-friendly Clothing Market: A Study of Willingness to Purchase Organic Cotton Clothing in Bangladesh. *Sustainability* **2022**, *14*, 4827. [CrossRef]
40. Souza, J.D.L.; Tondolo, V.A.G.; Sarquis, A.B.; Longaray, A.A.; Tondolo, R.D.R.P.; Costa, L.M.D. Effect of Perceived Value, Risk, Attitude and Environmental Consciousness on the Purchase Intention. *Int. J. Bus. Environ.* **2020**, *11*, 11–31. [CrossRef]
41. Szabo, S.; Webster, J. Perceived Greenwashing: The Effects of Green Marketing on Environmental and Product Perceptions. *J. Bus. Ethics* **2021**, *171*, 719–739. [CrossRef]
42. Zeithaml, V.A. Consumer Perceptions of Price, Quality, and Value: A Means-end Model and Synthesis of Evidence. *J. Mark.* **1988**, *52*, 2–22. [CrossRef]
43. Sheth, J.N.; Newman, B.I.; Gross, B.L. Why We Buy What We Buy: A Theory of Consumption Values. *J. Bus. Res.* **1991**, *22*, 159–170. [CrossRef]
44. Bielawska, K.; Grebosz-Krawczyk, M. Consumers' Choice Behaviour toward Green Clothing. *Eur. Res. Stud. J.* **2021**, *24*, 238–256. [CrossRef]
45. Baek, E.; Oh, G.E.G. Diverse Values of Fashion rental Service and Contamination Concern of Consumers. *J. Bus. Res.* **2021**, *123*, 165–175. [CrossRef] [PubMed]
46. Chun, E.; Jiang, W.; Yu, J.; Ko, E. Perceived Consumption Value, Pro-environmental Belief, Attitude, eWOM, and Purchase Intention toward Upcycling Fashion Products. *Fash. Text. Res. J.* **2018**, *20*, 177–190. [CrossRef]
47. Chi, T.; Ganak, J.; Summers, L.; Adesanya, O.; McCoy, L.; Liu, H.; Tai, Y. Understanding Perceived Value and Purchase Intention toward Eco-friendly Athleisure Apparel: Insights from US Millennials. *Sustainability* **2021**, *13*, 7946. [CrossRef]
48. Dowling, G.R.; Staelin, R. A Model of Perceived Risk and Intended Risk-handling Activity. *J. Consum. Res.* **1994**, *21*, 119–134. [CrossRef]
49. Featherman, M.S.; Pavlou, P.A. Predicting E-services Adoption: A Perceived Risk Facets Perspective. *Int. J. Hum. Comput. Stud.* **2003**, *59*, 451–474. [CrossRef]
50. Kim, D.J.; Ferrin, D.L.; Rao, H.R. A Trust-based Consumer Decision-making Model in Electronic Commerce: The Role of Trust, Perceived Risk, and Their Antecedents. *Decis. Support Syst.* **2008**, *44*, 544–564. [CrossRef]

51. Jacoby, J.; Kaplan, L.B. The Components of Perceived Risk. In *SV—Proceedings of the Annual Conference of the Association for Consumer Research*; Venkatesan, M., Ed.; Association for Consumer Research: Chicago, IL, USA, 1972; pp. 382–393.
52. Mitchell, V.W. Consumer Perceived Risk: Conceptualisations and Models. *Eur. J. Mark.* **1999**, *33*, 163–195. [CrossRef]
53. Peter, J.P.; Tarpey Sr, L.X. A Comparative Analysis of Three Consumer Decision Strategies. *J. Consum. Res.* **1975**, *2*, 29–37. [CrossRef]
54. Pires, G.; Stanton, J.; Eckford, A. Influences on the Perceived Risk of Purchasing Online. *J. Consum. Behav. Int. Res. Rev.* **2004**, *4*, 118–131. [CrossRef]
55. Qalati, S.A.; Vela, E.G.; Li, W.; Dakhan, S.A.; Hong Thuy, T.T.; Merani, S.H. Effects of Perceived Service Quality, Website Quality, and Reputation on Purchase Intention: The Mediating and Moderating Roles of Trust and Perceived Risk in Online Shopping. *Cogent Bus. Manag.* **2021**, *8*, 1869363. [CrossRef]
56. Yoo, F.; Jung, H.J.; Oh, K.W. Motivators and Barriers for Buying Intention of Upcycled Fashion Products in China. *Sustainability* **2021**, *13*, 2584. [CrossRef]
57. Armstrong, C.M.; Niinimäki, K.; Lang, C.; Kujala, S. A Use-oriented Clothing Economy? Preliminary Affirmation for Sustainable Clothing Consumption Alternatives. *Sustain. Dev.* **2016**, *24*, 18–31. [CrossRef]
58. Brydges, T.; Retamal, M.; Hanlon, M. Will COVID-19 Support the Transition to a More Sustainable Fashion Industry? *Sustain. Sci. Pract. Policy* **2020**, *16*, 298–308. [CrossRef]
59. Charnley, F.; Knecht, F.; Muenkel, H.; Pletosu, D.; Rickard, V.; Sambonet, C.; Schneider, M.; Zhang, C. Can Digital Technologies Increase Consumer Acceptance of Circular Business Models? The Case of Second Hand Fashion. *Sustainability* **2022**, *14*, 4589. [CrossRef]
60. Park, H.H.; Choo, T.G. The Influence of Perceived Risk of Up-cycling Fashion Product on Trust, Purchase Intention and Recommendation Intention. *Fash. Text. Res. J.* **2015**, *17*, 216–226. [CrossRef]
61. Mannheim, K. The Problem of Generations. *Psychoanal. Revie* **1970**, *57*, 378–404.
62. Alwin, D.F.; McCammon, R.J. Generations, Cohorts, and Social Change. In *Handbook of the Life Course*; Mortimer, J.T., Shanahan, M.J., Eds.; Springer: Boston, MA, USA, 2003; pp. 23–49. [CrossRef]
63. Edmunds, J.; Turner, B.S. *Generations, Culture and Society*; Open University Press: Buckingham, UK, 2002.
64. McCrindle, M.; Wolfinger, E. *The ABC of XYZ: Understanding the Global Generations*; University of New South Wales: Sydney, NSW, Australia, 2009.
65. Bolin, G. Generational Analysis as a Methodological Approach to Study Mediatised Social Change. In *Digital Technologies and Generational Identity*; Taipale, S., Wilska, T.-A., Gilleard, C., Eds.; Routledge: London, UK, 2017; pp. 23–36.
66. Dimock, M. Defining Generations: Where Millennials End and Generation Z Begins. *Pew Res. Cent.* **2019**, *17*, 1–7.
67. Dorie, A.; Loranger, D. The Multi-generation: Generational Differences in Channel Activity. *Int. J. Retail Distrib. Manag.* **2020**, *48*, 395–416. [CrossRef]
68. Thangavel, P.; Pathak, P.; Chandra, B. Consumer Decision-making Style of Gen Z: A Generational Cohort Analysis. *Glob. Bus. Rev.* **2022**, *23*, 710–728. [CrossRef]
69. Ladhari, R.; Gonthier, J.; Lajante, M. Generation Y and Online Fashion Shopping: Orientations and Profiles. *J. Retail. Consum. Serv.* **2019**, *48*, 113–121. [CrossRef]
70. Lissitsa, S.; Kol, O. Four Generational Cohorts and Hedonic M-shopping: Association between Personality Traits and Purchase Intention. *Electron. Commer. Res.* **2021**, *21*, 545–570. [CrossRef]
71. Sharma, S.; Singh, G.; Pratt, S. Modeling the Multi-dimensional Facets of Perceived Risk in Purchasing Travel Online: A Generational Analysis. *J. Qual. Assur. Hosp. Tour.* **2022**, *23*, 539–567. [CrossRef]
72. Goldring, D.; Azab, C. New Rules of Social Media Shopping: Personality Differences of US Gen Z versus Gen X Market Mavens. *J. Consum. Behav.* **2020**, *20*, 884–897. [CrossRef]
73. Çera, G.; Pagria, I.; Khan, K.A.; Muaremi, L. Mobile Banking Usage and Gamification: The Moderating Effect of Generational Cohorts. *J. Syst. Inf. Technol.* **2020**, *12*, 243–263. [CrossRef]
74. Ismail, A.R.; Nguyen, B.; Chen, J.; Melewar, T.C.; Mohamad, B. Brand Engagement in Self-concept (BESC), Value Consciousness and Brand Loyalty: A Study of Generation Z Consumers in Malaysia. *Young Consum.* **2021**, *22*, 112–130. [CrossRef]
75. Yamane, T.; Kaneko, S. Is the Younger Generation a Driving Force toward Achieving the Sustainable Development Goals? Survey Experiments. *J. Clean. Prod.* **2021**, *292*, 125932. [CrossRef]
76. Kim, J. Luxury Fashion Goods Ownership and Collecting Behavior in an Omni-channel Retail Environment: Empirical Findings from Affluent Consumers in the US. *Res. J. Text. Appar.* **2019**, *23*, 212–231. [CrossRef]
77. Liang, J.; Xu, Y. Second-hand Clothing Consumption: A Generational Cohort Analysis of the Chinese Market. *Int. J. Consum. Stud.* **2018**, *42*, 120–130. [CrossRef]
78. Kyriazos, T.A. Applied psychometrics: Sample size and sample power considerations in factor analysis (EFA, CFA) and SEM in general. *Psychology* **2018**, *9*, 2207. [CrossRef]
79. Bryant, F.B.; Yarnold, P.R.; Michelson, E.A. Statistical Methodology: VIII. Using Confirmatory Factor Analysis (CFA) in Emergency Medicine Research. *Acad. Emerg. Med.* **1999**, *6*, 54–66. [CrossRef] [PubMed]
80. Hair, J.F., Jr.; Sarstedt, M.; Hopkins, L.; Kuppelwieser, V.G. Partial least squares structural equation modeling (PLS-SEM): An emerging tool in business research. *Eur. Bus. Rev.* **2014**, *26*, 106–121. [CrossRef]

81. Hair, J.F.; Anderson, R.E.; Babin, B.J.; Black, W.C. *Multivariate Data Analysis: A Global Perspective*; Pearson: Upper Saddle River, NJ, USA, 2010.
82. Cronbach, L.J. Coefficient Alpha and the Internal Structure of Tests. *Psychometrika* **1951**, *16*, 297–334. [CrossRef]
83. Hu, L.T.; Bentler, P.M. Cutoff Criteria for Fit Indexes in Covariance Structure Analysis: Conventional Criteria Versus New Alternatives. *Struct. Equ. Modeling Multidiscip. J.* **1999**, *6*, 1–55. [CrossRef]

sustainability MDPI

Article

The Revitalization Design of Regional Ethnic Cultural Capital in a Sustainable Perspective: The Case of Traditional Chinese Garment Yunjian

Han Chen [1], Han Xu [1], Yudian Zhang [2] and Leilei Jia [1,*]

1 School of Design, Jiangnan University, Wuxi 214122, China; chenhanisaac@163.com (H.C.); xh502443490@foxmail.com (H.X.)
2 School of Fashion Design and Engineering, Zhejiang Sci-Tech University, Hangzhou 310018, China; zoey0712@126.com
* Correspondence: jialeilei@jiangnan.edu.cn

Abstract: The formation of the globalization pattern of regional ethnic culture brings new opportunities for the sustainable development of cultural capital. In order to solve the problem of mismatch between the high development demand of regional ethnic costumes and their traditional design methods, this study proposes an innovative design method for regional ethnic traditional costumes, taking the Yunjian, a cloud-shaped sash-like shoulder garment as an example. Firstly, we deeply explored the historical and cultural connotation of the Yunjian through fieldwork and literature research. Then we sorted out and refined its elements, colors and structural features. We classified the patterns according to the composition forms and clarified the path of pattern creation based on fractal theory. We combined the representative elements of Yunjian with the fractal algorithm and proposed the design model of Yunjian fractal pattern. Finally we verified the feasibility of the proposed design method through design practice. The results show that the Yunjian garment created by the method of this paper can better reflect the characteristics of fashion art and ethnic culture.

Keywords: sustainability; digital design; regional ethnic culture; cultural capital; revitalization; Yunjian; traditional garment

Citation: Chen, H.; Xu, H.; Zhang, Y.; Jia, L. The Revitalization Design of Regional Ethnic Cultural Capital in a Sustainable Perspective: The Case of Traditional Chinese Garment Yunjian. *Sustainability* **2022**, *14*, 8090. https://doi.org/10.3390/su14138090

Academic Editors: Hafeezullah Memon, Xiaoke Jin, Tian Wei and Chengyan Zhu

Received: 20 April 2022
Accepted: 28 June 2022
Published: 1 July 2022

1. Introduction

Ethnic traditional costumes are not only intangible cultural heritage carrying the spirit and centripetal force of national culture, but also cultural capital and wealth containing great revitalization value [1]. With the formation of the globalization of regional ethnic culture, Chinese regional fashion has become part of the global popular style [2]. The Yunjian is a unique costume item placed on the shoulder in ancient China, which has a rich art form and profound cultural heritage, and has a prominent position and high research value in the Han Chinese costume culture. The culture of traditional Yunjian garments is constantly being explored and moving towards the world fashion stage. This market-oriented fashion phenomenon has transformed the regional ethnic "cultural heritage" into "cultural capital" that can be transferred, and brought opportunities for sustainable development to the regional ethnic traditional culture, which was facing a crisis of discontinuity. The worldwide popularity of Chinese traditional costume styles has put forward higher demands on the design and development methods. With the development of artistic aesthetics and value experience, the emerging culture of ethnic fashion has changed the clothing consumption behavior of modern people. More scientific and efficient innovative design methods that take into account ethnic characteristics and modern fashion need to be researched and developed.

The rapid development of science and technology can give new possibilities to artistic expression. Research on the innovative design of costumes is not uncommon, but most of the studies focus on material and functional innovations. Few researchers have introduced computer algorithms into the innovative design of ethnic costumes with remarkable results. Nowadays, fractal algorithms have received a lot of attention from researchers. Mathematician Mandelbrot was the first to use the term fractal to describe a figure or phenomenon with similar properties. Fractal patterns contain self-similarity between the whole and the parts, with the possibility of infinite iteration and repetition. Currently, there is a lot of literature applying fractal theory to modern design. A Laubsted explored the mathematical logic behind fractal geometry and fractal art [3]. JZ Zhang et al. used fractal geometric methods to analyze traditional furniture shapes and decorative arts, and concluded a design method applicable to furniture shapes in practice [4]. JC Driscoll used fractal dimension and Vollendorf method to create unique fractal compositions and tried to apply them in architectural design development [5]. However, fewer research results have applied fractal theory to the design of regional ethnic traditional costumes with fractal art features and concluded a proven systematic approach. The Yunjian costumes created by ancient Chinese ancestors are not lacking in fractal art features, but have not been formed into a scientific and systematic pattern design system based on fractal theory. This study proposes an innovative method for Yunjian costumes based on the Newtonian function model, aiming to expand a new way of inheritance, innovation and application of national costumes.

2. Research Background

Yunjian is draped over the front and back of the shoulder. In Chinese, the meaning of "Yunjian" is "a colorful cloud on the shoulder", because people wear it as beautiful as a colorful cloud. Yunjian was developed during the Five Dynasties and Ten Kingdoms period (907–979 AD). At that time, it had no decorative meaning. It was mainly used to protect the collar and prevent the hair from staining the clothes. During the Qing dynasty, the decorative meaning of the Yunjian was greater than its practical function, and it became an indispensable decorative garment for women as part of their wedding ceremony. The Yunjian was rich in color, beautifully crafted, and had a very distinctive structure and design. In this section, we will introduce and analyze the pattern themes, color rules and structural features of the Yunjian.

2.1. Pattern Theme

Most of the patterns in traditional Chinese folk costumes come from life and are influenced by the background of the times, people's beliefs and local customs [6]. The common pattern images in Yunjian can be broadly divided into plant, animal, character, artifact and text categories. The common plant themes are pomegranates, grapes, pumpkins, loquats, and so on. The Chinese pronunciation of "many seeds" is the same as "many sons", so the Chinese people give pomegranate an auspicious meaning of many sons and many blessings. People also like the round shape and golden skin of pumpkins, which they believe symbolizes completeness and wealth. The long, curved and tough vines of the pumpkin symbolize longevity and family prosperity. In addition to the melon and fruit patterns, the lotus flower, which symbolizes harmony between husband and wife, and the peony flower, which symbolizes wealth and elegance, are also common motifs in the Yunjian. Animal motifs include bats, butterflies, Chinese unicorn, cranes, dragons and phoenixes. The Chinese pronunciation of "bat" is the same as "happiness", so bat patterns are loved by Chinese people as a symbol of happiness. The Chinese unicorn is a mythological animal in China. In mythology, the appearance of the Chinese unicorn usually means that good things are about to happen. As a symbol of hope and good luck, the Chinese unicorn pattern is often found in the clothing of Chinese nobility. Ruyi, coral, pearl, vase, ivory and other objects are also common subjects for auspicious patterns. Word patterns are usually derived from the deformation of words with auspicious meanings and good wishes, such as Shou (longevity) word patterns, Wan (ten thousand) word patterns and Xi (joyful) word patterns. Character patterns are mainly chosen from the subjects of maids, babies, couples and so on. These characters usually form lively scenes of life, such as farming, marriage, boat trips, flower viewing, children playing, etc. The common patterns in the five main themes and their line drawing extracts are shown in Figure 1.

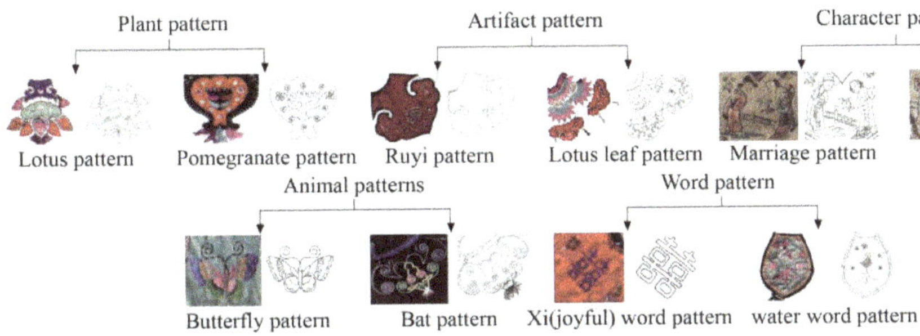

Figure 1. Representative Yunjian pattern themes and line drawings.

2.2. Color Rules

The color scheme of the Yunjian garment follows the two traditional color rules of "Homochromatic system" and "Contrast color system" in ancient costumes. Yunjian costumes commonly use five colors (white, cyan, black, red, yellow) as the main colors for large areas, while using related average colors for auxiliary embellishment. The same color series of Yunjian focuses on the harmony and unity of colors. Contrasting colors are used to adjust the color according to brightness, purity and area size to achieve a lively visual effect. There has been related literature that classifies Yunjian colors into three categories according to the rules of color usage, which are primary colors, auxiliary colors and decorative colors. Primary colors of Yunjian mainly refers to the fabric base color. Auxiliary colors refer to the embroidery, applique and other process colors to meet the overall or local modeling needs of the garment. Auxiliary colors usually have no obvious effect on the visual effect of Yunjian colors. The decorative colors mainly refer to the complicated, small block of gorgeous colors in the Yunjian costume.

2.3. Structural Features

Although the Yunjian is used as a costume accessory most of the time, it is worn in the visual center of the body. Therefore, it plays an important role in the overall structure of the costume. The modeling of Yunjian has unique artistic charm and regularity, its structure can be divided into hierarchical structure and external structure [7].

The hierarchical structure of the Yunjian can be divided into four categories: the single-piece structure, the multi-layer structure, the connected structure and the mixed structure. Different structure of Yunjian gives people completely different style feeling. The earliest Yunjians were basically symmetrical single-piece structures that were simple, lightweight, and suitable for children to wear. Modern children's bibs are developed from this. The multi-layer structure are single pieces of shaped Yunjians laminated together, this kind of design is ordered and looks richer. The connected structure Yunjian is light, it consists of the overall shape breaking down and cutting off the excess padding parts to form the skeleton, and then putting it together in an orderly manner. The connected structure Yunjian is more complex, it not only requires fine single piece, but also pays more attention to the overall arrangement effect. The mixed structured Yunjian combines all of the above forms and is more elegant and refined.

The appearance structure of the Yunjian can be divided into two types: centrifugal structure and centripetal structure. The centrifugal structure is centered on the neckline and spreads out in a straight line to lead the sight outward. The centripetal structure is a contraction from the periphery to the center, and the visual center is more focused. When the Yunjian is worn on the body, its shape outline can be divided into crescent-shaped, semi-circular and T-shaped, as shown in Figure 2.

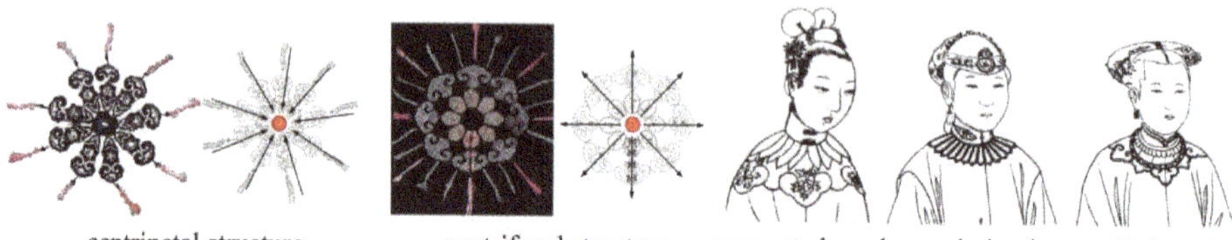

Figure 2. Cloud shoulder structure and shape.

Analyzed from the overall visual effect, the Yunjian has rich formal aesthetic characteristics. The modeling of the Yunjian reflects an orderly aesthetic, with a sense of symmetry and balance between the individual pieces. The same single-piece shape and form are used repeatedly, while the decorative elements are varied to give it a sense of rhythm.

3. Research Methods

3.1. Elemental Derivation Methods

This study focuses on the derivative design of pattern elements through the morphological displacement method. The morphological displacement method can summarize the derivation process of the elements into a regular design language [8]. The basic elements of the Yunjian pattern are moved in a directional way by the morphological displacement method, and the size of the elements and the logic of the displacement trajectory can be changed during the movement. We take the classical element of the Yunjian pattern, Ru Yi pattern, as an example, and show 8 predefined rules of morphological displacement derivation in Figure 3. The combined use of different rules and their derivation results are shown in Figure 4.

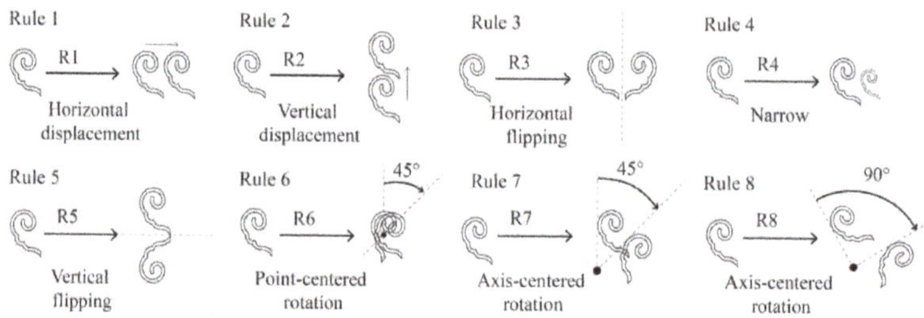

Figure 3. Elemental derivation rules.

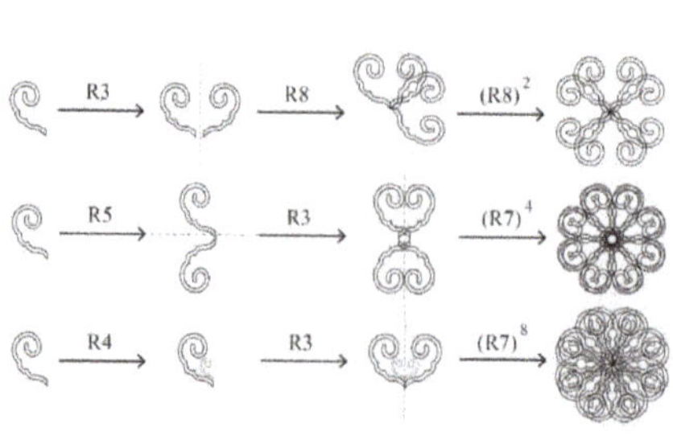

Figure 4. Example of the application of derivative rules.

3.2. Color Extraction Method

3.2.1. K-Means Clustering Algorithm

Color design is one of the most direct ways to convey the aesthetics of a garment [9]. Clustering algorithms can merge and extract colors from existing patterns based on a preset range of color fields. k-means is an iterative solving algorithm for cluster analysis [10]. It uses a circular positioning technique to translate different intervals to improve the quality of interval delineation. The algorithm selects K objects as initial clustering centers. The algorithm assigns each object to the closest cluster center and calculates the distance between each object and the center. When all objects are assigned, the cluster centers are recalculated based on the existing objects in the class. Applying the k-means clustering algorithm to Yunjian pattern color extraction and making color cards can analyze the color distribution more accurately and quantitatively [11,12].

3.2.2. Color Analysis Method Based on HSB Model and L*a*b* Model

Compared with the common RGB and CMYK color models, L*a*b* is a description model with a wider range of colors [13]. Considering that Yunjian garment focuses more on subjective color imagery feeling in artistic expression, we introduced HSB model, which is closer to human sensory intuition based on L*a*b* color model. In this research, the two color models L*a*b* and HSB are combined to analyze color. We extracted the RGB values of the colors in Adobe photoshop, and obtained the L*a*b* color axes by converting the RGB values through the XYZ intermediate parameters. The specific calculation is shown in Equations (1) and (2).

Converts RGB values to XYZ values:

$$\begin{bmatrix} X \\ Y \\ Z \end{bmatrix} = \begin{bmatrix} 0.412\ 453 & 0.357\ 580 & 0.180\ 423 \\ 0.212\ 671 & 0.715\ 160 & 0.072\ 169 \\ 0.019\ 334 & 0.119\ 193 & 0.950\ 227 \end{bmatrix} \begin{bmatrix} R \\ G \\ B \end{bmatrix} \tag{1}$$

Conversion of XYZ values to L*a*b* values.

$$\begin{aligned} L &= 116f\left(\frac{Y}{Y_n}\right) - 16 \\ a &= 500\left[f\left(\frac{X}{X_n}\right) - f\left(\frac{Y}{Y_n}\right)\right] \\ b &= 200\left[f\left(\frac{Y}{Y_n}\right) - f\left(\frac{Z}{Z_n}\right)\right] \end{aligned} \tag{2}$$

$X_n = 95.047$, $Y_n = 100.000$, $Z_n = 108.883$. The a and b values of the L*a*b* model parameters are used as the horizontal and vertical coordinates to plot the color distribution scatter diagram. In addition, we extracted the HSB values of the colors through Adobe photoshop and made a histogram for statistical analysis [14]. L*a*b* color distribution scatter diagram and HSB histogram can show the color usage rules of the Yunjian garment more intuitively.

3.3. Pattern Design Method

The overall structure of the Yunjian garment is symmetrical, and the combination of each single piece has regularity and repetition. Such a symmetrical structure makes most of the Yunjian patterns have the fractal characteristics of self-similarity and repetition iteration. Based on this feature, this study uses the fractal art principle of escape time algorithm to innovate the design of Yunjian pattern. The escape time algorithm is a computer graphics generation algorithm based on iterative function f_n. Assuming that f is a variational iterative function, $f_0(x) = x$, $f_1(x) = f(x)$, $f_2(x) = f(f(x))$, $f_{n+1}(x) = f(f_n(x))$, $n = 0, 1, 2 \cdots$. The computer establishes a coordinate system in the graphical determination region, brings the pixel coordinates in the region into the iteration function in turn, and judges whether the pixel coordinates converge under the given iteration function. Because of the different number of iterations for different convergence point divergence points, we can get the pattern with rich structure and conform to the preset parameter fractal characteristics. Based on the symmetrical and divergent skeletal structure of the Yunjian pattern, we chose the Newton iterative fractal prototype that fits this feature. The structure of the Newton iterative fractal prototype can be expressed by the formula as shown in Equation (3).

$$Z_{n+1} = Z_n - \frac{F(Z_n)}{F'(Z_n)}, (n = 1, 2, \cdots) \tag{3}$$

$F(z)$ is a complex function, $z = x + yi$, x, y are variables, and i is an imaginary number. In the specific calculation, generally $F(z) = z^p - 1$, and p is a positive integer greater than 2, so the iterative formula can be rewritten as Equation (4).

$$Z_{n+1} = \frac{(p-1)z_n^p + 1}{p z_n^{p-1}}, (n = 1, 2, \cdots) \tag{4}$$

We can change the p-value in Equation (4) according to the specific design requirements to obtain different evolutionary results based on Newton's iterative fractal prototype to be used as the key design material for the Yunjian garment pattern.

3.4. Evaluation Methodology

This study uses a rank sum algorithm to determine the weights of each indicator in the evaluation system. We invited an expert panel of senior related practitioners to score the importance of the evaluation indicators [15]. Before calculating the weights, we conducted consistency tests on the expert opinions, as shown in Equations (5)–(7).

$$X^2 = m(n-1)w \tag{5}$$

$$w = S/\left[(12)^{-1} m^2 \left(n^3 - n\right)\right] \tag{6}$$

$$S = \sum R_j^2 - \left(\sum R_j\right)^2 / n \tag{7}$$

X^2 is the statistic, R_j is the rank sum of the jth indicator, m is the number of experts, and n is the number of indicators. By substituting Equations (6) and (7) into Equation (5), we can check whether the experts' opinions are consistent by calculating and comparing.

The specific calculation of the evaluation index weights is shown in Equation (8).

$$a_j = 2\left[m(1+n) - R_j\right] / [mn(1+n)] \tag{8}$$

After determining the evaluation indexes, evaluation item weights and evaluation scales, we adopt a questionnaire survey conducted both online and offline to conduct a comprehensive evaluation of the design practice results [16].

4. Design Practice

4.1. Yunjian Pattern Derivation

We extracted several common elements from the Yunjians and combined them into a new pattern. Then we apply the morphological displacement method to mirror and rotate the elements to generate a new pattern with prototypical features.The pattern can be used as a new basic element unit. We have completed several representative Yunjian element units by this method, as shown in Figure 5. These element units will be further used in the subsequent design.

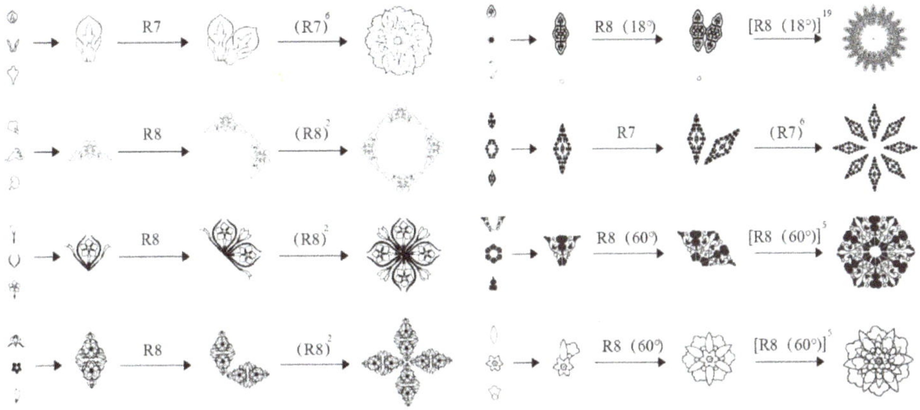

Figure 5. Basic pattern unit generation process.

4.2. *Yunjian Color Design*

In this research, five color intervals were created using the K-means clustering algorithm, and the clustered colors were batch normalized and averaged by the color distance value ΔE. The image cluster partition of the 5 groups of images is set to consist of all colors within $\Delta E \leq 5$ pixels. To achieve the desired effect, we obtain the color share of the Yunjian pattern by presetting the color separation value with the distance value and multiple clustering calculations. Then we clarify the color collection area and size by image cluster partitioning, as shown in Table 1. HSB represents hue, saturation, brightness and RGB represents the colors of red, green and blue channels [17]. The RGB color extraction results of each group of Yunjian patterns are shown in Table 2.

Table 1. Yunjian color extraction.

Main Color	Pattern	Image Cluster Partitioning	Color Ratio (%)
White series		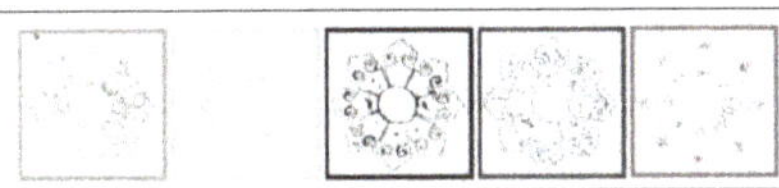	25.7 24.6 20.7 15.5 13.5
Cyan series			23.9 22.4 21.3 20.0 12.4
Black series		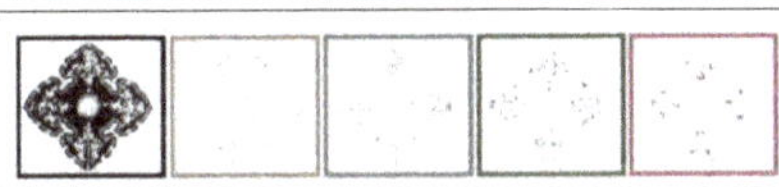	81.5 5.3 5.1 4.7 3.4
Red series	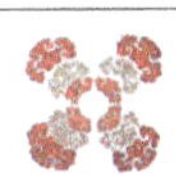	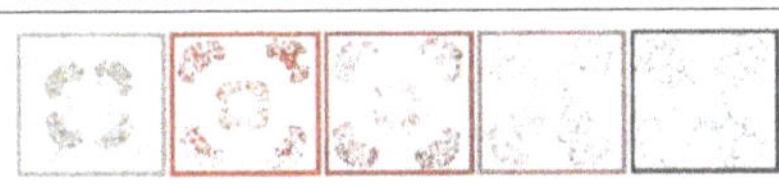	31.2 22.2 19.4 17.3 9.9
Yellow series			33.1 22.8 18.4 17.9 7.8

Table 2. HSB values and RGB values of colors.

White Series					
RGB	194/177/169	227/216/208	44/41/46	77/79/77	145/124/123
HSB	19/13/76	25/8/89	276/11/18	120/3/31	3/15/57
Cyan series					
RGB	116/138/132	150/177/165	47/36/40	156/123/117	102/85/89
HSB	164/16/54	153/15/69	338/13/28	9/25/61	346/17/40
Black series					
RGB	12/7/8	165/142/121	122/129/128	87/106/73	172/96/121
HSB	348/42/5	29/27/65	171/5/51	95/31/42	340/44/67
Red series					
RGB	191/176/166	208/75/70	181/88/87	180/128/127	94/92/93
HSB	24/13/75	2/66/82	1/52/71	1/29/71	330/2/37
Yellow series					
RGB	234/185/138	178/139/112	60/43/62	113/90/83	209/86/77
HSB	29/41/92	25/37/70	294/31/24	14/27/44	4/63/82

From Figure 6, it can be seen that most of the extracted colors are warm colors, and a small amount of colors are distributed in the neutral and cold color regions. In terms of saturation, most of the colors are low-saturation colors, a small number of medium-saturation colors, and no high-saturation colors appear. In terms of brightness, the color distribution is relatively uniform, and there are few low brightness colors.

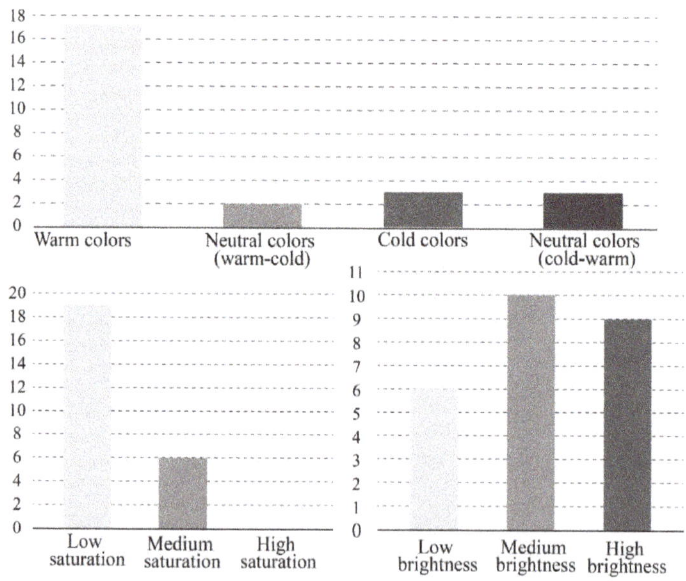

Figure 6. Number of different hues, saturation and lightness in the color extraction results.

According to Equation (2), we plotted the scatter diagram of color distribution of five color families with the a value as the horizontal coordinate and the b value as the vertical coordinate, as shown in Figure 7.

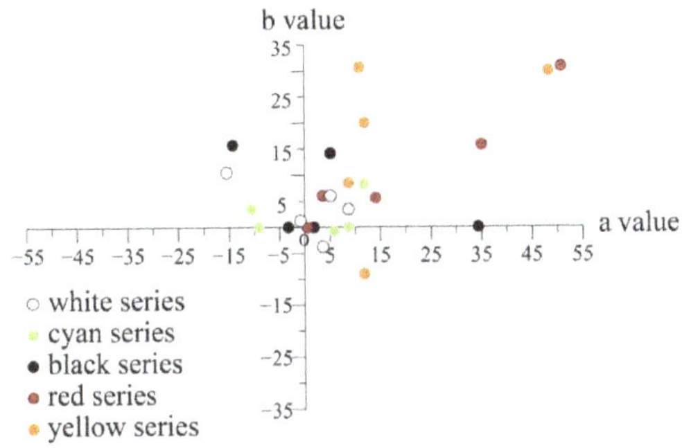

Figure 7. Distribution of color dots for the five color series.

The L-axis represents brightness, with values from 0 to 100 indicating brightness from black to white. The a-axis represents the relative colors of red and green, and the b-axis represents the relative colors of yellow and blue.

As can be seen from Figure 7, the distances between the dots of the white and cyan series are relatively close to each other and are concentrated around the origin L value. The dots of the black series are not far from the origin, but their positions are scattered and their colors differ in hue. The dots of the red series are all located in the first quadrant and distributed in a diagonal line, but the distance between the dots is farther. The yellow series is similar to the red series, the a-value gap between each of their dots is small and the b-value is relatively scattered. In general, the dots of the five color series are basically distributed in the first quadrant and concentrated near the L value of the origin, with a small number of dots scattered in more distant locations.

The histogram statistics of the HSB color model and the scatter plot of the L*a*b* model both confirm that the Yunjian garments usually choose warm colors, low saturation colors, and evenly distributed color brightness.

4.3. Yunjian Fractal Pattern Designcolor Design

According to Equation (4), when $p = 4, 6, 8, 20$, we can obtain the Newton iterative fractal diagram when the "arms" of the fractal are equal to the corresponding values, as shown in Figure 8. We can know that the value of the real part of the exponent p is proportional to the number of "arms" of the Newtonian fractal, and the area of the central intersection circle increases gradually as the value of p increases. We rewrite the Exponent (Re) index in the primary Newton fractal in the Ultra Fractal software platform in version 6.04 to make the model match the pattern structure according to the design characteristics of the Yunjian. When the exponent is 4, 6, 8, and 20, different models and their mapping patterns are obtained. This research lists four series of Yunjian fractal pattern processes produced by different exponent real parts.

When we set the Exponent (Re) index to 4, we get two new models with completely different forms and effects by adding Inverse mapping, Glass Hemisphere mapping and Triangle Inequality Average external mapping respectively. On top of that, we can further superimpose new mappings or formulas to pursue more complex and exquisite pattern effects. After selecting the mapping, we add a color factor with the Yunjian style in the Adjustment Gradient window in the model and change the opacity to achieve the desired effect. Similarly, when the Exponent (Re) index is 6 or 8 or 20, the corresponding fractal pattern can be obtained by the same procedure. It can be seen that the final fractal patterns obtained are extremely different due to the different settings of indices. We made 8 fractal patterns according to the model. These patterns conform to the basic composition form of the Yunjian pattern with infinitely subdivided fractal features and brilliant colors, which can reflect the exquisite and complicated style features of the Yunjian to a certain extent.

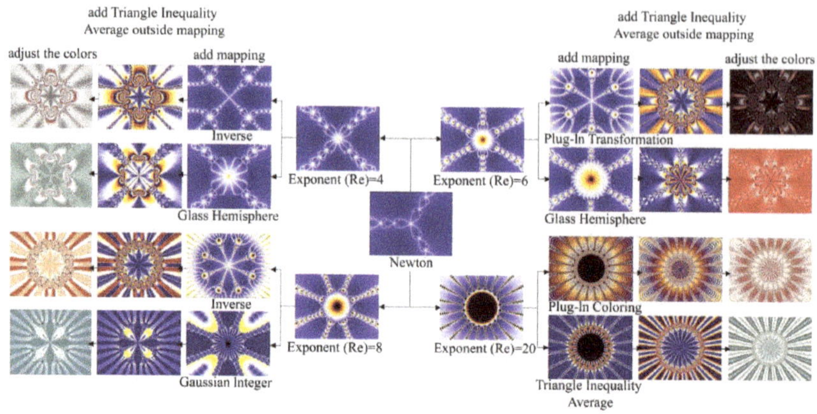

Figure 8. Yunjian fractal pattern design process.

4.4. Yunjian Clothing Design

In order to make the fractal pattern not lose the original characteristics of the Yunjian, we integrated the representative element unit of the Yunjian into the fractal pattern and made the design processing of mirroring or rotating according to the structural characteristics of the fractal pattern. We subjectively processed the elemental units to make them blend naturally, and finally finished the fractal pattern. We combined the typical outer outline structure of the Yunjian with the fractal pattern and removed the outer pattern to obtain 8 tiling style figure of the fractal Yunjian. The production process is shown in Figure 9.

Figure 9. Cloud shoulder shape design process.

This research focuses on 8 styles of Yunjians, combined with modern clothing to complete a series of Yunjian clothing design, as shown in Figure 10. The series of garments are simple in design, and the overall style is casual and comfortable. In the color selection of the series, we retained the traditional color custom of using the same color scheme as the overall clothing, and only added a small amount of contrasting colors in the details to enrich the color layers. The styles of this collection are mostly skirts, mixing Yunjians with modern casual style blazers, loose shirts and halter skirts brings us a new fashion feeling. And the Yunjian became the highlight of the whole set.

Figure 10. Series of Yunjian clothing design development.

4.5. Yunjian Clothing Design Evaluation

This research determined the presentation effect evaluation index U according to the characteristics of the Yunjian garment, U = (u_1, u_2, u_3, u_4) = (shape, structure, pattern, color, aesthetics). A five-level scale criterion was used, V = (v_1, v_2, v_3, v_4, v_5) = (very bad, bad, mediocre, good, very good). A team of expert judges, consisting of eight relevant senior practitioners, ranked the importance of the five indicators for recovery effects, and the results obtained are shown in Table 3.

Table 3. Evaluation index importance ranking.

Rank	Shape	Structure	Pattern	Color	Aesthetics
P_1	1	2	4	3	5
P_2	2	1	3	4	5
P_3	1	2	3	4	5
P_4	2	1	4	5	3
P_5	1	2	4	3	5
P_6	1	2	3	4	5
P_7	1	4	2	3	5
P_8	2	1	5	4	3
Rank Sum (R_j)	11	15	28	30	36

Assuming that the 8 judges did not agree on the importance of the 6 recovery effect indicators, we calculated $X^2 = 22.300$, took the significance level $\alpha = 0.01$, and the degree of freedom df = 4, checked the table to get $X^2_{0.01}(4) = 13.277$. Since $X^2 = 22.300 > 13.277$, we can assume that the 8 judges' opinions are significantly consistent.

The weights of each indicator were calculated from Equation (8) as: $a_1 = 0.3083$, $a_2 = 0.2750$, $a_3 = 0.1667$, $a_4 = 0.1500$, $a_5 = 0.1000$, yielding the weight of each indicator A = (0.3083, 0.2750, 0.1667, 0.1500, 0.1000).

After identifying the evaluation items, evaluation scales and evaluation item weights, we took online and offline questionnaires at the same time. We set the survey target as university students aged 18 to 25 who have knowledge of Yunjian clothing, and we collected a total of 146 questionnaires, of which 139 were valid, resulting in a valid rate as 95.2%. The data of our survey questionnaires are shown in Table 4.

Table 4. Statistics of questionnaire results.

Evaluation Indicators	Very Bad	Bad	Mediocre	Good	Very Good
Shape	0.00	0.01	0.12	0.38	0.49
Structure	0.00	0.03	0.16	0.28	0.53
Pattern	0.02	0.12	0.05	0.10	0.71
Color	0.00	0.03	0.10	0.18	0.69
Aesthetics	0.01	0.08	0.09	0.28	0.54

We can get the matrices from this table:

$$R = \begin{bmatrix} 0.00 & 0.01 & 0.12 & 0.38 & 0.49 \\ 0.00 & 0.03 & 0.16 & 0.28 & 0.53 \\ 0.02 & 0.12 & 0.05 & 0.10 & 0.71 \\ 0.00 & 0.03 & 0.10 & 0.18 & 0.69 \\ 0.01 & 0.08 & 0.09 & 0.28 & 0.54 \end{bmatrix}$$

Using equation B = A·R for the calculation:

$$B = \begin{bmatrix} 0.3083 & 0.2750 & 0.1667 & 0.1500 & 0.1000 \end{bmatrix} \cdot \begin{bmatrix} 0.00 & 0.01 & 0.12 & 0.38 & 0.49 \\ 0.00 & 0.03 & 0.16 & 0.28 & 0.53 \\ 0.02 & 0.12 & 0.05 & 0.10 & 0.71 \\ 0.00 & 0.03 & 0.10 & 0.18 & 0.69 \\ 0.01 & 0.08 & 0.09 & 0.28 & 0.54 \end{bmatrix}$$

$$= \begin{bmatrix} 0.0043 & 0.0438 & 0.1133 & 0.2658 & 0.5727 \end{bmatrix}$$

The comprehensive evaluation results show that 11% of the evaluators think the design effect of the Yunjian garment is "mediocre", 26% of the evaluators think the design effect is "good", and 57% of the evaluators think the design effect is "very good". According to the principle of maximum affiliation, the effect of this series of Yunjian apparel design was evaluated as "very good". Therefore, the Yunjian fractal pattern design model proposed in this study can be better used and practiced in clothing products.

5. Conclusions

Yunjian garment is an important intangible cultural heritage of the Han Chinese, with artistic expression and profound cultural connotation of oriental intentions. Under the globalization of regional ethnic culture, the Yunjian has also become a sustainable cultural capital with great revitalization value. By analyzing the motifs, color principles and structural rules of the Yunjian, this study proposes a digital innovation design method for regional ethnic costumes based on the Newtonian fractal model and carries out design practice. The comprehensive evaluation results show that the Yunjian costume designed by the design method proposed in this study has a good combination of ethnic cultural core and modern fashion communication characteristics. Taking the Yunjian garment as an example, this study tries to explore the capital revitalization method of regional ethnic fashion culture based on digital design method, aiming to provide theoretical and practical reference for the sustainable design of regional characteristic fashion culture.

Author Contributions: Conceptualization, H.C. and L.J.; Data curation, H.X. and Y.Z.; Formal analysis, H.X.; Investigation, H.X.; Methodology, H.C. and L.J.; Resources, H.C. and Y.Z.; Supervision, L.J. All authors have read and agreed to the published version of the manuscript.

Funding: This work was supported by the Jiangsu Province Social Science Foundation Youth Project (grant number 22YSC001), Jiangsu Province Culture and Tourism Research General Project (grant number 22YB30), Jiangsu Province Universities Philosophy and Social Science General Project (grant number 2021SJA0866). The authors acknowledge the above financial support.

Institutional Review Board Statement: Not applicable.

Informed Consent Statement: Not applicable.

Data Availability Statement: Not applicable.

Acknowledgments: The garment photos for this article were obtained from the Jiangnan University Folk Costume Heritage Museum. The authors appreciate the support.

Conflicts of Interest: The authors declare no conflict of interest.

References

1. Delgado, M.J.B.L.; Albuquerque, M. The contribution of regional costume in fashion. *Procedia Manuf.* **2015**, *3*, 6380–6387. [CrossRef]
2. Thompson, C.J.; Haytko, D.L. Speaking of fashion: Consumers' uses of fashion discourses and the appropriation of countervailing cultural meanings. *J. Consum. Res.* **1997**, *1*, 15–42. [CrossRef]
3. Rian, I.M. Fractal-based algorithmic design of Chinese ice-ray lattices. *Front. Archit. Res.* **2022**, *11*, 324–339. [CrossRef]
4. Zhang, J.Z.; Xu, J.; Bai, W.D. Furniture design and fractal geometry-the fractal art in traditional furniture. *Adv. Mater. Res.* **2014**, *933*, 655–661. [CrossRef]
5. Driscoll, J.C. Fractal patternsas fitness criteria in genetic algorithms applied as a design tool in architecture. *Nexus Netw. J.* **2021**, *23*, 21–37. [CrossRef]
6. Gong, Y.X.; Qiao, C.Q.; Zhong, B.C.; Zhong, J.R.; Gong., D.C. Analysis and characterization of materials used in heritage theatrical figurines. *Herit. Sci.* **2020**, *8*, 13. [CrossRef]
7. Jin, X.; Yin, S.; Liu, N.N.; Li, X.D.; Zhao, G.; Ge, S.M. Color image encryption in non-RGB color spaces. *Multimed. Tools Appl.* **2018**, *77*, 15851–15873. [CrossRef]
8. Liu, L.; Zhao, Z.M. Morphological image processing in the speckle metrology. *Imaging Sci. J.* **2011**, *59*, 303–310. [CrossRef]
9. Ding, M.; Dong, W. Product color emotional design considering color layout. *Color Res. Appl.* **2019**, *44*, 285–295.
10. Liu, J.L.; He, R.; Tian, W.; Xia, S.F.; Zhu, C.Y. An overview of color system development and its application in Dai brocade design. *J. Silk* **2021**, *58*, 117–125.
11. Wang, S.H.; Li, X.M.; Jin, X.K.; Tian, W.; Zhu, C.Y. Design and application of jacquard fabric patterns based on the geomorphologic color features. *J. Silk* **2022**, *59*, 101–110.
12. Jin, X.K.; Zhang, S.C.; Li, Q.Z.; Du, L.; Zhu, C.Y. Development of color difference formula and its application in fabric color evaluation. *J. Silk* **2013**, *50*, 33–38.
13. Miao, P.Y.; Fang, Z.Q.; Sun, T.; Jin, X.K.; Zhu, C.Y.; Tian, W. Primary color extraction method for brocade. *J. Text. Res.* **2021**, *42*, 120–125.
14. Huang, S.M. A study of hue identification in the hue circle of the HSB color space. *Percept. Motor Skills* **2005**, *100*, 1143–1154. [CrossRef] [PubMed]
15. Zhu, L. Research and application of AHP-fuzzy comprehensive evaluation model. *Evol. Intell.* **2020**, *13*, 1–7. [CrossRef]
16. Chen, H.; Shen, L.; Wang, M.M.; Tang, Y. Man-algorithm Cooperation Intelligent Design of Clothing Products in Multi Links. *FIBRES TEXTILES East. Eur.* **2022**, *30*, 59–66. [CrossRef]
17. Chen, L.; Halepoto, H.; Liu, C.; Kumari, N.; Yan, X.; Du, Q.; Memon, H. Relationship Analysis among Apparel Brand Image, Self-Congruity, and Consumers' Purchase Intention. *Sustainability* **2021**, *13*, 12770. [CrossRef]

sustainability MDPI

Article

Sinicized Exploration of Sustainable Digital Fashion: Chinese Game Players' Intention to Purchase Traditional Costume Skins

Yawen Fu and Hui'e Liang *

College of Textile Science and Engineering, Jiangnan University, Wuxi 214122, China; 7213007003@stu.jiangnan.edu.cn
* Correspondence: lianghe@jiangnan.edu.cn

Abstract: Digitalization is not only blurring boundaries between the real world and virtual space, but, since COVID-19, it has also made the traditional fashion industry less reliant on physical materials. In this context, digital fashion and virtual design have emerged. Although China has started to digitally distribute cultural products based on the digitization of museums, there is still a gap in the field of digital fashion. In order to achieve the sustainable development of Chinese traditional costume culture and explore the Sinicization of digital fashion, this article proposes to transfer Chinese traditional costumes into a series of digitalized commodities, serving as computer game skins for online sales. This research involved the construction of a hypothetical model based on the technology acceptance model (TAM), including perceived usefulness, purchase price perceptions, perceived playfulness, cultural elements authenticity, satisfaction, and purchase intention through the related literature. In total, the survey data of 219 Chinese game players were collected, and a structural equation model (SEM) was constructed to verify these research hypotheses. The results showed that perceived playfulness and cultural elements authenticity had a positive effect on the players' satisfaction and purchase intentions, whereas perceived usefulness had a negative influence on the players' satisfaction and purchase intentions. In addition, purchase price perceptions had no significant negative effect on the players' satisfaction or their purchase intentions. Overall, this study contributes to the co-development of digital fashion and traditional costume culture in China, as well as the improvement of product design and marketing management for game companies.

Keywords: digital fashion; digitalized collection; online game; Chinese traditional costume; virtual game items; structural equation modeling

Citation: Fu, Y.; Liang, H. Sinicized Exploration of Sustainable Digital Fashion: Chinese Game Players' Intention to Purchase Traditional Costume Skins. *Sustainability* **2022**, *14*, 7877. https://doi.org/10.3390/su14137877

Academic Editors: Hafeezullah Memon, Xiaoke Jin, Tian Wei and Chengyan Zhu

Received: 13 May 2022
Accepted: 17 June 2022
Published: 28 June 2022

1. Introduction

Global political and economic changes, coupled with the ongoing COVID-19 pandemic, have further instilled the necessity for rapid digital innovation in fashion [1]. Fashion designers and brands are attempting to enter the virtual space and computer game world in new ways [2]. The first virtual fashion brand, Tribute Brand, allows consumers to upload a photograph of themselves without consideration of size or gender; the consumers then receive a digital garment with zero pollution. The Fabricant, a Dutch fashion company, came under the spotlight after they auctioned a digital-only dress called Iridescence for cryptocurrency worth USD 9500 in May 2019 [3]. Furthermore, various luxury brands have collaborated with game companies to launch game character clothing. Gucci was the first brand to partner with Drest, a game in which players make themselves up as hyper-realistic supermodels. More recently, more than 200 other luxury brands, such as Prada, Off-White, and Christian Louboutin, have joined this trend [4]. Digital fashion refers not only to a processing tool for the development and visualization of virtual goods, but also as a marketing or educational tool for online shops and virtual museums to manufacture digital-only end goods for virtual use [5]. Recent examples of cooperation between fashion brands and game companies show that digital fashion might become an essential part of business management and fashion design practice [2].

In the post-epidemic period, China has formed a new creative cultural model, namely, "museums + digital collections" based on the digitization of museums [6]. To bequeath Chinese traditional culture, the National Museum of China, the most representative museum in China, has launched a series of digitalized collections through digitalization technology [6]. However, compared with foreign countries, China has paid little attention to digital fashion. Therefore, in order to achieve the sustainable development of Chinese traditional costume culture and explore the Sinicization of digital fashion, this paper proposes to transfer Chinese traditional costumes, dating from hundreds of years ago and now collected in museums, into digitalized commodities, serving as game character skins for sale. The paper also investigates the factors in Chinese game players' intentions to purchase traditional costume skins by constructing theoretical models and hypotheses.

Prior studies centered on the positive effects of games. For example, online games bring not only leisure and pleasure to players [7], but they have also enhanced their cognitive abilities [8], learning capabilities [9], prosocial thoughts [10] and cooperation [11], aiming at specific problems or specific groups. However, further research about the intention to purchase virtual goods in computer games needs to be conducted. In their research about Habbo Hotel, an online game, Lehdonvirta et al. [12] found that even nonrealistic digital clothing is regarded as a commodity, and users are willing to pay for virtual goods. Therefore, the exploration of the factors influencing the purchase of virtual goods in computer games can reveal players' intentions of use, and improve the service quality of game companies and the operating performance of online games. Laura Belmond, a fashion expert from a sustainability nonprofit, claims that this virtual clothing could help replace fashion-focused pieces that are worn just once, potentially reducing the amount of waste that this industry generates [4]. Furthermore, the combination of online games and traditional clothing can be considered as an attempt to lead modern fashion consumption and the sustainable development of Chinese traditional costumes in a digitalized manner [13].

The technology acceptance model (TAM) is currently the most influential model for explaining and predicting people's behavior in their use of products and information technologies [14]. Considering the virtual items' specificity, this paper presents some adjustments based on the TAM. Gefen and Straub classified the TAM according to the purpose of use, and argued that when the information technology or product is used as a tool to achieve other purposes, the perceived ease of use has no significant influence on the intention of use [15]. In this paper, after an interview with five players using traditional costume skins to play games, it was found that using game skins is not operationally difficult. Therefore, the perceived ease of use was removed from the research model in this paper.

2. Literature Review and Hypothesis Deduction

2.1. Perceived Usefulness

Perceived usefulness is the degree to which users think that using a particular service or product would enhance their job performance [16]. In the TAM, perceived usefulness is considered to be the main factor influencing users' attitudes and intentions, both of which ultimately determine their behaviors. The positive effect between perceived usefulness and users' attitudes was verified in previous studies, such as the willingness to use mobile networks [17], educational technologies [18], MOOCs [14] and virtual-reality simulations [19]. In the study of online game platforms, it was pointed out that if players perceive a game platform as useful, they will continue to use it, which will also positively affect their purchase intentions toward virtual goods on the platform [20]. Similarly, if a virtual item is perceived as useful, game players will be more likely to acquire positive emotions when using it and tend to purchase it. Therefore, the following hypotheses are proposed:

Hypothesis 1 (H1). *Perceived usefulness has a positive correlation with the Chinese players' satisfaction from using traditional costume skins.*

Hypothesis 2 (H2). *Perceived usefulness has a positive correlation with Chinese players' intentions of purchasing traditional costume skins.*

2.2. Purchase Price Perceptions

From the consumers' perspective, the purchase price is what is given up or sacrificed to obtain a service or a product [21]. Purchase price perceptions refer to the consumers' consideration of the price to evaluate something's worthiness or utility [22]. Previous research indicated that the higher purchase price perceptions negatively influenced the consumers' perceived value and purchase intentions, whereas lower purchase price perceptions were positively related to the consumers' perceived value and purchase intentions. Li et al. [23] concluded that perceived price perceptions negatively affected the users' willingness to pay for an online Q&A platform. Hsiao [24] found that perceived price perceptions had a negative influence on perceived value in the research of the willingness to pay for social networking services. Some game players may consider skins, a virtual good, as a low practical value commodity compared with a physical good. Others may think that they can obtain a similar game experience using some free skins, thus it is not necessary to pay for them. Therefore, the following hypotheses are proposed:

Hypothesis 3 (H3). *Purchase price perceptions have a remarkably negative correlation with the Chinese players' satisfaction from using traditional costume skins.*

Hypothesis 4 (H4). *Purchase price perceptions have a remarkably negative correlation with Chinese players' intentions of purchasing traditional costume skins.*

2.3. Perceived Playfulness

Perceived playfulness was used in the study of child behavior first. Barnett [25] defined it as the ability to draw satisfaction and fulfill one's intrinsic motives interacting with an activity or a behavior. Previous research showed that if users had a greater ability to perceive the playfulness trait, they would achieve more positive emotions and a higher level of satisfaction. In the study of computer usage, it was found that perceived playfulness positively influenced users' consumption behaviors. Martocchio et al. [26] concluded that perceived playfulness made users show a higher performance and higher emotional responses when performing computer training tasks. Moon et al. [27] found that perceived playfulness can increase individuals' perceptions about concentration, curiosity and enjoyment while using the World Wide Web, and it had a much more significant influence on use intentions than perceived usefulness and perceived ease of use. Game skins, a visually stimulating element in the game design, offer not only a specific image to game characters, but they have also provided a visual experience for game players. For players, game skins can bring them more excitement and pleasure, resulting in buying more game skins. Therefore, the following hypotheses are proposed:

Hypothesis 5 (H5). *Perceived playfulness has a positive correlation with the Chinese players' satisfaction from using traditional costume skins.*

Hypothesis 6 (H6). *Perceived playfulness has a positive correlation with Chinese players' intentions of purchasing traditional costume skins.*

2.4. Cultural Elements Authenticity

Authenticity is often used to appeal to consumers in beverage and food advertisements because it is regarded as a criterion for evaluating cultural products [28]. Littrell et al. [29] noted that authenticity included "uniqueness and originality, workmanship, cultural and historical integrity, aesthetics, function and use, craftspeople and materials, as well as shopping experience and geniuses". Laroche et al. [30] first proposed the term cultural elements authenticity and introduced it into the study of purchase intentions toward international brands using Chinese elements. According to Lee et al. [31], consumers' attitudes toward traditional cultural products are divided into four levels, which are cultural/historical integrity, uniqueness, manufacturing properties and aesthetic properties. Cultural/historical integrity refers to conventional cultural products with authentic historical value and traditional cultural connotation, originated from the original culture. Aesthetic properties refer to the materialistic features of the pattern, color and design that conform to traditional aesthetic satisfaction. In addition, Lee et al. concluded that Chinese consumers pay more attention to cultural/historical integrity and aesthetic properties of cultural products, and both of them positively influence their purchase intentions.

Laroche et al. [30] borrowed Lee et al.'s [31] definition of traditional cultural product attitudes (TCPs) and defined cultural elements authenticity as cultural uniqueness originating from the original culture and consisting of integrity and aesthetic features. Laroche et al. [30] came to a conclusion that for Chinese consumers and Chinese immigrants, cultural elements authenticity had a positive impact on purchasing international brand products with Chinese cultural elements. Traditional costume skins with ancient Chinese clothing elements, such as traditional colors, patterns and styles, meet Chinese consumers' aesthetic requirements and are consistent with cultural and historical values in modern society. Therefore, the following hypotheses are proposed:

Hypothesis 7 (H7). *Cultural elements authenticity has a significantly positive correlation with the Chinese players' satisfaction from using traditional costume skins.*

Hypothesis 8 (H8). *Cultural elements authenticity has a significantly positive correlation with Chinese players' intentions of purchasing traditional costume skins.*

2.5. Satisfaction and Purchase Intention

Consumer attitudes and purchase intentions have been studied in many models, such as the theory of planned behavior (TPB) [32] and technology acceptance model (TAM) [33]. Satisfaction is one of the consumer attitudes, referring to the consumers' assessment of the service and product characteristics to meet their expectations [34]. Purchase intention is a tendency of consumers' behaviors [32]. Previous studies [35,36] showed that the consumers' satisfaction positively affected their purchase intentions. Therefore, the following hypothesis is proposed:

Hypothesis 9 (H9). *Satisfaction has a significantly positive correlation with Chinese players' intentions of purchasing traditional costume skins.*

On the basis of the above theoretical discussions and research hypotheses, this research proposes the following model (see Figure 1). This theoretical model includes six constructs and nine relevant research hypotheses.

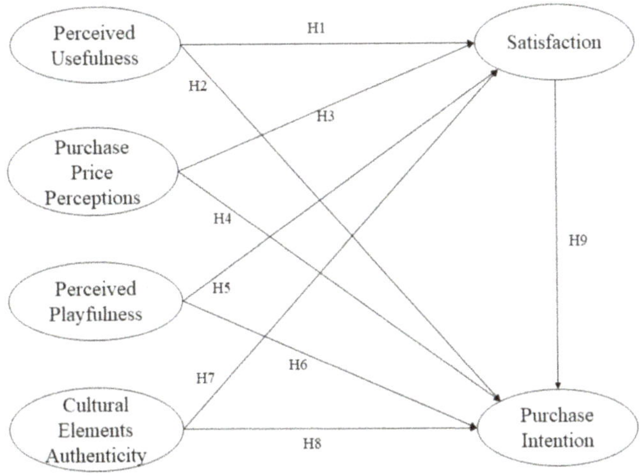

Figure 1. The theoretical model in this study.

3. Research Methods

3.1. Data Collection

This research adopted a questionnaire survey to collect data via WENJUANXING, which is a professional online questionnaire survey platform in China (https://www.wjx.cn/?source=baidu&plan=%E9%97%AE%E5%8D%B7%E6%98%9F%EF%BC%88%E6%AD%A3%E5%B8%B8%EF%BC%89PC&keyword2=%E9%97%AE%E5%8D%B7%E6%98%9F%E5%93%81%E4%B8%93%E6%A0%87%E9%A2%98, accessed on 17 March 2022). The targeted respondents in this research were Chinese players who have played online games before. Before distributing questionnaires, we conducted a predictive test at first and then modified the questionnaire items according to the feedback. To acquire more authentic and credible research results, we set the first question as, "Have you played online games before?" If participants had not, the questionnaire would come to an end. A total of 251 questionnaires were collected in this survey. After removing 32 invalid samples, the number of valid ones remaining were 219, yielding an effective recovery rate of 87.25 percent. The analysis results of the sample characteristics are displayed in Table 1.

Table 1. Demographic characteristics of respondents.

Sample	Category	Number	Percentage
Gender	Male	77	35.2%
	Female	142	64.8%
Age	Under 18	5	2.3%
	18–25	144	65.8%
	26–30	67	30.6%
	31–40	2	0.9%
	Above 40	1	0.4%
Education	Below bachelor's degree	35	16.0%
	Bachelor's degree	107	48.9%
	Master's degree	57	26.0%
	Doctoral degree	20	9.1%
Income (RMB)	Below 1000	20	9.1%
	1000–3000	116	53.0%
	3000–5000	20	9.1%
	5000–8000	24	11.0%
	8000–10,000	17	7.8%
	Above 10,000	22	10.0%
Occupation	Fashion industry	79	36.1%
	Non-fashion industry	140	63.9%
Experience of buying chinoiserie skins	Yes	147	67.1%
	No	72	32.9%

3.2. *Variable Measurement*

This paper involved six measurement items, namely, perceived usefulness (PU), purchase price perceptions (PPP), perceived playfulness (PP), cultural elements authenticity (CEA), satisfaction (SAT) and purchase intention (PI). All constructs and items were derived from relevant mature scales. All questionnaire items and reference sources are shown in Appendix A (Table A1). Perceived usefulness drew on the scale developed by Davis [16], and three items were adopted, such as, "Using traditional costume skins enables me to accomplish tasks more quickly". Purchase price perceptions refer to the scale used in the research by Voss [37], for example, "The price I have to pay is very unreasonable". Three measurement items for perceived playfulness were adapted from the research by Moon [27], such as, "Using traditional costume skins stimulates my curiosity". The scales used in the study by Laroche [30] for cultural elements authenticity were referred to, such as, "Traditional costume skins look nice". The measurement scales for satisfaction from Oliver [38] and Hsu [39] were used, and, for purchase intention, from Perriens [40] and Cronin [41]. The questionnaire items were adjusted appropriately according to the context and respondents in this study. All items included in this paper used the Likert 7-point scale. Among them, "1" means "strongly disagree" and "7" means "strongly agree".

4. Empirical Analysis and Hypothesis Test Results

4.1. *Reliability and Validity*

SPSS 23.0 software (IBM, Armonk, NY, USA) was used to test the reliability of the measurements. As seen in Table 2, the Cronbach's α coefficient values of perceived usefulness, purchase price perceptions, perceived playfulness, cultural elements authenticity, satisfaction and purchase intention ranged from 0.898 to 0.964, which was well above the recommended acceptable level of 0.70, indicating high reliability of the measurements.

Table 2. Results of reliability analysis.

Construct	Item	Factor Loading	Cronbach's α	CR	AVE
Perceived Usefulness	PU1 PU2 PU3	0.851 0.862 0.912	0.908	0.908	0.766
Perceived Playfulness	PP1 PP2 PP3	0.883 0.932 0.887	0.927	0.928	0.812
Purchase Price Perceptions	PPP1 PPP2 PPP3	0.935 0.886 0.790	0.903	0.905	0.761
Cultural Elements Authenticity	CEA1 CEA2 CEA3	0.896 0.910 0.877	0.921	0.923	0.800
Satisfaction	SAT1 SAT2 SAT3	0.938 0.965 0.943	0.964	0.964	0.900
Purchase Intention	PI1 PI2 PI3	0.893 0.863 0.835	0.898	0.899	0.746

The validity test used AMOS 26.0 software to implement a confirmatory factor analysis to verify each construct validity. As presented in Table 2, the standardized factor load coefficients of each variable were between 0.790 and 0.965, which exceeded the reference value of 0.50. The combined reliability (CR) values of all constructs were greater than 0.80, manifesting a good convergence effect of the scales. Moreover, the average variance extracted (AVE) values of each variable ranged from 0.746 to 0.900, surpassing the judgment standard of 0.50. Furthermore, the square root of the AVE was more significant than the Pearson correlation coefficient between the variable and other variable (see Table 3), demonstrating that each construct has a high level of discrimination validity.

Table 3. Correlation matrix and AVE.

Construct	CEA	PP	PPP	PU	SAT	PI
CEA	0.894					
PP	0.534	0.901				
PPP	−0.146	−0.127	0.872			
PU	0.319	0.515	−0.381	0.875		
SAT	0.770	0.606	−0.247	0.446	0.949	
PI	0.749	0.703	−0.199	0.404	0.871	0.864

4.2. Model Fitting

The AMOS 26.0 software was used to assess the goodness-of-fit indices of this model. Table 4 presents the statistical indicators of the model fit test. The CMIN value of this study was 226.155, of which lower is better. The CMIN/DF value was 1.885, meeting the criteria of 1.0 and 3.0. The values of CFI, IFI, TLI and NFI were 0.972, 0.972, 0.964 and 0.942, respectively, all of which outstripped the acceptable level of 0.90. In addition, the RMSEA value was $0.064 < 0.08$ and the SRMR value was $0.036 < 0.05$, both of which showed that the fitting effect of this model is good.

Table 4. Main indicators of model fit test.

Model Fit Indices	Evaluation Index (Acceptable Level)	Values of the Model
CMIN/DF	1.0 < CMIN/DF < 3.0	1.885
CFI	>0.90	0.972
IFI	>0.90	0.972
TLI	>0.90	0.964
NFI	>0.90	0.942
SRMR	<0.05	0.036
RMSEA	<0.08	0.064

4.3. Hypothesis Test

The results of hypothesis testing are shown in Figure 2 and Table 5. Perceived usefulness had a negative effect on satisfaction ($\beta = 0.107$, $p = 0.069 > 0.05$) and purchase intention ($\beta = -0.071$, $p = 0.178 > 0.05$). Purchase price perceptions did not have a significantly negative bearing on satisfaction ($\beta = -0.091$, $p = 0.067 > 0.05$) and purchase intention ($\beta = -0.016$, $p = 0.716 > 0.05$). Perceived playfulness engendered a significantly positive influence on satisfaction ($\beta = 0.214$, $p < 0.001$) and purchase intention ($\beta = 0.289$, $p < 0.001$). Cultural elements authenticity produced a positive influence on satisfaction ($\beta = 0.608$, $p < 0.001$) and purchase intention ($\beta = 0.141$, $p = 0.037 < 0.05$). Satisfaction positively affected purchase intention ($\beta = 0.615$, $p < 0.001$). Therefore, the results support H5, H6, H7, H8 and H9, whereas H1, H2, H3 and H4 are invalid.

Table 5. Hypothesis testing results.

Hypothesis	Estimate	Std. Estimate	CR	*p*	Conclusion
H1 (PU → SAT)	0.098	0.107	1.818	0.069	Rejected
H2 (PU → PI)	−0.060	−0.071	−1.348	0.178	Rejected
H3 (PPP → SAT)	−0.090	−0.091	−1.828	0.067	Rejected
H4 (PPP → PI)	−0.015	−0.016	−0.363	0.716	Rejected
H5 (PP → SAT)	0.227	0.214	3.441	***	Supported
H6 (PP → PI)	0.286	0.289	5.022	***	Supported
H7 (CEA → SAT)	0.745	0.608	10.239	***	Supported
H8 (CEA → PI)	0.161	0.141	2.086	0.037	Supported
H9 (SAT→ PI)	0.573	0.615	8.133	***	Supported

Note: *** means *p*-value is less than 0.001.

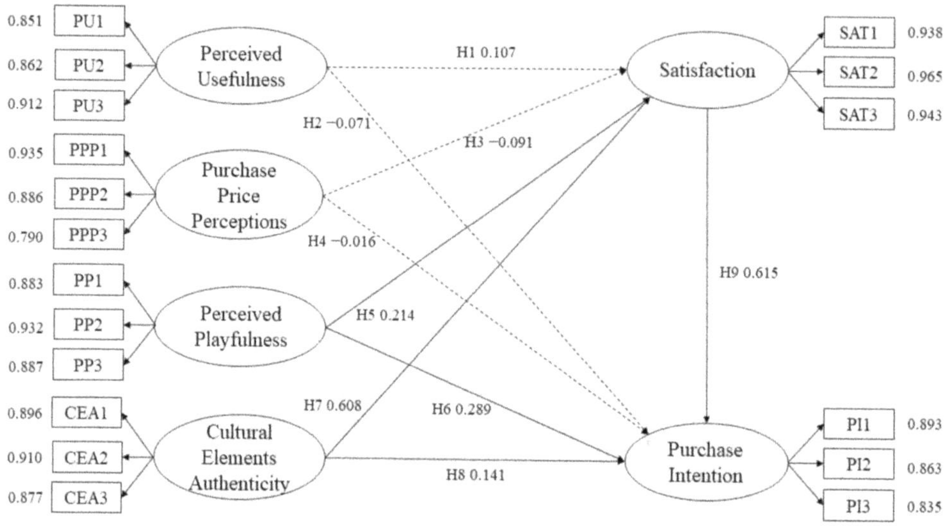

Figure 2. The influence between variables in the structural model.

4.4. Mediating Effect Test

According to the analysis above, it was found that satisfaction has a mediating effect between perceived playfulness and purchase intention, and between cultural elements authenticity and purchase intention. The bootstrap method was conducted to test the mediating effect of satisfaction. The number of bootstrap samples was 5000. The test results are illustrated in Table 6.

Table 6. Mediating effect of satisfaction.

Path		Effect Size	Boot Standard Error	BootCI Confidence (Lower Bound)	BootCI Confidence (Upper Bound)
PP → SAT → PI	Indirect effect	0.130	0.056	0.034	0.258
	Direct effect	0.286	0.073	0.153	0.442
	Total effect	0.416	0.087	0.251	0.593
CEA → SAT → PI	Indirect effect	0.427	0.104	0.235	0.644
	Direct effect	0.161	0.106	-0.035	0.376
	Total effect	0.588	0.096	0.416	0.792

The results presented that the bootstrap 95% confidence interval of the direct influence of perceived playfulness on purchase intention and the mediating influence of satisfaction were (0.153, 0.442) and (0.034, 0.258), and did not contain 0 between the upper and lower limits. This indicates that perceived playfulness can directly predict purchase intention and pass the mediating influence of satisfaction to indirectly predict purchase intention. The direct effect action value was 0.286, and the satisfaction mediation effect action value was 0.130, accounting for 68.75% and 31.25% of the total effect. Concerning the mediating role in the relationship between cultural elements authenticity and purchase intention, the bootstrap 95% confidence interval of mediating effect did not contain 0, but that of direct effect contained 0. This means that cultural elements authenticity only passes the mediating effect of satisfaction to indirectly predict purchase intention.

5. Discussion

This paper probed into the bearings of Chinese game players' purchase intentions toward traditional costume skins. According to the empirical analysis results, perceived playfulness and cultural elements authenticity positively impacted satisfaction and purchase intention, whereas perceived usefulness engendered a negative effect on satisfaction and purchase intention. Furthermore, purchase price perceptions did not have a significantly negative bearing on satisfaction and purchase intention. Finally, it was indicated that satisfaction played a partially mediating role in the link between perceived playfulness and purchase intention, and a full mediating role in the link between cultural elements authenticity and purchase intention.

First, H1 and H2 were rejected, as we demonstrated that using traditional costume skins did not improve game performance or accomplish tasks more quickly in the computer game, which is inconsistent with previous research [19,20]. Game items, designed to enhance the communication between online games and game players, provide functional, decorative and social values for game users in the virtual world [42]. The drivers of purchasing virtual items are classified into three categories, namely, functional attributes, hedonic attributes and social attributes [43]. Functional attributes refer to how virtual goods can provide performance advantages and new functionalities for game players, which is the same as perceived usefulness in this paper. Functional attributes of virtual items have a strong effect on users' purchase intentions, but many of the apparently highly desirable virtual goods lack these functionalities [43]. In this case, the goods' desirability can possibly be traced to the social or affective value that they contain. Traditional costume skins, a visual symbol, are the highly desirable virtual items that Lehdonvirta proposed. From the players' perspective, traditional costume skins neither help improve the game characters' performance and game players' concentration, nor advance the game process during the game. It is the reason why perceived usefulness has a negative correlation with the players' satisfaction and purchase intentions.

Second, the research findings indicated that purchase price perceptions did not have a significant negative influence on the players' satisfaction and purchase intentions, thus H3 and H4 were rejected, which is in alignment with the finding of adopting e-learning [44]. If the value consumers gain from a product or service is higher than the cost they pay, purchase intention will be formed. If the value is lower, users will reject this product or service. In the context of e-learning, when learners are truly glad to learn using a system, they are willing to trade costs to obtain a better learning performance [44]. Similarly, if game players truly like traditional costume skins, they are willing to pay for a better game experience. Before they purchase traditional costume skins, players will evaluate whether perceived values (playful and cultural attributes) of virtual items are higher than the cost or not. In fact, players value the more emotional and social values that traditional costume skins hold, and perceive risks brought by the expense relatively less.

Third, the research outcomes manifested that perceived playfulness exerted a significant positive effect on the users' satisfaction and purchase intentions, thus H5 and H6 were adopted. Perceived playfulness is the same as hedonic attributes mentioned in this paper, which are connected with aesthetic qualities. If the aesthetic aspects of virtual items are amply compelling, users may acquire hedonistic delight from experiencing them [43]. Game items centered on the affective value, such as game skins and hairstyles, can alter aesthetic properties and game scenes to make game users much more entertained [42]. Recent research on online games stressed the emotional value, demonstrating that games enable players to provoke increased pleasure and arousal [45,46]. Virtual items and augmented products in the game not only enhance the game characters' appearance and image [42], but also help players to regulate their negative attitudes into positive ones [47]. Therefore, traditional costume skins with a unique appearance can stimulate players' psychological arousal, making them feel pleased and excited, which eventually generates a positive influence.

Fourth, cultural elements authenticity produced a significant positive influence on the players' satisfaction and purchase intentions, in accordance with Laroche et al.'s finding [30]. Besides giving rise to individualistic hedonism, the visual appearance of traditional costume skins can also be considered as connected with their social value. The cultural elements authenticity reflects the same meaning as social attributes of virtual goods, but social attributes put more emphasis on cultural elements attached to virtual items compared with hedonic attributes [43]. Traditional costume skins, inspired by traditional costumes dating from hundreds of years ago and collected in museums, have Chinese traditional cultural elements. Authenticity makes not only virtual goods valuable, but it also offers the emotional resonance and cultural identity to game players. By purchasing and displaying traditional costume skins, game players can express their love and pride in Chinese culture, and establish an affective bond with other players, which also strengthens the emotional communication among Chinese people.

6. Conclusions

6.1. Theoretical Implications

This paper provides some academic implications. At first, the introduction of cultural elements authenticity into the exploration of purchase intention of digital fashion expands the research perspective and scope of digital fashion. Atacac (2016–), Carlings' digital-only collection (2017–), and The Fabricant (2018–), the earliest commercial examples of digital fashion, have put more emphasis on how to transfer traditional and material practices into new digitized ones using 3D software [5], but there has been a lack of introducing traditional costume elements into the realm of digital fashion. Digital fashion is a much more sustainable and inclusive method for fashion practice [5]. This paper proposed the combination of digital fashion and a digitalized collection, exploring the possibility of this combination for a mixture of sustainability and the collectible value [48]. Overall, digital fashion promotes the sustainable development of Chinese traditional costume culture and points out a direction suitable for Chinese digital fashion.

Furthermore, this paper expands the TAM and applies it to the realm of digital fashion. Now, many developed extensions of the TAM are more suitable for the information age [44,49]. However, in the current context, it still has some shortcomings due to the lack of consideration on how social and cultural factors influence users' acceptance of technology [44]. This study introduces cultural elements authenticity into the TAM to analyze the influence of virtual goods on the consumers' satisfaction and purchase intentions from cultural perspectives. We also found that the influence of cultural elements authenticity on satisfaction is more significant than that of perceived playfulness, which suggests that cultural elements authenticity should be given priority when game companies design Chinoiserie game skins. In addition, this study also demonstrates that the perceived usefulness of traditional costume skins has no significant effect on the consumers' satisfaction and purchase intentions. It indicates that consumers value hedonic and social attributes of virtual goods more than functional attributes of physical goods in the digitalization era. Therefore, this study provides a rich and varied map for researchers who use the TAM aimed at physical goods and virtual items.

6.2. Managerial Implications

First, this study argues that the expense is not a major factor to prevent consumers from purchasing virtual goods, and Chinese players are willing to pay for game skins due to the matched entertainment value and cultural value, which is coincident with the findings of the intention to purchase digital clothing for online games [12] and to use e-learning systems [44]. Game companies have to care about the balance between costs and values that game players gain. Indeed, players want to be protected, and they usually assess the risk rather than the fee before buying a game item. Therefore, when designing skins, game companies should give priority to meeting players' needs for perceived playfulness and cultural elements authenticity. Meanwhile, perceived values including perceived playfulness and cultural elements authenticity compensate for perceived risks such as purchase price perceptions, which stimulate players' desire to purchase and maximize benefits of virtual goods. Furthermore, cultural elements authenticity of traditional costume skins could enhance the Chinese game players' cultural pride and cultural self-confidence, leading the young generation to form correct cultural values and life values, and promote the high-quality development of the digital culture industry.

Second, in order to enhance the players' experience, game companies should take into account the comprehensive utilization of various game objects with Chinese cultural elements. Narrative transportation theory can be applied to form an explanation. Narrative transportation refers to a distinct mental process, an integrative melding of the attention, imagery and feeling. When individuals are absorbed into a story, they will be immersed in events occurring in the story and feel some aspects of the world that are inaccessible [50]. When using Chinese cultural elements, game companies could adopt matching game characters, scenes and plots that conform to Chinese historical and cultural values. It could attract people to play the game and become lost in the narrative game, stimulating their curiosity, imagery and exploration, and in turn, increasing players' continuous intentions of playing the game [50]. Furthermore, using traditional costume skins could build affective bonds between game users and game characters, and enhance their cultural identity in the real world and well-being in the virtual world. Previous research showed that in order to increase game users' exposure to pleasurable stimuli, game companies should take aesthetic and emotional decorative game goods, for instance, character skins, into consideration [42]. Such innovative game items could provide value for game players, increasing their purchase intentions [47]. Therefore, this study reveals that game companies should take into account the game scene and game plot design that match character skins to evoke game users' positive emotions through narrative transportation.

6.3. Limitations and Future Research

There are still several limitations in this paper. First, prior research on digital fashion focused on three categories: communication and marketing, design and production, and culture and society [51]. The combination of digital collections in the museum and game skins proposed in this paper is connected with design and production, but does not fully show the digital fashion practice in China. Therefore, communication and marketing and culture and society should also be taken into account in terms of exploring the sinicized digital fashion. With the rapid development of digital fashion and the rapid reflection of the fashion industry in China, many Chinese fashion brands have stepped into the digital world. Taking Balabala, the best children's wear brand in China, as an example, it recently released the first digital girl in the world, called Gu Yu, and plans to have interactions with real models in the upcoming Metaverse show [52], which provides a digitally innovative experience for Chinese families and children.

Second, this study did not separately explore the respondents' purchase intentions toward traditional costume skins who have not purchased chinoiserie skins before. Analyzing the psychology and behavior of these potential customers could help game companies to design items that are suitable for more players.

Finally, this study did not consider the influence mechanism of some moderating variables on game players' purchase intentions. For example, it could be taken into consideration whether players' positive emotions stimulated by cultural pride or happiness could lead to higher purchase intentions when they see game items with Chinese elements. For future investigations, cultural pride and cultural identity developed by purchasing cultural products with Chinese elements are expected to be further analyzed.

Author Contributions: Conceptualization, Y.F.; formal analysis, Y.F.; data curation, Y.F.; writing—original draft, Y.F.; writing—review and editing, Y.F. and H.L. All authors have read and agreed to the published version of the manuscript.

Funding: This research was funded by the National Social Science Foundation of China (No. 21BG142).

Institutional Review Board Statement: Not applicable.

Informed Consent Statement: Not applicable.

Data Availability Statement: The data presented in this study are available on request from the corresponding author. The data are not publicly available due to confidentiality assurance of each participant's information.

Conflicts of Interest: The authors declare no conflict of interest.

Appendix A.

Table A1. Measurement items.

Construct	Items		References
Perceived Usefulness	PU1	Using traditional costume skins enables me to accomplish tasks more quickly.	Davis [16]
	PU2	Using traditional costume skins allows me to accomplish more missions.	
	PU3	I find the traditional costume skins useful while I play online games.	
Purchase Price Perceptions	PPP1	The price I have to pay is a rip-off.	Voss [37]
	PPP2	The price I have to pay is very unreasonable.	
	PPP3	I would not be pleased to pay this price.	
Perceived Playfulness	PP1	Using traditional costume skins stimulates my curiosity.	Moon [27]
	PP2	Using traditional costume skins leads to my exploration.	
	PP3	Using traditional costume skins arouses my imagination.	
Cultural Elements Authenticity	CEA1	Traditional costume skins look nice.	Laroche [30]
	CEA2	Traditional costume skins have traditional elements such as colors and patterns.	
	CEA3	Traditional costume skins conform to my cultural knowledge.	
Satisfaction	SAT1	I am satisfied with my decision to use traditional costume skins.	Oliver [38] Hsu [39]
	SAT2	I am pleased with the experience of using traditional costume skins.	
	SAT3	My choice to use traditional costume skins was a wise one.	
Purchase Intention	PI1	I would be tempted to try traditional costume skins.	Perrien [40] Cronin [41]
	PI2	I would be tempted to look actively for traditional costume skins.	
	PI3	I would recommend traditional costume skins to my friends.	

References

1. Silva, E.S.; Bonetti, F. Digital humans in fashion: Will consumers interact? *J. Retail. Consum. Serv.* **2021**, *60*, 102430. [CrossRef]
2. Makryniotis, T. Fashion and Costume Design in Electronic Entertainment-Bridging the Gap between Character and Fashion Design. *Fash. Pract. J. Des. Creat. Process Fash. Ind.* **2018**, *10*, 99–118. [CrossRef]
3. Renwick, F. In the Future Your Clothes Will Be Made Outof Pixels: Inside the 'Wild West' World of Digital Fashion. Available online: https://www.esquire.com/uk/style/fashion/a30150947/digital-fashion-the-fabricant-menswear-future-trends/ (accessed on 13 April 2022).
4. Segran, E. Would You Spend $10,000 on a Virtual Dress? Gucci Is Betting on It. Available online: https://www.fastcompany.com/90546878/would-you-spend-10000-on-a-virtual-dress-gucci-is-betting-on-it (accessed on 10 April 2022).
5. Särmäkari, N. Digital 3D Fashion Designers: Cases of Atacac and The Fabricant. *Fash. Theory* **2021**, *25*, 1–30. [CrossRef]

6. Antchain Launches "Treasure Project" to Protect Traditional Culture Using Trusted Digital Technology. Available online: http://ent.people.com.cn/n1/2021/1021/c1012-32260535.html (accessed on 10 April 2022).
7. Dunn, E.W.; Gilbert, D.T.; Wilson, T.D. If money doesn't make you happy, then you probably aren't spending it right. *J. Consum. Psychol.* **2011**, *21*, 115–125. [CrossRef]
8. Spence, I.; Feng, J. Video Games and Spatial Cognition. *Rev. Gen. Psychol.* **2010**, *14*, 92–104. [CrossRef]
9. Mayer, R.E. Computer Games in Education. *Annu. Rev. Psychol.* **2019**, *70*, 531–549. [CrossRef]
10. Greitemeyer, T.; Osswald, S. Playing Prosocial Video Games Increases the Accessibility of Prosocial Thoughts. *J. Soc. Psychol.* **2011**, *151*, 121–128. [CrossRef]
11. Childress, M.D.; Braswell, R. Using Massively Multiplayer Online Role Playing Games for Online Learning. *Distance Educ.* **2006**, *27*, 187–196. [CrossRef]
12. Lehdonvirta, V.; Wilska, T.A.; Johnson, M. Virtual Consumerism: Case Habbo Hotel. *Inf. Commun. Soc.* **2009**, *12*, 1059–1079. [CrossRef]
13. Pan, Y.; Roedl, D.; Blevis, E.; Thomas, J.C. Fashion Thinking: Fashion Practices and Sustainable Interaction Design. *Int. J. Des.* **2015**, *9*, 53–66.
14. Wu, B.; Chen, X.H. Continuance intention to use MOOCs: Integrating the technology acceptance model (TAM) and task technology fit (TTF) model. *Comput. Hum. Behav.* **2017**, *67*, 221–232. [CrossRef]
15. David, G.; Straub, D.W. The Relative Importance of Perceived Ease of Use in IS Adoption: A Study of E-Commerce Adoption. *J. Assoc. Inf. Syst.* **2000**, *1*, 1–30.
16. Davis, F.D. Perceived Usefulness, Perceived Ease of Use, and User Acceptance of Information Technology. *Mis Q.* **1989**, *13*, 319–340. [CrossRef]
17. Alalwan, A.A.; Baabdullah, A.M.; Rana, N.P.; Tamilmani, K.; Dwivedi, Y.K. Examining adoption of mobile internet in Saudi Arabia: Extending TAM with perceived enjoyment, innovativeness and trust. *Technol. Soc.* **2018**, *55*, 100–110. [CrossRef]
18. Scherer, R.; Siddiq, F.; Tondeur, J. The technology acceptance model (TAM): A meta-analytic structural equation modeling approach to explaining teachers' adoption of digital technology in education. *Comput. Educ.* **2019**, *128*, 13–35. [CrossRef]
19. Fagan, M.; Kilmon, C.; Pandey, V. Exploring the adoption of a virtual reality simulation: The role of perceived ease of use, perceived usefulness and personal innovativeness. *Campus-Wide Inf. Syst.* **2012**, *29*, 117–127. [CrossRef]
20. Mantymaki, M.; Salo, J. Teenagers in social virtual worlds: Continuous use and purchasing behavior in Habbo Hotel. *Comput. Hum. Behav.* **2011**, *27*, 2088–2097. [CrossRef]
21. Zeithaml, V.A. Consumer Perceptions of Price, Quality, and Value: A Means-End Model and Synthesis of Evidence. *J. Mark.* **1988**, *52*, 2–22. [CrossRef]
22. Liang, L.J.; Choi, H.S.C.; Joppe, M. Understanding repurchase intention of Airbnb consumers: Perceived authenticity, electronic word-of-mouth, and price sensitivity. *J. Travel Tour. Mark.* **2018**, *35*, 73–89. [CrossRef]
23. Wu, L.; Pengya, A.; Rong, X. Users' Willingness of Paying for Online Q&A Platform Based on the Perceived Value. *Doc. Inf. Knowl.* **2018**, *4*, 4–14.
24. Hsiao, K.L. Why internet users are willing to pay for social networking services. *Online Inf. Rev.* **2011**, *35*, 770–788. [CrossRef]
25. Barnett, L.A. The playful child: Measurement of a disposition to play. *Play Cult.* **1991**, *4*, 51–74.
26. Martocchio, J.J.; Webster, J. Effects of Feedback and Cognitive Playfulness on Peerformanc in Microcomputer Software Training. *Pers. Psychol.* **1992**, *45*, 553–578. [CrossRef]
27. Moon, J.W.; Kim, Y.G. Extending the TAM for a World-Wide-Web context. *Inf. Manag.* **2001**, *38*, 217–230. [CrossRef]
28. Cohen, E. Authenticity and commoditization in tourism. *Ann. Tour. Res.* **1988**, *15*, 371–386. [CrossRef]
29. Littrell, M.A.; Anderson, L.F.; Brown, P.J. What makes a craft souvenir authentic? *Ann. Tour.* **1993**, *21*, 197–215. [CrossRef]
30. Laroche, M.; Li, R.; Richard, M.O.; Shao, M.X. Understanding Chinese consumers' and Chinese immigrants' purchase intentions toward global brands with Chinese elements. *J. Prod. Brand Manag.* **2021**, *30*, 1077–1093. [CrossRef]
31. Lee, S.; Ko, E.; Chae, H.; Minami, C. A study of the authenticity of traditional cultural products: Focus on Korean, Chinese, and Japanese consumers. *J. Glob. Sch. Mark. Sci.* **2017**, *27*, 93–110. [CrossRef]
32. Fishbein, M.; Ajzen, I. Belief, Attitude, Intention and Behaviour: An Introduction to Theory and Research. Addison-Wesley, Reading MA. *Philos. Rhetor.* **1977**, *41*, 842–844.
33. Fred, D.; Davis, R.P.B.; Paul, R. Warshaw, User Acceptance of Computer Technology: A Comparison of Two Theoretical Models. *Manag. Sci.* **1989**, *35*, 982–1003.
34. Sasono, I.; Jubaedi, A.D.; Novitasari, D.; Wiyono, N.; Riyanto, R.; Oktabrianto, O.; Jainuri, J.; Waruwu, H. The Impact of E-Service Quality and Satisfaction on Customer Loyalty: Empirical Evidence from Internet Banking Users in Indonesia. *J. Asian Financ. Econ. Bus.* **2021**, *8*, 465–473.
35. García-Salirrosas, E.E.; Acevedo-Duque, Á. PERVAINCONSA Scale to Measure the Consumer Behavior of Online Stores of MSMEs Engaged in the Sale of Clothing. *Sustainability* **2022**, *14*, 2638. [CrossRef]
36. Yang, C.; Tu, J.-C.; Jiang, Q. The Influential Factors of Consumers' Sustainable Consumption: A Case on Electric Vehicles in China. *Sustainability* **2020**, *12*, 3496. [CrossRef]
37. Voss, G.B.; Parasuraman, A.; Grewal, D. The Roles of Price, Performance, and Expectations in Determining Satisfaction in Service Exchanges. *J. Mark.* **1998**, *62*, 46–61. [CrossRef]

38. Oliver, R.L. A cognitive model of the antecedents and consequences of satisfaction decisions: A suggested framework and research propositions. *Am. Mark. Assoc.* **1989**, *17*, 460–469.
39. Hsu, M.H.; Chiu, C.M. Predicting electronic service continuance with a decomposed theory of planned behaviour. *Behav. Inf. Technol.* **2004**, *23*, 359–373. [CrossRef]
40. Perrien, J.; Paul, D.F. Advertisers and the Factual Content of Advertising. *J. Advert.* **1985**, *14*, 30–35, 53. [CrossRef]
41. Cronin, J.J., Jr.; Bradyb, M.K.; Hult, G.T.M. Assessing the effects of quality, value, and customer satisfaction on consumer behavioral intentions in service environments. *J. Retail.* **2000**, *76*, 193–218. [CrossRef]
42. Bae, J.; Kim, S.J.; Kim, K.H.; Koo, D.M. Affective value of game items: A mood management and selective exposure approach. *Internet Res.* **2019**, *29*, 315–328. [CrossRef]
43. Lehdonvirta, V. Virtual item sales as a revenue model: Identifying attributes that drive purchase decisions. *Electron. Commer. Res.* **2009**, *9*, 97–113. [CrossRef]
44. Liao, Y.K.; Wu, W.Y.; Le, T.Q.; Phung, T.T.T. The Integration of the Technology Acceptance Model and Value-Based Adoption Model to Study the Adoption of E-Learning: The Moderating Role of e-WOM. *Sustainability* **2022**, *14*, 815. [CrossRef]
45. Bae, J.; Koo, D.M.; Mattila, P. Affective motives to play online games. *J. Glob. Sch. Mark. Sci.* **2016**, *26*, 174–184. [CrossRef]
46. Li, K.; Nguyen, H.V.; Cheng, T.C.E.; Teng, C.I. How do avatar characteristics affect avatar friendliness and online gamer loyalty? Perspective of the theory of embodied cognition. *Internet Res.* **2018**, *28*, 1103–1121. [CrossRef]
47. Kim, S.J.; Kim, K.H.; Choi, J. The Role of Design Innovation in Understanding Purchase Behavior of Augmented Products. *J. Bus. Res.* **2019**, *99*, 354–362. [CrossRef]
48. Crane, D. Fashion and Artification in the French Luxury Fashion Industry. *Cult. Sociol.* **2019**, *13*, 293–304. [CrossRef]
49. Bae, Y.; Choi, J.; Gantumur, M.; Kim, N. Technology-Based Strategies for Online Secondhand Platforms Promoting Sustainable Retailing. *Sustainability* **2022**, *14*, 3259. [CrossRef]
50. Green, M.C.; Brock, T.C. The Role of Transportation in the Persuasiveness of Public Narratives. *J. Personal. Soc. Psychol.* **2000**, *79*, 701–721. [CrossRef]
51. Nobile, T.H.; Noris, A.; Kalbaska, N.; Cantoni, L. A review of digital fashion research: Before and beyond communication and marketing. *Int. J. Fash. Des. Technol. Educ.* **2021**, *14*, 293–301. [CrossRef]
52. Balabala: The First Digital Girl in the World, Called Gu Yu, Is Coming. Available online: https://new.qq.com/omn/20220427/20220427A0CBQL00.html (accessed on 10 April 2022).

sustainability

MDPI

Article

Research on Influencing Mechanism of Fashion Brand Image Value Creation Based on Consumer Value Co-Creation and Experiential Value Perception Theory

Lihong Chen [1], Habiba Halepoto [2], Chunhong Liu [1,*], Xinfeng Yan [3] and Lijun Qiu [1]

1 Shanghai International Fashion Science and Innovation Center, Donghua University, Shanghai 200051, China; lhckxyy@dhu.edu.cn (L.C.); qiulijun20180916@163.com (L.Q.)
2 Engineering Research Center of Digitized Textile and Fashion Technology (Ministry of Education), Donghua University, Shanghai 201620, China; 317111@mail.dhu.edu.cn
3 International Cultural Exchange School, Donghua University, Shanghai 200051, China; yanxf@dhu.edu.cn
* Correspondence: chliu@dhu.edu.cn

Abstract: In view of the current lack of fashion brand competitiveness and innovation in China, this paper puts forward the concept of fashion brand image value creation and analyzes it from five dimensions: fashion brand image design, image publicity, brand aesthetics, brand charm, and brand function. This paper explores the relationship between fashion brand image value creation, customer participation behavior, experience value perception, intention, trust, and loyalty based on consumer value co-creation and experience value perception theories. On this basis, the structural equation model is used to test the research hypothesis empirically. An online survey questionnaire was subsequently developed and conducted to verify validity and reliability by statistical analysis. The results show that the value creation of fashion brand image will positively impact brand loyalty. Customer participation behavior and experience value perception play an intermediary and chain intermediary role, and customer participation willingness and fashion brand trust play a regulatory role. This study provides new ideas and references for the value creation of fashion brand image and provides quantitative scientific data for fashion enterprises to grasp the direction of brand image value creation and implement brand construction and marketing strategies.

Keywords: value creation; brand loyalty; customer participation behavior; experience value perception; structural equation model

Citation: Chen, L.; Halepoto, H.; Liu, C.; Yan, X.; Qiu, L. Research on Influencing Mechanism of Fashion Brand Image Value Creation Based on Consumer Value Co-Creation and Experiential Value Perception Theory. *Sustainability* **2022**, *14*, 7524. https://doi.org/10.3390/su14137524

Academic Editor: Lester Johnson

Received: 16 May 2022
Accepted: 14 June 2022
Published: 21 June 2022

1. Introduction

With the development of technology and the transformation of market competition, garment enterprises have realized the importance of branding. However, while the garment industry is developing towards "branding", it faces some problems, such as low competitiveness, insufficient innovation and creativity, and insufficient brand value promotion space utilization [1,2]. Therefore, it is in the interests of individual garment businesses to enhance brand competitiveness, innovation and creativity and create higher brand value [3,4]. Currently, fashion brand image value creates a direction for fashion enterprises to address these issues [5]. Fashion brand image value creation can meet the new needs of consumers and enable customers to identify the differences between brands and product categories to improve the competitiveness of brands, and finally bring more significant economic benefits to enterprises, prolong their life and increase the value of brands [6]. Therefore, to enhance brand competitiveness and brand value, the creation of fashion brand image value has been a key topic of common concern in business and academia [7–9]. At the same time, enterprises or brands will also encounter many challenges when creating brand image value [10]. For example, brands do not pay enough attention to brand aesthetics, which leads to low brand image value innovation ability and reducing customer stickiness.

In the process of brand image value creation, due to the lack of a correct direction of value creation activities, fashion brand image value creation is poor, customers' perception of brand image value creation is low, and finally reduces brand loyalty [11–13].

Previous research on fashion brand value creation focused on the use-value of brand products, but the space for value creation of product use-value was limited. In order to create a more significant value space, fashion brand value creation should be quickly transferred to more personalized service provision, customer experience, and co-creation. Therefore, this paper explores the influencing mechanism of fashion brand image value creation on brand loyalty based on consumer value co-creation and experiential value perception theory to provide quantitative scientific data for fashion enterprises to conduct the brand image value creation and marketing strategies. Keeping this idea in mind following questions were explored through this research.

1. What effect does the value creation of fashion brand image, customer engagement behavior, and perceived value of fashion brand experience have on brand loyalty?
2. What impact does fashion brand image value creation have on customer participation?
3. What impact does customer participation behavior have on the perceived value of fashion brand experience?
4. What role do the customer participation behavior and apparel brand experience value perception play in influencing brand image value creation and brand loyalty?
5. What role does the customer participation intention play in influencing brand image value creation on customer participation behavior?
6. What role does the fashion brand trust play in influencing customer engagement behavior and the perceived value of fashion brand experience on brand loyalty?

This research would be beneficial for textile and fashion marketers to understand the influencing mechanism of image value creation of Chinese fashion brands based on consumer value co-creation and experiential value perception theory. Also, it would be helpful for academia to understand its influence more deeply.

This introduction has shown that from the point of view of fashion brand image, increasing fashion brand image value creation is preferable, while knowing which factors and how to influence fashion brand image value creation is also important. In the next sections, we have reviewed the value creation of fashion brand image, experiential value perception, consumer value co-creation, and participation behavior in order to see which factors could contribute to achieving a larger portion of fashion brand image value creation. Furthermore, the research hypotheses and theoretical model have been established in Section 3, to explain and influence the mechanism of fashion brand image value creation and brand loyalty in Section 4, as empirical tests. The article concludes with a discussion of the findings in Section 5, their implications and suggestion, and future research directions in Section 6.

2. Literature Review

2.1. Value Creation of Fashion Brand Image

Brand value creation is an important research content in the field of fashion brands. High-value brands can reduce marketing costs and bring premium income to enterprises, so enterprises are increasingly concerned about brand value creation. France et al. believe that brand value creation is to satisfy customers' pursuit of higher value with new brand value [14]. Duan and Qu believe that brand value creation results from enterprises' brand investment [15]. Jayaraman and Luo proposed that brand value creation is a means for enterprises to establish a competitive advantage through brand power, brand image, and reputation [16]. Since the ultimate goal of brand value creation is to enable customers to form a unique brand experience. The content of brand value creation includes products, services, innovation, brand image, brand relationship, etc. [17]. Based on scholars' research on brand value creation, it can be found that the connotation of brand value creation has not been clearly defined. Therefore, fashion brand image value creation refers to a series of innovative and creative activities produced by enterprises that aim to create additional brand value based on the original limited value space to meet customers' higher goal value pursuit and maximize brand image value.

Brand image is indispensable in creating brand value, so this paper believes that the connotation of value creation of fashion brand image can be defined with the help of the connotation of brand value creation. The connotation of the fashion brand value creation can be defined as the fashion brand or enterprise using all aspects of innovation to create brand image, to let the brand grow based on original finite value space, to create additional brand value, to satisfy customers with the brand image value's pursuit of a higher goal, to maximize the value of the brand image.

2.2. Consumer Value Co-Creation

At present, many scholars believe that in value co-creation, customers are the co-creators of value, and what they co-create with enterprises is the experience value. Prahalad and Ramaswamy [18,19] point out that there are multiple interaction points between consumers and enterprises, enabling them to achieve value co-creation and a personalized experience. Zaborek and Mazur [20] believe that value co-creation is a positive interactive process between consumers and enterprises. Consumers actively contribute their wisdom and labor and cooperate with enterprises to invent, design, and provide valuable products, services, and experiences for other consumers. Andreu, Sánchez et al. [21] believe that consumer value co-creation broadly means consumers lead value creation activities.

Similarly, Pongsakornrungsilp and Schroeder [22] point out that the dominant player in value co-creation is consumers, and value is reflected in the experience of consumers. Consumer value co-creation can also be understood as customer participation. For example, Nambisan and Baron [23] point out that customers can satisfy their needs for information, emotion, and other content by participating in product innovation to realize value co-creation. Auh et al. [24] believe that customer participation value co-creation is an active customer participation behavior. Customers contribute knowledge, experience, and resources in participation, which can provide value to themselves and enterprises at the same time. Payne et al. [25] put forward that customer participation is the critical factor in value co-creation, reflecting the contribution degree of customers in value co-creation. To sum up, the connotation of consumer value co-creation includes the concepts of "consumers as value co-creators" and "customer participation". Therefore, the text defines consumer value co-creation as consumers, as the subject of value creation, who participate in value creation activities and create value together with enterprises.

Marcos and his peers [26] believed that engagement is the source of experience and the primary way to realize value co-creation. Zhang et al. [27] show that the primary expression mode of value creation is interaction, and both the interaction between consumers and enterprises and among consumers can bring good experience value to customers. Luo and coworkers [28] proposed that customer participation value co-creation is the premise for customers to obtain co-creation experience value. Based on the viewpoints of the above scholars, it can be known that the essence of co-creation is experience value, and the premise of producing experience value is customer participation or interaction. Based on this, Mandlik and Kadirov [29] proposed a mechanism model of consumer value co-creation, including anamorphic, process, and result. Antecedent variables are composed of factors that affect customer participation value co-creation. These variables will influence the experience value by influencing customer value co-creation behavior and finally impacting customer loyalty. Moise et al. [30] pointed out that customer participation in value co-creation, perceived risk, unique needs, control desire, and organizational support as motivation variables will affect customer participation in value co-creation behavior. Finally, customers can gain unique consumption experiences or value perception in value co-creation, and enterprises can gain brand loyalty.

Based on the studies of the above scholars, the internal mechanism of consumer value co-creation conforms to the "motivation-process-result" model. Among them, customer participation in value co-creation varies according to different research fields, and the motivation will impact customer participation in value co-creation (customer participation behavior). Experiential value or perceived value can be either a process variable or a result variable of customer participation in value co-creation, but it is always produced by value co-creation. Customer and brand loyalty are the direct result variables of consumer value co-creation.

2.3. *Customer Participation Behavior*

Existing research on consumer value co-creation primarily focuses on the level of customer participation behavior. Dai and Gu [31] define the connotation of customer participation from different research fields: Consumer participation refers to the specific behavior in which customers help create value in participating in products and services based on the traditional consumption field. Based on virtual community, customer participation behavior is a dynamic behavior of forwarding, sharing, and leaving comments. However, both emphasize the initiative of consumers or users in the process of participation. Growth [32] believes that customer participation is customers' behavior in service production and delivery, also called customer cooperative production behavior. Bove [33] states that customer participation behavior is a necessary consumer value co-creation behavior, and enterprises expect this behavior. Based on the above analysis, the connotation of customer participation behavior is defined as the behavior that consumers must take to successfully realize value creation in producing and delivering products or services.

Scholars have studied the constitutive dimensions of customer participation behavior. Ennew and Binks [34] proposed that customer participation is essentially the behavior and process of customer participation in value creation, mainly including three dimensions: information sharing, responsible behavior, and interpersonal interaction. Yi and Gong [35] believed that information sharing should be based on information search and thus divided into four dimensions of consumer participation behavior: information search, information sharing, responsible behavior, and interpersonal interaction. Bu Qingjuan et al. [36] divided customer participation behaviors in virtual communities into help-seeking, interpersonal interaction, feedback, and advocacy. Wu and Chen [37] divided customer participation behavior into three dimensions: information sharing, cooperation, and joint decision making according to the degree and process of customer participation. Participation behavior in consumer value co-creation behavior focuses on the interaction behavior between consumers and enterprises. Therefore, this paper chooses responsibility behavior and interpersonal interaction as the dimensions of customer participation behavior. In addition, the willingness of consumers to participate in value co-creation is the prerequisite for the occurrence of participation behavior, and it is influenced by the information provided by brands to consumers. Therefore, information search and information sharing are also considered the dimensions of customer participation behavior in this study.

2.4. *Experiential Value Perception*

Experiential value perception is a new concept that combines customer value perception, experiential value theory, and customer consumption behavior. The experience value perception comes from value co-creation and interaction between customers and brands [38]. Consumers' perceived experience value refers to consumers' perceived preference and evaluation of branded clothes and products [39]. The experience value is the cognition and evaluation of customers on enterprise services and advertisements based on their perception [40], which tends to be a psychological feeling formed by the interaction of experience and feeling on many vital points. This study proposes that experience value perception is consumers' comprehensive feeling and evaluation of many experience elements in value creation or other activities.

Scholars in different research fields have different classifications of perceived dimensions of experiential value. Kim and Oh [41] point out that the corresponding dimensions of experiential value perception are functionality, emotion, and social and experiential value perception in social networks. For mobile data services, experience value perception is divided into practicality and hedonic experience value perception. Huang [42] believes that customers' experience value perception includes functional experience value perception and emotional experience value perception in the social field of the mobile short video. This paper holds that the experiential value perception under consumer value co-creation manifests as practical, functional, and social experiential value perception. The connotation of emotional experience value perception can be defined as an inner feeling generated by consumers' processing and analysis of perceived information through thinking activities such as association based on their own experience and experience. Specifically, it refers to the value of feelings such as "fun" and "relaxation" obtained in the process. Functional experience value perception refers to consumers' perception of functional utility, shown by the brand's inherent attributes and primary performance. Social experience value refers to the customer's perception of the brand image value or symbolic value during the experience, such as the brand's social status, social reputation, and other content.

3. Hypotheses and Model Establishment

3.1. Research Hypotheses

The high quality of garments impacts the buyer's choice, particularly knowing the raw material source [1]. A high-value brand image can bring higher consumer loyalty and brand premium to a brand. Chen and Zhang [43] believe that customer-centered and value-oriented brand image value creation has successfully created brand awareness, reputation, and loyalty as brand value creation can affect consumers' cognition, trust, and loyalty to brands through value co-creation. Joshi and Sharma [44] believed that brand value creation based on consumer participation would directly affect customers' repeated purchase behavior. Dai et al. [45] point out that value creation can help brands attract more customer groups and gain higher brand loyalty. Brand value creation includes many aspects, among which the value creation of the brand image is one of them, and it can bring positive effects to enterprises, including price premium, customer satisfaction, and brand loyalty [46]. In conclusion, brand image value creation has a positive impact on brand loyalty; therefore, Hypothesis H1 is proposed in this paper:

Hypothesis H1. *The value creation of fashion brand image has a positive effect on brand loyalty.*

Enterprises will transmit various resource information to consumers in value creation, including corporate values and corporate culture connotation [47]. This will have an impact on customer participation in product development and upgrading. Consumers began to actively participate in the process of value creation through the internet [48], during which the value creation of enterprises' products or services would promote the participation of consumers, and such customer participation behavior is reflected in the contribution of customers' knowledge, skills, and experience in the consumption field. Wu and Chen [37] pointed out that consumers play value creators, and value co-creation with the brand is inseparable from the brand's offerings. Brand elements' design and production activities and other aspects will affect the consumer's participation behavior [1].

Similarly, in the process of creating the value of fashion brand image, the realization method of creating the value of fashion brand image will increase the impression of the brand image in consumers' minds, enhance consumers' perception of the brand image, and thus promote the creation of consumers' value co-creation behavior. For example, differentiated marketing can meet the unique needs of customers. The customer's unique demand is an essential motivation for participating in value co-creation and significantly influences customers participating in value co-creation.

To sum up, value creation will influence customers' participation behavior or value co-creation behavior to contribute their knowledge and experience through value transmission. Accordingly, Hypothesis H2 is proposed in this paper:

Hypothesis H2. *Fashion brand image value creation has a positive impact on customer participation.*

When analyzing the relationship between customer participation behavior and brand loyalty, customer participation behavior in participating in value co-creation will enhance customers' understanding of the enterprise, thus promoting brand loyalty [49]. Based on the co-creation theory of consumer value, Hoyer and coworkers proposed that customer engagement enhances brand loyalty by improving the brand experience [50]. Etgar [51] points out that consumers can have a high level of interaction with enterprises through cooperative behavior and information sharing (cooperative behavior and information sharing are the manifestations of customer participation behavior). Consumers enjoy more autonomy and control during this period to satisfy their personalized needs and preferences, thus enhancing brand loyalty [52]. Experience-based consumer value co-creation is a high level of customer participation mode, providing innovative thinking for enterprises and positively influencing consumer loyalty [16,19,27]. Thus, customer participation in consumer value co-creation will promote brand loyalty, and in the process of participation, it is continuously strengthened and consolidated. Accordingly, Hypothesis H3 is proposed in this paper:

Hypothesis H3. *Customer engagement behavior has a positive impact on brand loyalty.*

Zhao [53] believes that experience value is essentially the value of customers' subjective perception and is affected by customers' participation behavior. In the research field of value co-creation, Yang [54] proposed that customers have initiative, and their active participation can enable them to obtain unique consumption experiences and promote the generation of experience value perception. Customers' participation in value co-creation can enhance their sense of identity with enterprise value, thus enhancing experience value perception, including emotional, social, and functional experience value perception. My-Quyen [55] believes that experience value perception can be obtained from the function, efficiency, and other physical attributes of customer participation in value co-creation activities or creating certain feelings. Zhang and Chen [56] believe that customers' participation in value co-creation will generate deeper emotions for brands, namely, improving the value perception of emotional experience. Vahdat [57] and Tang [58] proposed that luxury goods are not available in luxury marketing, so customer participation behavior has a more significant impact on emotion and social experience value perception.

In conclusion, under the value co-creation mode, customer participation behavior will affect the perceived value of the fashion brand experience. Therefore, Hypothesis H4 is proposed in this paper:

Hypothesis H4. *Customer participation behavior has a positive impact on the perceived value of fashion brand experience.*

There are many driving factors of brand loyalty, among which is customer perceived experience value. Paulose and Shakeel 2021 [59] believe consumers are loyal to brands with high perceived experience value. From the perspective of customer participation value co-creation, experience value perception is affected by customer participation behavior, affecting customer loyalty to the brand. Zhao [53] points out that when customers participate in value co-creation, their emotional experience value perception and recreational experience value perception are strengthened, and their feelings towards products are deepened. As a result, they have a higher brand commitment and loyalty and actively promote and recommend brands and products to others. Zhang and coworkers [60] pointed out that customers' perceived emotional and functional experience value would promote their trust and loyalty to the brand. Guan [61] points out that the value of the social experience will significantly meet consumers' psychological and spiritual needs and then promote consumers to have a strong cognition, association, and loyalty to the brand. It can be seen from the above that the perceived value of fashion brand experience will have a positive impact on brand loyalty. Therefore, Hypothesis H5 is proposed in this paper:

Hypothesis H5. *The perceived value of fashion brand experience has a positive impact on brand loyalty.*

In the process of customer participation in value co-creation, enterprises or brands make full use of the valuable resources brought by consumers through the participation of consumers, complete the process of value creation and delivery, and obtain economic returns and customer satisfaction loyalty. Nardi et al. [62] proposed that brand value creation could affect consumers' cognition and attitude towards the brand through customer participation, including customer satisfaction, trust, and loyalty. Therefore, it can be inferred that customer participation during value co-creation plays an essential role in the relationship between value creation and brand loyalty. Chen [63] points out that brands will guide brands to grasp the consumption trend through accurate customer information exchange, information sharing, and other participating behaviors conducive to improving customers' brand recognition, association, satisfaction, and loyalty in value creation activities. Focusing on introducing new materials in the garment industry or proposing logistics models for the supply chain has been practiced for a long time [64,65]. The enterprise or the brand value creation pattern changed under value-creating behavior or participation behavior [27]. They maintained that primary performance for our customers to participate in the brand product research and development, production, logistics, and marketing leads to value creation concerns from use-value to the experience value transfer, thereby enabling customers to form a higher quality of brand loyalty. In conclusion, brand value creation will promote brand loyalty under the influence of value co-creation behavior or customer participation behavior. Therefore, Hypothesis H6 is proposed in this paper:

Hypothesis H6. *Customer participation plays a mediating role in the relationship between the value creation of fashion brand image and brand loyalty.*

With the coming of the era of the experience economy, value creation is rapidly shifting to customer experience, and customer participation in the behavior of value co-creation is the premise of customer experience value perception. Teng and Tsai put forward from tourism management that the value creation of tourism brands will affect tourists' perception of tourism experience value through their participation behavior [66]. For example, when tourists join the brand's interactive social activities, their participation behaviors such as information exchange and sharing will impact the value perception of social experience. Brand value creation under customer participation behavior would impact experiential value perception; that is, customer participation behavior has a specific mediating effect on the influence relationship between brand value creation and experiential value perception [16,19,67]. From experiential marketing, Yoo et al. [68] point out that brand value creation must rely on certain material carriers and bring different experiential value perceptions to customers through their participation behavior. Therefore, in creating the image value of a fashion brand, the carriers of its experience value are image design, image publicity, brand aesthetics, brand charm, and brand function, which will impact the perception of experience value through the customer's participation behavior.

Experiential value perception significantly affects the relationship between participation behavior, "information exchange and interpersonal interaction", and tourist loyalty. Since customer participation in value co-creation can impact customer loyalty through experience value perception. Fang et al. [69] believe that user participation in the context of value co-creation can benefit both customers and brands. Customers' participation behaviors (comments, inquiries, and interpersonal interactions with other consumers) in the online shopping environment can improve customers' perception of experience value, thus improving their repeated purchases, recommendation, and other behaviors [6,70]. In conclusion, customer participation behavior positively impacts brand loyalty, and the perceived value of fashion brand experience plays a significant role in this influencing relationship. In other words, customer participation behavior can positively affect brand loyalty through the perceived value of the fashion brand experience.

Therefore, the image value creation of fashion brands can make customers perceive the experience value of fashion brands through customer participation behavior and positively impact brand loyalty. Therefore, Hypothesis H7 is proposed in this paper:

Hypothesis H7. *Customer participation behavior and apparel brand experience value perception play an intermediary chain role in brand image value creation affecting brand loyalty.*

The willingness of customers to participate in this paper refers to the willingness of consumers to participate in value co-creation, which originates from consumers' cognition of value creation activities to a certain extent and is the basis for participating in value co-creation [71]. Venkatesh and Davis [72] believe that consumers' willingness to participate directly influences actual participation behavior. In the context of customer participation, it should be noted that the "willingness" is customers' attitude toward their participation in product creation and shows in her research that willingness can regulate consumers' behavior of participating in value creation. When consumers are highly willing to participate, they want to participate in value-creation activities. The customers usually gradually generate the willingness for value co-creation in the interactive atmosphere created by enterprises and then produce value co-creation behavior. Based on the above analysis, the willingness of customers to participate in value co-creation can promote customer participation behavior. Therefore, Hypothesis H8 is proposed in this paper:

Hypothesis H8. *Customer participation intention plays a moderating role in the influence of brand image value creation on customer participation behavior.*

Gavilan [73] points out that customers' trust in hotels is mainly based on other consumers' online participation behaviors. For example, online reviews of hotels will affect customers' trust in the hotel brand to some extent. Brand trust is a process of regulating the influence of customer participatory value co-creation on brand loyalty through the level of customer perceived risk and perceived value. When customers' perceived risk or value is higher, their desire to participate in the value co-creation will be stronger, leading to more active participation behavior of customers to enhance brand loyalty. After the occurrence of participatory behavior (value co-creation behavior), customers will have a psychological evaluation of the brand, namely brand trust, and thus affect consumer loyalty [35,38]. Azize [74] believes that users' shared and interactive participation in the brand community will increase brand loyalty, and users' trust in the community plays a moderating role. Thus, consumers' participation in value co-creation can deepen their understanding of enterprises and help them establish emotional links with enterprises. Moreover, under the influence of brand trust, consumers' repeat purchase behavior and recommendation intention will be further enhanced.

Based on the above analysis, fashion brand trust plays a facilitating role in the relationship between customer participation behavior and brand loyalty. Therefore, Hypothesis H9 is proposed in this paper:

Hypothesis H9. *Fashion brand trust plays a moderating role in influencing customer engagement behavior on brand loyalty.*

Consumer experience value perception results from a comprehensive feeling, which is essential for understanding the customer-brand relationship. For example, in the virtual brand community, only when consumers first form trust in the brand community will they take the initiative to deeply understand the brand products and services in the community to obtain the perception of experience value and finally form brand loyalty. Elena and Jose [12] believe that value perception is influenced by trust, and customers with high brand trust will get more perceived benefits and value than other consumers, and it is easier to form brand loyalty. The experience value perception generated by customer value co-creation behavior will positively impact brand loyalty through brand trust. In addition, the level of consumer trust is positively correlated with the value perception, and the interaction between the two will further determine brand loyalty. The consumers' level of trust in brands will affect their access to brand resource information, thus affecting their perception of experience value and ultimately influencing consumers' brand attachment [45].

In conclusion, the relationship between perceived experience value and brand loyalty is disrupted by brand trust, and when consumers form brand trust, they are more likely to generate brand loyalty. Therefore, Hypothesis H10 is proposed in this paper:

Hypothesis H10. *Fashion brand trust plays a moderating role in influencing the perceived value of fashion brand experience on brand loyalty.*

3.2. Theoretical Model Establishment

Taken together, this theoretical model is established based on consumer value co-creation theory and experiential value perception theory. This is achieved by combining the above research hypotheses, this paper takes the customer participation behavior and the perceived value of apparel brand experience as the intermediary variable and the customer participation willingness and brand trust as the moderating variable and constructs a theoretical model of the impact of apparel brand image value creation on brand loyalty, as shown in Figure 1.

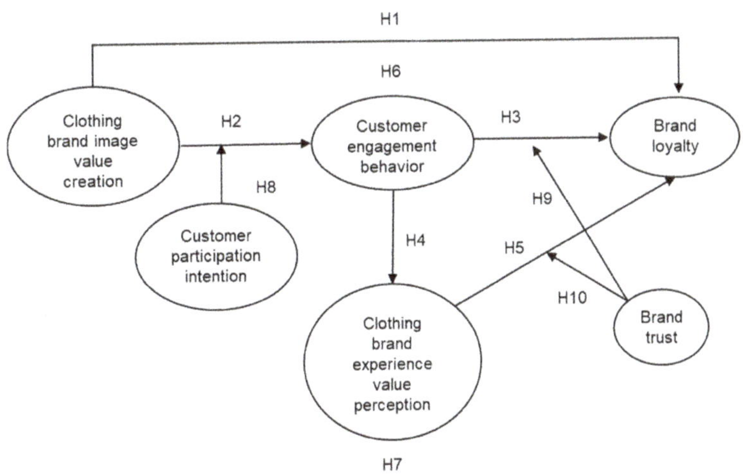

Figure 1. The theoretical model for this study.

4. Empirical Test and Data Analysis

4.1. Item Development

The measurement scale in this paper is based on the relevant mature scale to be screened and modified. Among them, customer participation behavior mainly borrows from the scale of Yi and Gong [35]. The perceived value of fashion brand experience is mainly based on the scale of Shin [75] and Hennigs [76]. Brand loyalty is mainly based on the measurement scale of Guan [61] and Apenes Solem [11]. Customer participation intention is mainly based on the scale of [53]. Brand trust is mainly based on the scale of Elena and Jose [12].

There is no mature fashion brand image value creation scale, so it is necessary to develop a measurement scale. This paper constructs an index system of fashion brand image value creation based on the grounded theory, which includes five dimensions, including image design, image publicity, brand aesthetics, brand charm, and brand function, and

29 indexes. Based on much literature and previous studies, a suitable measurement scale was formed through modification and adjustment, as shown in Table 1.

Table 1. Measurement scale of fashion brand image value creation.

Dimension	Number	Secondary Index	Item	Reference
Image design	ID1	Product image design	Product image design novel.	Cai [4] Marine [7] Cai and Cheng [77]
	ID2	Store image design	Store image design is unique.	
	ID3	Brand identity image design	Brand identity image design has identification.	
	ID4	Packaging design	The packaging design is simple and generous.	
	ID5	Price image design	Reasonable price image design.	
Image publicity	IP1	Designer image publicity	The brand designer image is consistent with the designer image in my mind.	Fu and Cao [78]
	IP2	Advertising image promotion	Brand ads usually get my attention.	
	IP3	Service image publicity	Branded service usually satisfies me.	
	IP4	Corporate image publicity	The brand that belongs to the enterprise has a good reputation.	
	IP5	Customer image promotion	Customers of brands usually have a good personal image.	
	IP6	Promotion image publicity	The promotional activities of the brand can continually deepen my impression of the brand.	
Brand Aesthetics	BA1	Brand style	The brand style is usually popular with the public.	Godey [79] Cai [4]
	BA2	Brand symbol	Brand symbols are usually aesthetic.	
	BA3	Brand color	The use of brand color is usually in line with the public aesthetic concept.	
	BA4	Brand design	Brand patterns are usually representative	
	BA5	Brand poster	Brand posters are usually unique and original.	
	BA6	Brand store atmosphere	The store atmosphere is usually pleasant and comfortable.	
Brand Charm	BC1	Brand concept	The brand concept is in line with the public concept	Koenigsberg [80]
	BC2	Brand spirit	The brand spirit is positive	
	BC3	Brand culture	Profound brand culture	
	BC4	Brand reputation	Good brand reputation	
	BC5	Brand personality	Strong brand personality	
	BC6	Brand added value	The brand has added brand value.	
	BC7	Brand value orientation	The public can recognize brand value orientation	
Brand function	BF1	Recognition function	This brand makes it easy for me to distinguish myself from other brands.	Ma [81] Jiao [8] Gu and Xu [82]
	BF2	Quality commitment and assurance function	The brand's products are of reliable quality.	
	BF3	Communication and shopping guide function	The brand can convey the brand or product message to me.	
	BF4	Competition function	The brand has a highly competitive advantage	
	BF5	Value chain function	The brand will usually cooperate with well-known brands	

4.2. Data Collection

The questionnaire survey was used to collect data, and online questionnaires were distributed through the "Questionnaire Star" software. As a principal target to investigate the perceptions of fashion brand image value creation by diverse types of consumers, all consumers who had ever purchased branded clothing via physical or internet shops were the subject of our research. For the purpose of matching the subjects' characteristics, we asked consumers to choose what brands they were concerned about. A total of 642 questionnaires were issued, of which 567 were valid, with an effective rate of 88.3%. In order to ensure the reliability and validity of the questionnaire, this study first conducted a pre-survey within a small group, and this questionnaire was partially modified based on the pre-survey results. The descriptive statistical results of the final survey samples collected are shown in Table 2. It can be seen from the table that all ages, occupations, regions, and monthly incomes were involved in the survey, indicating that the selection of survey groups was reasonable.

Table 2. The basic information of survey samples.

Statistical Variables		Sample Size	Proportion (%)
Gender	Male	270	47.62%
	Female	297	52.38%
Age	≤18	19	3.41%
	19–25	185	32.68%
	26–30	157	27.63%
	31–40	110	19.34%
	41–50	70	12.35%
	≥51	26	4.59%
Occupation	Party and government officials	41	7.23%
	Professional and technical personnel, doctors	53	9.35%
	Teachers	63	11.11%
	Company management cadre	30	5.29%
	Company worker	115	20.28%
	Students	184	32.45%
	Business service worker	42	7.41%
	Self-employed	36	6.35%
	Other	3	0.53%
Education background	High school and below	55	9.65%
	Junior college	84	14.88%
	Undergraduate	245	43.26%
	Master's degree or above	183	32.21%
Monthly income	≤1500 RMB	167	29.45%
	1501–3000 RMB	71	12.52%
	3001–5000 RMB	114	20.11%
	5001–8000 RMB	137	24.16%
	8001–12,000 RMB	54	9.52%
	≥12,001 RMB	24	4.24%
Place of abode	First-tier cities	173	30.51%
	Second-tier cities	222	39.15%
	Third line cities	104	18.34%
	Fourth tier cities or other areas	68	12.00%

4.3. Reliability and Validity Test

Data testing was carried out on the measurement scale of value creation of fashion brand image, mainly including reliability and validity test, factor test, and goodness of fit test of the model. In the exploratory factor molecule, the cumulative variance contribution rate of the scale was 73.061% > 60%, and the overall KMO and Cronbach's α values were 0.946 and 0.965, respectively, both greater than 0.7. Meanwhile, the KMO and Cronbach's α values of each dimension also reached the acceptance standard. In the confirmative factor analysis, the standardized factor loading (EFA) was between 0.648 and 0.912, meeting the criteria of greater than 0.5, and the combinatorial reliability (CR) and square extraction variance (AVE) also met the criteria of 0.7 and 0.5. In the goodness of fit test of the model, the chi-square to the degree of freedom ratio (X2/DF) is 2.898, which is less than 3, AGFI and GFI are more significant than 0.8, and NFI, CFI, TLI, and IFI are all greater than 0.9. The measurement scale has good reliability and validity based on the above analysis.

SPSS25.0 was used for reliability and validity tests and confirmative factor analysis in this study. The specific results are shown in Table 3. Cronbach's α coefficient and KMO value of each measurement variable meet the accepted standard of the lowest 0.5, and Cronbach's α coefficient and KMO value of the whole model are more significant than 0.7, which can be considered good reliability and validity of the model. In addition, the combined reliability and mean-variance of all the measured variables were within the acceptable range, and the confirmatory factor analysis results were good.

Table 3. The test of model reliability and validity and confirmatory factor analysis.

Dimension		Measurement Item	CITC	α After Deleting	α	KMO	Bartlett's Test of Sphericity			EFA	CR	AVE
							Chi-Square	Degrees of Freedom	Significant			
Fashion brand image value creation		ID	0.730	0.864	0.888	0.888	1274.11	10	0.000	0.775	0.899	0.642
		IP	0.727	0.863						0.811		
		BA	0.716	0.866						0.859		
		BC	0.768	0.854						0.832		
		BF	0.700	0.870						0.723		
Customer engagement behavior		CPB1	0.745	0.913	0.922	0.851	2291.85	15	0.000	0.794	0.923	0.666
		CPB2	0.813	0.903						0.852		
		CPB3	0.815	0.903						0.865		
		CPB4	0.771	0.909						0.818		
		CPB5	0.760	0.910						0.777		
		CPB6	0.757	0.911						0.785		
Fashion brand experience value perception	Emotional experience value perception	AEVP1	0.694	0.871	0.875	0.694	831.85	3	0.000	0.767	0.883	0.716
		AEVP2	0.840	0.750						0.916		
		AEVP3	0.752	0.832						0.849		
	Functional experience sense of value	FEVP1	0.802	0.846	0.892	0.805	1193.63	6	0.000	0.851	0.904	0.701
		FEVP2	0.748	0.866						0.801		
		FEVP3	0.802	0.844						0.901		
		FEVP4	0.697	0.883						0.792		
	Value perception of social experience	SEVP1	0.686	0.863	0.873	0.695	820.47	3	0.000	0.758	0.881	0.713
		SEVP2	0.834	0.749						0.913		
		SEVP3	0.755	0.823						0.854		

Table 3. *Cont.*

Dimension	Measurement Item	CITC	α After Deleting	α	KMO	Bartlett's Test of Sphericity			EFA	CR	AVE
						Chi-Square	Degrees of Freedom	Significant			
Brand loyalty	BL1	0.695	0.861	0.882	0.814	1394.81	10	0.000	0.756	0.884	0.604
	BL2	0.697	0.861						0.753		
	BL3	0.765	0.844						0.808		
	BL4	0.774	0.843						0.828		
	BL5	0.652	0.871						0.736		
Customer participation intention	CPI1	0.780	0.903	0.910	0.747	1081.33	3	0.000	0.829	0.911	0.774
	CPI2	0.838	0.852						0.899		
	CPI3	0.837	0.853						0.909		
Brand trust	BT1	0.799	0.850	0.894	0.813	1210.69	6	0.000	0.731	0.850	0.588
	BT2	0.730	0.875						0.712		
	BT3	0.829	0.837						0.893		
	BT4	0.703	0.886						0.716		

Overall KMO = 0.927, Bartlett test chi-square value = 10,153.84, Sig value = 0.000, Cumulative variance contribution rate = 74.78%, overall Cronbach's α coefficient = 0.954.

4.4. Goodness of Fit Test of Model

AMOS22.0 was used for the goodness of fit test of the model in this study, and the specific results are shown in Table 4. Since when X2/DF is less than 3, RMR is less than 0.05, RMSEA is less than 0.08, AGFI and GFI are greater than 0.8, and other indicators are greater than 0.9, it indicates that the model has a good degree of the fitting. It can be seen from the table that all indicators of each measurement variable meet the requirements, so it can be considered that the relevant results of each measurement model are relatively ideal.

Table 4. The test of model goodness of fit index.

Fitting Index	Fashion Brand Image Value Creation	Customer Engagement Behavior	Fashion Brand Experience Value Perception	Brand Loyalty	Customer Participation Intention	Brand Trust
CMIN	71.613	17.928	76.321	55.384	5.274	31.216
X2/DF	2.170	2.988	2.544	2.408	2.637	1.951
RMR	0.020	0.029	0.033	0.012	0.039	0.026
RMSEA	0.049	0.063	0.056	0.053	0.033	0.044
GFI	0.939	0.988	0.972	0.945	0.918	0.944
AGFI	0.911	0.958	0.948	0.927	0.859	0.908
NFI	0.953	0.992	0.977	0.946	0.942	0.913
IFI	0.956	0.995	0.966	0.948	0.945	0.937
CFI	0.956	0.995	0.966	0.948	0.945	0.937

4.5. Hypothesis Testing

The test results of the theoretical model are shown in Table 5. H1 path coefficient is 0.602, and the T value is 6.324, which meets the significant standard, indicating that the value creation of fashion brand image has a significant positive impact on brand loyalty. Therefore, Hypothesis H1 is established. H2 path coefficient is 0.863, T value is 16.182, reaching the significant standard, indicating that the value creation of fashion brand image has a significant positive impact on customer participation behavior, so the assumption of

H2 is valid. The diameter coefficient of H3 is 0.500, and the T value is 4.475, which reaches the significant standard, indicating that customer participation behavior has a significant positive impact on brand loyalty. Hypothesis H3 is established. The path coefficient of H4 was 0.876, and the T value was 18.436, which reached the significant standard, indicating that customer participation behavior had a significant positive impact on the perceived value of fashion brand experience. Hypothesis H4 was established. The path coefficient of H5 was 0.481, and the T value was 4.287, which reached the significant standard. Based on the above analysis, the perceived value of fashion brand experience significantly impacts brand loyalty, so hypothesis H5 is established.

Table 5. The result of a hypothesis test.

Hypothesis	Path Coefficient	S.E.	T Value	*p*-Value	Conclusion
H1	0.602	0.107	6.324	***	Support
H2	0.863	0.067	16.182	***	Support
H3	0.500	0.116	4.475	***	Support
H4	0.876	0.039	18.436	***	Support
H5	0.481	0.123	4.287	***	Support

Note: *** indicates that p is significant below 0.001.

4.6. Mediation Effect Test

In this study, the Bootstrap method was used to test the mediating effect of Hypothesis H6 customer participation behavior, the chain mediating effect of Hypothesis H7 customer participation behavior, and the perceived value of fashion brand experience. When testing Hypothesis H6, a simple mediation model (Model4) is used to analyze the mediation effect of customer participation behavior under controlling the basic information of the survey objects. The test results are shown in Table 6. The upper and lower limits of the bootstrap95% confidence interval do not contain 0, the direct effect value is 0.451, and the indirect effect value is 0.322, accounting for 58.34% and 41.66% of the total effect, respectively. It shows that customer participation has a significant (partial) mediating effect on the relationship between the value creation of fashion brand image and brand loyalty, and hypothesis H6 has been verified. When testing Hypothesis H7, Model6 analyzes the chain mediating effect between customer participation behavior and fashion brand experience value perception under the same control of basic information. The specific results are shown in Table 7. The upper and lower limit of the BootCI confidence interval does not contain 0, the direct effect value is 0.463, and the indirect effect value is 0.180 and 0.158, respectively. This indicates that customer participation behavior and fashion brand experience value perception have an intermediary chain role (partial intermediary) in the impact of fashion brand image value creation on brand loyalty. Hypothesis H7 is verified.

Table 6. Customer Participation Behavior Mediation Effect Analysis Results.

Path	Effect Value	Boot Standard Error	BootCI Lower Limit of the Confidence Interval	BootCI Upper Limit of the Confidence Interval	Effect Proportion
Total effect	0.773	0.0455	0.7506	0.8993	-
Direct effect	0.451	0.0515	0.5610	0.7633	58.34%
Indirect effect (fashion brand image value creation → customer participation behavior → brand loyalty)	0.322	0.0393	0.0593	0.2163	41.66%

Table 7. Chain Effect Analysis Results.

Path	Effect Value	Boot Standard Error	BootCI Lower Limit of the Confidence Interval	BootCI Upper Limit of the Confidence Interval
Total effect	0.801	0.0347	0.7323	0.8684
Direct effect	0.463	0.0455	0.3523	0.6130
Indirect effect (fashion brand image value creation → customer participation behavior → brand loyalty)	0.180	0.0434	0.0928	0.2638
Indirect effect (fashion brand image value creation → customer participation behavior → fashion brand experience value perception → brand loyalty)	0.158	0.0391	0.0195	0.1715

4.7. Moderation Effect Test

This study used a hierarchical regression model to examine the moderating effect of customer participation intention and brand trust. Under the condition of controlling the basic information of the survey objects, the interaction items of "fashion brand image value creation × customer participation willingness", "customer participation behavior × brand trust" and "fashion brand experience value perception × brand trust" were respectively introduced for regression analysis. The specific results are shown in Table 8. After introducing the interaction term, the value of the adjusted R square gradually increases, indicating that the model is better fitted. In addition, the normalization coefficient was 1.320, 1.321, and 1.472, and the T value was 7.034, 5.903, and 5.923, respectively, and the p was significant at the level of 0.01. This indicates that interaction items have a significant impact on customer participation behavior or brand loyalty. Customer participation willingness has a significant moderating effect on the impact of fashion brand image value creation on customer participation behavior, and brand trust has a significant moderating effect on customer participation behavior on brand loyalty. At the same time, the perceived value of fashion brand experience also has a significant regulating effect on the influence of brand loyalty. Therefore, Hypotheses H8, H9, and H10 have been verified.

Table 8. The results of moderating effect.

Variable	Customer Participation Behavior		Variable	Brand Loyalty		Variable	Brand Loyalty	
	Standardization Coefficient	T Value		Standardization Coefficient	T Value		Standardization Coefficient	T Value
Gender	0.060	2.098	Gender	−0.065	−1.797	Gender	−0.026	−0.786
Age	−0.013	−0.430	Age	0.050	1.365	Age	0.104	3.059
Occupation	0.003	0.103	Occupation	0.018	0.510	Occupation	0.007	0.219
Education	−0.014	−0.484	Education	0.000	0.002	Education	−0.008	−0.251
Income	−0.046	−1.560	Income	0.050	1.351	Income	0.026	0.757
Residence	−0.055	−1.908	Residence	0.031	0.868	Residence	0.016	−0.472
Fashion brand image value creation	1.201 ***	16.368	Customer engagement behavior	0.561 **	2.939	Fashion brand experience value perception	0.745 ***	4.234
Customer participation intention	1.038 ***	6.442	Brand trust	0.983 ***	5.721	Brand trust	0.140 **	3.142
Fashion brand image value creation × customer participation willingness	1.320 ***	7.034	Customer engagement behavior × brand trust	1.321 ***	5.903	Fashion brand experience value perception × brand trust	1.472 ***	5.923
Adjust R squared change	−0.002→0.557→0.597		Adjust R squared change	−0.006→0.369→0.378		Adjust R squared change	−0.006→0.466→0.477	
F value change	0.811→78.665→82.524		F value change	0.517→37.398→33.189		F value change	0.517→55.206→48.94	

Note: *** means p is significant at the level of 0.001, ** means p is significant at the level of 0.01.

5. Discussions

The empirical results show that fashion enterprises can further form brand loyalty by promoting consumers' value co-creation and enhancing experience value perception when carrying out value creation activities. The SEM model was developed to establish the relationships among variables as follows in Figure 2.

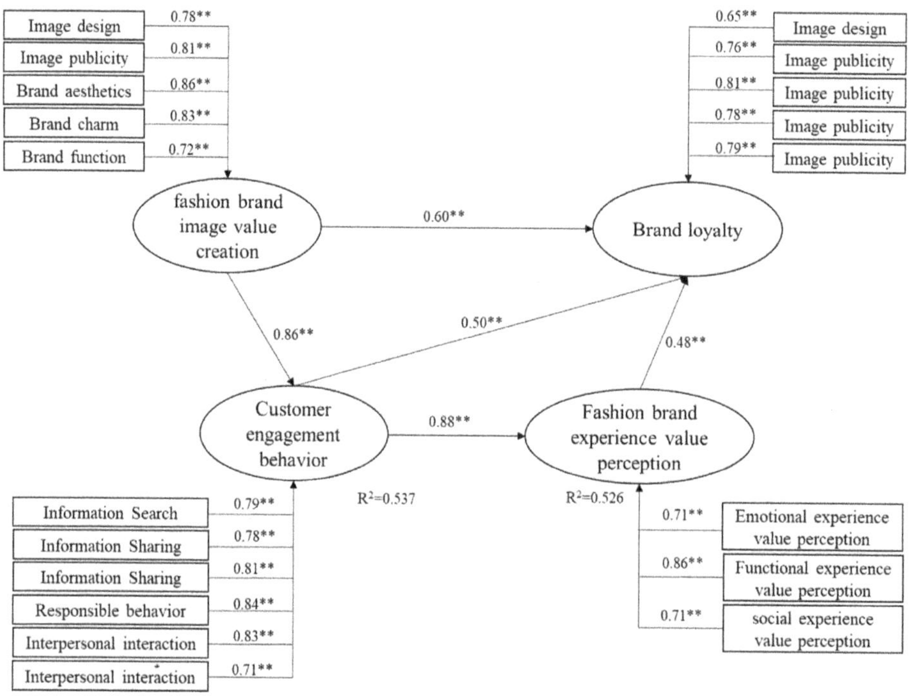

Figure 2. The graphical representation of the SEM model. ** means p is significant at the level of 0.01.

Assuming the establishment of H1 indicates that creating fashion brand image value will directly affect brand loyalty when fashion enterprises carry out value creation activities, consumers' perception of various elements of fashion brand image can be improved, thus enhancing brand loyalty. Hypotheses H2, H3, and H6 show that customer participation plays an intermediary role in the relationship between the value creation of fashion brand image and brand loyalty. When consumers' perception of various elements of the value creation of fashion brand image is high, it will stimulate consumers' interest in participating in the co-creation and then produce customer participation behavior, impacting brand loyalty. Furthermore, customer participation behavior belongs to creating the necessary link of consumer value. It can make consumers and brand both sides establish good relations of cooperation, to a certain extent, also affect the customer to the enterprise's trust, so that customers have a positive attitude orientation and enhance customer repeat purchase desire, to enhance the brand loyalty further.

It is assumed that H2, H4, H5, and H7 are established, indicating that the creation of fashion brand image value will first affect consumers' customer participation behavior and then influence brand loyalty through the role of perceived experience value, that is, customer participation behavior and perceived experience value of fashion brand have an intermediary chain role. The fashion brand experience value perception needs to be based on consumer value co-creation in fashion brand image value creation. According to the theory of consumer value co-creation, consumers' positive perception of fashion brands will promote them to participate in the value co-creation, and the inevitable result of participating in the value co-creation is to produce a unique consumption experience, during which consumers will generate value perception of experience elements, namely, experience value perception. In addition, this perception usually leads to a favorable orientation; consumers will have a more profound emotional attachment to the brand, thus have a higher brand commitment and loyalty to the brand, and will actively promote and recommend the brand and products to the brand.

The establishment of Hypothesis H8 indicates that creating fashion brand image value on customer participation behavior is regulated by customer participation willingness. Consumers will evaluate and judge the value creation of fashion brand image based on their perception. When consumers are interested in the value creation of fashion brand image, they will hope to better experience fashion brand products or services by participating in value creation. Therefore, the degree of consumers' interest in the value creation of fashion brand image and the degree of consumers' desire to participate in the value creation is customers' willingness to participate in this influence relationship. Furthermore, the more interested they are, the stronger their hope is, their willingness to participate, and easier to generate customer participation behavior.

It is assumed that the establishment of H9 and H10 indicates that the process of customer participation behavior on brand loyalty and the process of apparel brand experience value perception on brand loyalty are both regulated by apparel brand trust. From the perspective of consumer value co-creation, consumers' trust in fashion brands will affect their behavior in value co-creation and further affect their purchasing decisions. When consumers have high trust in fashion brands, they are more active in value co-creation; Through such frequent brand interaction, consumers will have a sense of dependence on fashion brands and thus form brand loyalty. According to the theory of experiential value perception, brand trust strongly correlates with consumers' experiential value perception. When consumers trust a brand, it indicates that they have a high psychological evaluation of the brand and will actively acquire more information related to the brand or product, resulting in a positive perception of the brand, and thus promoting the generation of brand loyalty.

6. Conclusions

6.1. Theoretical Implications

This research provides a comprehensive view of the factors that influence brand loyalty: customer participation behavior, perceived experience value, brand trust, and fashion brand image value creation attributes. Customer participation causes good relations between consumers and the brand and develops brand trust through positive attitude orientation by enhancing customer repeat purchase desire and brand loyalty. Thus, brand value creation activities improve fashion brand image and enhance brand loyalty. The customer participation behavior is developed before brand loyalty through the role of perceived experience value. The value creation of a fashion brand image depends on consumers' perceptions. The consumers' interest or desire (i.e., participation willingness) of the value creation of fashion brand image is related to their hope (i.e., trust) with brands. Fashion brand image value will directly affect brand loyalty. Customer value co-creation and participation behavior on brand loyalty and purchasing decisions are regulated by apparel brand trust. Brand trust strongly correlates with consumers' experiential value perception. More brand interaction results in more brand or product information, causing higher brand psychological evaluation, which ultimately causes higher brand positive perception, higher brand loyalty, and trust in brand trust.

6.2. Practical Implications

In terms of promoting consumer value co-creation, it can be achieved by enhancing the interaction between customers and brands based on empirical results. First, fashion enterprises need to open interactive social platforms on websites such as Sina Weibo,

Xiaohongshu, Facebook, and WeChat to establish a social network connection between brands and consumers. Secondly, fashion enterprises should set appropriate usernames and other essential information, including corporate brand logos and related pictures in social media accounts, which is conducive to gaining the attention and trust of other users. Third, fashion enterprises need to establish an effective information feedback mechanism to obtain the consumption needs of online consumers, which requires enterprises to enhance their proactive awareness and actively release topics and activities to enhance interaction and realize value co-creation. In addition, fashion enterprises can also set up an influencers business department, focus on content marketing, and in the form of a network broadcast and soft text push to trigger the discussion among users and brands.

The primary way to increase offline interaction among consumers and brands is to use the star effect. Fashion enterprises can invite stars to participate in offline activities to attract consumers to experience the scope of brand activity atmosphere and take the opportunity to promote consumers to participate in value co-creation. Secondly, fashion enterprises can enhance the interaction with consumers by opening offline experience stores to experience the category of fashion brand concept. For example, fashion enterprises can transform fashion shows into offline experience stores with different themes in different regions. Consumers can experience the design concept and development concept of brand clothes and jointly discuss their views on the concepts to achieve value co-creation.

In terms of enhancing the perception of experience value, it can be achieved by enhancing the brand information symmetry perceived by consumers. Fashion enterprises can choose to join the virtual brand community to solve asymmetric brand information perceived by consumers. Releasing many high-quality contents related to the brand in the virtual community and setting keywords and title classification information can make consumers better understand the brand and improve it. Secondly, to enhance information symmetry, fashion enterprises should standardize the transmission mechanism of brand information and actively send "screening" information to consumers to realize the transformation of information to consumers' demands. For example, when consumers have low information perception of brand connotation, spirit, concept, and culture, fashion enterprises should promote the spiritual content of fashion brands and reduce the output of material content such as products, posters, and stores to reduce information asymmetry. For example, training offline sales personnel to explain the brand story and history can effectively improve customers' understanding of the brand connotation.

6.3. Limitations and Future Research

The present study has a few limitations. First, this study is limited to Chinese clothing consumers in a particular geographical region. Hence, further testing of other objects across different geographical regions is required before these results can be generalized. Second, the participants filling out questionnaires based solely on the understanding of mass consumers is not comprehensive enough and more survey samples of occupations for apparel enterprise practitioners can be added to further ensure the accuracy of the research. Third, Fashion brand image value creation is currently a relatively new concept in the field of public cognition. Through the form of questionnaires, especially in the form of online distribution, it is easy to cause some respondents to have a partially thorough understanding of the research topic and cause the conclusions to be inconsistent with the real ideas. It can also be considered to use scientific and technological means to improve the accuracy of the recovered data, or to try to use mathematical analysis methods and experimental methods to break through the limitations of research methods. Therefore, we invite future research in this area. Finally, different target consumer perceptions of fashion brand image value creation may lead to inconsistent results, thus it is recommended that other consumers from different countries be selected as subjects for investigation in the future. Further research emphasizing different subjects should be undertaken in other countries for different market-oriented targets.

Author Contributions: Conceptualization, L.C. and H.H.; Data curation, L.Q.; Formal analysis, L.C. and L.Q.; Funding acquisition, X.Y.; Investigation, L.C. and L.Q.; Methodology, L.C. and C.L.; Project administration, C.L.; Resources, C.L. and X.Y.; Software, L.C.; Supervision, C.L.; Validation, L.C. and H.H.; Visualization, L.C.; Writing—original draft, L.C. and X.Y.; Writing—review & editing, H.H. All authors have read and agreed to the published version of the manuscript.

Funding: The authors received General Projects of National Social Science Foundation (education) "internationalization development and overseas communication strategy of university brand image driven by education of foreign students in China" (BGA200057) for this research.

Institutional Review Board Statement: Not applicable.

Informed Consent Statement: Not applicable.

Data Availability Statement: This data can be provided by corresponding author on request.

Acknowledgments: The authors would like to extend our sincere thanks to anonymous reviewers for providing helpful comments and suggestions on earlier drafts of the manuscript.

Conflicts of Interest: The author declared no potential conflicts of interest with respect to the research, authorship, and/or publication of this article.

References

1. Khanzada, H.; Khan, M.Q.; Kayani, S. Cotton Based Clothing. In *Cotton Science and Processing Technology: Gene, Ginning, Garment and Green Recycling*; Wang, H., Memon, H., Eds.; Springer: Singapore, 2020; pp. 377–391.
2. Domingos, M.; Vale, V.T.; Faria, S. Slow Fashion Consumer Behavior: A Literature Review. *Sustainability* **2022**, *14*, 2860. [CrossRef]
3. Chen, L.; Qie, K.; Memon, H.; Yesuf, H.M. The Empirical Analysis of Green Innovation for Fashion Brands, Perceived Value and Green Purchase Intention—Mediating and Moderating Effects. *Sustainability* **2021**, *13*, 4238. [CrossRef]
4. Cai, R.; Tang, C.; Sun, W. The Connotation, Scale Development and Inspection of Product Design Innovation. *Soft Science* **2019**, *33*, 138–143.
5. Zameer, H.; Wang, Y.; Yasmeen, H. Reinforcing green competitive advantage through green production, creativity and green brand image: Implications for cleaner production in China. *J. Clean. Prod.* **2020**, *247*, 119119. [CrossRef]
6. Chen, L.; Halepoto, H.; Liu, C.; Kumari, N.; Yan, X.; Du, Q.; Memon, H. Relationship Analysis among Apparel Brand Image, Self-Congruity, and Consumers' Purchase Intention. *Sustainability* **2021**, *13*, 12770. [CrossRef]
7. Aghekyan-Simonian, M.; Forsythe, S.; Kwon, W.S.; Chattaraman, V. The role of product brand image and online store image on perceived risks and online purchase intentions for apparel. *J. Retail. Consum. Serv.* **2012**, *19*, 325–331. [CrossRef]
8. Xuan, J.; Jianhong, L.; Yiwen, C. The factor structure of brand image system. *Acta Psychol. Sin.* **2004**, *36*, 359.
9. Jaiprakash, A.T. A conceptual research on the association between celebrity endorsement, brand image and brand equity. *J. Mark. Manag.* **2008**, *7*, 54–64.
10. Chenting, Z.; Yue, W. New Changes of Digital Marketing in the Outbreak. *Furnit. Inter. Des.* **2020**, *3*, 16–17. [CrossRef]
11. Apenes Solem, B.A. Influences of customer participation and customer brand engagement on brand loyalty. *J. Consum. Mark.* **2016**, *33*, 332–342. [CrossRef]
12. Delgado-Ballester, E.; Luis Munuera-Aleman, J. Brand trust in the context of consumer loyalty. *Eur. J. Market.* **2001**, *35*, 1238–1258. [CrossRef]
13. Atulkar, S. Brand trust and brand loyalty in mall shoppers. *Mark. Intell. Plan.* **2020**, *38*, 559–572. [CrossRef]
14. France, C.; Grace, D.; Iacono, J.L.; Carlini, J. Exploring the interplay between customer perceived brand value and customer brand co-creation behaviour dimensions. *J. Brand Manag.* **2020**, *27*, 466–480. [CrossRef]
15. Zhuoran, D.; Hongjian, Q. Research on relationship between competitive advantage and value creation of garment industrial cluster. *J. Silk* **2020**, *57*, 47–54. [CrossRef]
16. Jayaraman, V.; Luo, Y. Creating competitive advantages through new value creation: A reverse logistics perspective. *Acad. Manag. Perspect.* **2007**, *21*, 56–73. [CrossRef]
17. Lin, L.; Lihong, C. Content Marketing Index System of Apparel Brand Cultural Image. *J. Silk* **2020**, *57*, 71–77.
18. Prahalad, C.K.; Ramaswamy, V. Co-opting customer competence. *Harv. Bus. Rev.* **2000**, *78*, 79–90.
19. Prahalad, C.K.; Ramaswamy, V. Co-creation experiences: The next practice in value creation. *J. Interact. Mark.* **2004**, *18*, 5–14. [CrossRef]
20. Zaborek, P.; Mazur, J. Enabling value co-creation with consumers as a driver of business performance: A dual perspective of Polish manufacturing and service SMEs. *J. Bus. Res.* **2019**, *104*, 541–551. [CrossRef]
21. Andreu, L.; Sánchez, I.; Mele, C. Value co-creation among retailers and consumers: New insights into the furniture market. *J. Retail. Consum. Serv.* **2010**, *17*, 241–250. [CrossRef]
22. Pongsakornrungsilp, S.; Schroeder, J.E. Understanding value co-creation in a co-consuming brand community. *Mark. Theory* **2011**, *11*, 303–324. [CrossRef]

23. Nambisan, S.; Baron, R.A. Different roles, different strokes: Organizing virtual customer environments to promote two types of customer contributions. *Organ. Sci.* **2010**, *21*, 554–572. [CrossRef]
24. Auh, S.; Bell, S.J.; McLeod, C.S.; Shih, E. Co-production and customer loyalty in financial services. *J. Retail.* **2007**, *83*, 359–370. [CrossRef]
25. Payne, A.F.; Storbacka, K.; Frow, P. Managing the co-creation of value. *J. Acad. Mark. Sci.* **2008**, *36*, 83–96. [CrossRef]
26. Marcos-Cuevas, J.; Nätti, S.; Palo, T.; Baumann, J. Value co-creation practices and capabilities: Sustained purposeful engagement across B2B systems. *Ind. Mark. Manage.* **2016**, *56*, 97–107. [CrossRef]
27. Zhang, M.; Guo, L.; Hu, M.; Liu, W. Influence of customer engagement with company social networks on stickiness: Mediating effect of customer value creation. *Int. J. Inf. Manag.* **2017**, *37*, 229–240. [CrossRef]
28. Mingyu, L.; Qian, F.; Shouzhong, H. Research on the influencing factors of adoption of "Shared Customization" operation mode in garment industry. *J. Silk* **2021**, *58*, 80–85. [CrossRef]
29. Mandlik, M.A.; Kadirov, D. Towards a theory of integrated empowerment: A service ecosystems agenda for future. *Int. J. Qual. Res. Serv.* **2020**, *4*, 56–76. [CrossRef]
30. Moise, M.S.; Gil-Saura, I.; Ruiz-Molina, M.-E. Implications of Value Co-Creation in Green Hotels: The Moderating Effect of Trip Purpose and Generational Cohort. *Sustainability* **2020**, *12*, 9866. [CrossRef]
31. Dai, D.; Gu, X. User participation behavior, perceived value and customer loyalty analysis based on mobile short video social application. *Consum. Econ.* **2017**, *33*, 58–65.
32. Groth, M. Customers as good soldiers: Examining citizenship behaviors in internet service deliveries. *J. Manag.* **2005**, *31*, 7–27. [CrossRef]
33. Bove, L.L.; Pervan, S.J.; Beatty, S.E.; Shiu, E. Service worker role in encouraging customer organizational citizenship behaviors. *J. Bus. Res.* **2009**, *62*, 698–705. [CrossRef]
34. Ennew, C.T.; Binks, M.R. Impact of participative service relationships on quality, satisfaction and retention: An exploratory study. *J. Bus. Res.* **1999**, *46*, 121–132. [CrossRef]
35. Yi, Y.; Gong, T. Customer value co-creation behavior: Scale development and validation. *J. Bus. Res.* **2013**, *66*, 1279–1284. [CrossRef]
36. Qingjuan, B.; Yongsheng, J.; Zhaohui, L. Does Interactive Behavior Certainly Create Value? The Effect of Customer Value Co-creation Interactive Behavior on Customer Value. *Foreign Econ. Manag.* **2016**, *38*, 21–37.
37. Wenzhen, W.; Qijie, C. Impacts of Customer Participation Behavior on Customer Satisfaction and Behavioral Intentions Based on Co-creation Perspective. *Manag. Rev.* **2017**, *29*, 167.
38. Yen, C.-H.; Teng, H.-Y.; Tzeng, J.-C. Innovativeness and customer value co-creation behaviors: Mediating role of customer engagement. *Int. J. Hosp. Manag.* **2020**, *88*, 102514. [CrossRef]
39. Kim, I.; Jung, H.J.; Lee, Y. Consumers' Value and Risk Perceptions of Circular Fashion: Comparison between Secondhand, Upcycled, and Recycled Clothing. *Sustainability* **2021**, *13*, 1208. [CrossRef]
40. Ying, L. Research on Brand Image Perception of Historical and Cultural Blocks in Guangzhou Based on Brand Concept Map (BCM). *Furnit. Inter. Des.* **2021**, *11*, 123–127.
41. Kim, B.; Oh, J. The difference of determinants of acceptance and continuance of mobile data services: A value perspective. *Expert Syst. Appl.* **2011**, *38*, 1798–1804. [CrossRef]
42. Huang, L.; Mou, J.; See-To, E.W.; Kim, J. Consumer perceived value preferences for mobile marketing in China: A mixed method approach. *J. Retail. Consum. Serv.* **2019**, *48*, 70–86. [CrossRef]
43. Huili, C.; Jingqiong, Z. Research on the Chinese goods activities of relevant clothing industry guildin modern Shanghai. *J. Silk* **2020**, *57*, 80–87.
44. Joshi, A.W.; Sharma, S. Customer knowledge development: Antecedents and impact on new product performance. *J. Mark.* **2004**, *68*, 47–59. [CrossRef]
45. Yuqian, D.; Yuyilan, C.; Xiaogang, L. The mechanism of the effect of clothing brand color salience on brand loyalty. *J. Silk* **2021**, *58*, 80–85.
46. Tran, V.D.; Vo, T.N.L.; Dinh, T.Q. The relationship between brand authenticity, brand equity and customer satisfaction. *J. Asian Financ. Econ. Bus.* **2020**, *7*, 213–221. [CrossRef]
47. Hlady-Rispal, M.; Servantie, V. Deconstructing the way in which value is created in the context of social entrepreneurship. *Int. J. Manag. Rev.* **2018**, *20*, 62–80. [CrossRef]
48. Lingyan, W. Research on Furniture Market Innovation under the Background of Internet. *Furnit. Inter. Des.* **2020**, *8*, 24–26. [CrossRef]
49. Yali, Z.; Hongjian, Q. Impact of experience marketing of clothing stores on customer loyalty:an analysis of double mediating effect based on perceived service quality and customer trust. *J. Silk* **2020**, *57*, 51–57. [CrossRef]
50. Hoyer, W.D.; Chandy, R.; Dorotic, M.; Krafft, M.; Singh, S.S. Consumer co-creation in new product development. *J. Serv. Res.* **2010**, *13*, 283–296. [CrossRef]
51. Etgar, M. A descriptive model of the consumer co-production process. *J. Acad. Mark. Sci.* **2008**, *36*, 97–108. [CrossRef]
52. Baiming, X. New Pattern of Customized Furniture Market after the Opportunity Dividends' disappearance. *Furnit. Inter. Des.* **2020**, *4*, 9–10. [CrossRef]

53. Zhao, Y.; Chen, Y.; Zhou, R.; Ci, Y. Factors influencing customers' willingness to participate in virtual brand community's value co-creation. *Online Inf. Rev.* **2019**, *43*, 440–461. [CrossRef]
54. Yang, H.; Vazquez, D.; Blazquez, M. Creative Art-based Initiatives Enabling Value Co-creation in the Luxury Fashion Industry. In *Creativity and Marketing: The Fuel for Success*; Pantano, E., Ed.; Emerald Publishing Limited: West Yorkshire, UK, 2021; pp. 133–147.
55. My-Quyen, M.T. Transforming customer brand engagement to co-creation value through participation energy and effort. *Serv. Bus.* **2021**, *15*, 493–514. [CrossRef]
56. Zhang, Y.; Chen, L. The impacts of website characteristics and customer participation on citizenship behaviors: The mediating role of co-creation experience in virtual brand communities. *Adv. Appl. Sociol.* **2017**, *7*, 151–164. [CrossRef]
57. Vahdat, A.; Hafezniya, H.; Jabarzadeh, Y.; Thaichon, P. Emotional brand attachment and attitude toward brand extension. *Serv. Mark. Q.* **2020**, *41*, 236–255. [CrossRef]
58. Xinling, T.; Jia, C.; Haijing, M. Luxury brand experience marketing and its influence effect: A case study of Gucci. *J. Silk* **2021**, *58*, 52–59.
59. Paulose, D.; Shakeel, A. Perceived Experience, Perceived Value and Customer Satisfaction as Antecedents to Loyalty among Hotel Guests. *J. Qual. Assur. Hosp. Tour.* **2021**, *23*, 447–481. [CrossRef]
60. Zhang, Y.; Zhao, H.; Hu, Y.; Yao, G. Smart Community Service Brand Functional Value and Sustainable Brand Relationship—The Mediating Role of Customer Emotional Cognition. *Sustainability* **2021**, *13*, 1833. [CrossRef]
61. Guan, J.; Wang, W.; Guo, Z.; Chan, J.H.; Qi, X. Customer experience and brand loyalty in the full-service hotel sector: The role of brand affect. *Int. J. Contemp. Hosp. Manag.* **2021**, *33*, 1620–1645. [CrossRef]
62. Nardi, V.A.M.; Jardim, W.C.; Ladeira, W.J.; Santini, F. A meta-analysis of the relationship between customer participation and brand outcomes. *J. Bus. Res.* **2020**, *117*, 450–460. [CrossRef]
63. Chen, C.-L. Value creation by SMEs participating in global value chains under industry 4.0 trend: Case study of textile industry in Taiwan. *J. Glob. Inf. Technol. Manag.* **2019**, *22*, 120–145. [CrossRef]
64. Memon, H.; Chaklie, E.B.; Yesuf, H.M.; Zhu, C. Study on Effect of Leather Rigidity and Thickness on Drapability of Sheep Garment Leather. *Materials* **2021**, *14*, 4553. [CrossRef] [PubMed]
65. Wang, H.; Memon, H.; Shah, S.H.H.; Shakhrukh, M. Development of a Quantitative Model for the Analysis of the Functioning of Integrated Textile Supply Chains. *Mathematics* **2019**, *7*, 929.
66. Teng, H.-Y.; Tsai, C.-H. Can tour leader likability enhance tourist value co-creation behaviors? The role of attachment. *J. Hosp. Tour. Manag.* **2020**, *45*, 285–294. [CrossRef]
67. Mustak, M.; Jaakkola, E.; Halinen, A. Customer participation and value creation: A systematic review and research implications. *Manag. Serv. Qual. An. Int. J.* **2013**, *23*, 341–359. [CrossRef]
68. Yoo, S.R.; Lee, S.W.; Jeon, H.M. The role of customer experience, food healthiness, and value for revisit intention in Grocerant. *Sustainability* **2020**, *12*, 2359. [CrossRef]
69. Qian, F.; Mingyu, L.; Shouzhong, H. Construction of new community platform of clothing based on value co-creation. *J. Silk* **2020**, *57*, 57–69. [CrossRef]
70. Zi-Xin, J.; Tie, J. A Customer Persona of Creative Products Based on Purchasing Behavior. *Packag. Eng.* **2021**, *42*, 218–224. [CrossRef]
71. Ramaswamy, V. Co-creation of value—towards an expanded paradigm of value creation. *Mark. Rev. St. Gallen* **2009**, *26*, 11–17. [CrossRef]
72. Venkatesh, V.; Davis, F.D. A theoretical extension of the technology acceptance model: Four longitudinal field studies. *Manag. Sci.* **2000**, *46*, 186–204. [CrossRef]
73. Gavilan, D.; Avello, M.; Martinez-Navarro, G. The influence of online ratings and reviews on hotel booking consideration. *Tour. Manag.* **2018**, *66*, 53–61. [CrossRef]
74. Azize, Ş.; Cemal, Z.; Hakan, K. Does brand communication increase brand trust? The empirical research on global mobile phone brands. *Procedia-Soc. Behav. Sci.* **2012**, *58*, 1361–1369. [CrossRef]
75. Shin, K.; Peng, X.; Qin, P. Research on the effect of customer-to-customer interaction of virtual brand community on customer participation in value co-creation by taking experiential value as mediating variable. *Chin. J. Manag.* **2016**, *13*, 1808–1816.
76. Hennigs, N.; Wiedmann, K.-P.; Klarmann, C. Consumer Value Perception of Luxury Goods: A Cross-Cultural and Cross-Industry Comparison. In *Luxury Marketing*; Springer: Berlin/Heidelberg, Germany, 2013; pp. 77–99.
77. Xuejiao, C.; Qiaoming, C. Research on Operation Process and Market Behavior of Design Project. *Furnit. Inter. Des.* **2020**, *3*, 56–58. [CrossRef]
78. Fu, Y.; Cao, N. Scale Development of Online Public Opinion on Tourism Image Communication Based on Grounded Theory. *Stat. Decis.* **2016**, *20*, 65–68.
79. Godey, B.; Lagier, J.; Pederzoli, D. A measurement scale of "aesthetic style" applied to luxury goods stores. *Int. J. Retail. Distrib. Manag.* **2009**, *37*, 527–537. [CrossRef]
80. Koenigsberg, O.; Kohli, R.; Montoya, R. The Design of Durable Goods. *Mark. Sci.* **2011**, *30*, 111–122. [CrossRef]
81. Ma, J.; Diao, Y.; Shan, J. The influence of Internet Brand Perceived Fit on Extended Evaluation: Based on the Moderating Effect of Parent Brand Functional Image. *Enterp. Econ.* **2015**, *19*, 85–90.
82. Jiawei, G.; Baiming, X. Research on Furniture Brand Building and Communication Strategy Based on IKEA Cases. *Furnit. Inter. Des.* **2021**, *8*, 76–79. [CrossRef]

sustainability

MDPI

Article

Study of the Enhancements of Porous Structures of Activated Carbons Produced from Durian Husk Wastes

Tongsai Jamnongkan [1,*], Nitchanan Intaramongkol [1], Nattharika Kanjanaphong [1], Kemmika Ponjaroen [1], Wasana Sriwiset [1], Rattanaphol Mongkholrattanasit [2], Piyada Wongwachirakorn [3], Kun-Yi Andrew Lin [4,*] and Chih-Feng Huang [5,*]

1 Department of Fundamental Science and Physical Education, Faculty of Science at Sriracha, Kasetsart University, Chonburi 20230, Thailand; nitchanan.int@ku.th (N.I.); nattharika.k@ku.th (N.K.); kemmika.p@ku.th (K.P.); wasana.sriw@ku.th (W.S.)

2 Department of Textile Chemistry Technology, Faculty of Industrial Textiles and Fashion Design, Rajamangala University of Technology Phra Nakhon, Bangkok 10110, Thailand; rattanaphol.m@rmutp.ac.th

3 Department of Environmental Science, Faculty of Science and Technology, Pibulsongkarm Rajabhat University, Phitsanulok 65000, Thailand; piyada.w@psru.ac.th

4 Department of Environmental Engineering, Innovation and Development Center of Sustainable Agriculture & Research Center of Sustainable Energy and Nanotechnology, i-Center for Advanced Science and Technology (iCAST), National Chung Hsing University, Taichung 40227, Taiwan

5 Department of Chemical Engineering, i-Center for Advanced Science and Technology (iCAST), National Chung Hsing University, Taichung 40227, Taiwan

* Correspondence: jamnongkan.t@ku.ac.th (T.J.); linky@nchu.edu.tw (K.-Y.A.L.); huangcf@dragon.nchu.edu.tw (C.-F.H.)

Citation: Jamnongkan, T.; Intaramongkol, N.; Kanjanaphong, N.; Ponjaroen, K.; Sriwiset, W.; Mongkholrattanasit, R.; Wongwachirakorn, P.; Lin, K.-Y.A.; Huang, C.-F. Study of the Enhancements of Porous Structures of Activated Carbons Produced from Durian Husk Wastes. *Sustainability* **2022**, *14*, 5896. https://doi.org/10.3390/su14105896

Academic Editors: Hafeezullah Memon, Xiaoke Jin, Tian Wei and Chengyan Zhu

Received: 7 April 2022
Accepted: 9 May 2022
Published: 12 May 2022

Abstract: The idea of generating high-value practical materials, such as activated carbons, from agricultural wastes as a raw material has been a quite important trend recently due to its positive contributions to the environment and resource savings from biomass. In this paper, activated carbons prepared from durian husk waste by the KOH chemical activation method are studied. We focus on the effects of stages of the activating temperature on their properties. The optimum conditions for activation were a KOH/char ratio of 1:2 at the first and second activation process at the temperatures of 400 and 800 °C, respectively. The characterization results of Fourier transform infrared spectroscopy (FTIR), scanning electron microscopy (SEM), and Brunauer–Emmett–Teller (BET) surface area showed that the obtained activated carbons have a high surface area and small pore size. The adsorption/desorption isotherms of the obtained activated carbons showed type I and type II isotherms. The chemical structure of obtained activated carbons did not show any variation in the surface functional groups. A feasible method to produce the activated carbons with a high surface area and high adsorption capability from durian husk waste was eventually demonstrated.

Keywords: activated carbons; durian husk; activation; waste utilizations; adsorption

1. Introduction

Durian is a one of most famous fruits consumed locally and exported in the Southeast Asia region, especially in Malaysia and Thailand. A large quantity of durian fruit is produced per year for commercialization [1]. However, it is well known that not the entire durian fruit is edible. The outer part of the durian skin normally weighs more than half of the total of this fruit, and this cannot be eaten and is thrown away. Large quantities of durian husk waste are being produced, which is becoming an environmental problem. Therefore, it is worthwhile to find a solution to this problem. Generally, durian husk waste is usually discarded at a landfill site and ends up in open burning, which can release harmful pollution into the air, increasing the air pollution to the environment [2,3]. In the past decade, the idea of utilizing biomass from agricultural wastes as a raw material for the production of valuable material has attracted the interest of global researchers.

Activated carbons, carbonaceous material with a large surface area and high porosity, are one of the most commonly obtained materials produced from agricultural wastes. In recent times, this material has received much attention from the scientific community as it is renewable, low-cost, and environmentally friendly. The surface area and pore size distribution are important factors in the determination of activated carbon's performance, such as its adsorption and mechanical properties [4].

Recently, activated carbons with high surface area and pore volumes have been produced from various sources of agricultural wastes such as coal [5–7], coconut shell [8,9], sawdust [10,11], sugarcane bagasse [12], rice straw, and husk [13] plant materials [14–18], including agricultural industrial by-products such as coffee bean husk [19,20]. Among these precursors, durian husk waste is also a main source within agricultural industrial practice in the production of potential activated carbon due to its effectiveness and inexpensive cost as a raw material. In general, activated carbon is used in a wide range of environmental applications. In particular, it is a common material used as an adsorbent for adsorption or treatment in water purification, such as heavy metal removal in groundwater [21–23], wastewater treatment [24,25], and for the removal of volatile organic compounds (VOCs) [26], carbon dioxide [27], pigments, and odors [28]. It is well known that the potential efficiency of activated carbons depends on their surface area and active surfaces. Thus, it is very important to prepare activated carbon with a high surface area and high porosity in order to enhance its adsorption capacity due to these characteristics, which are particularly suitable for specific applications [29–31]. Currently, activated carbon is manufactured via physical or chemical stimulation processes, such as the physical activation of pyrolysis to produce char followed by steam gasification [32,33], and chemical activation, such as using solutions of zinc chloride [34], phosphoric acid [35], potassium hydroxide [36], sodium hydroxide [37], and so on. Besides this, the nature of raw materials has also commonly affected the pore structure and pore size distribution of activated carbon. A functional activated carbon will be obtained through a two-step process: (i) hydrothermal carbonization of the raw materials in an atmosphere without oxygen followed by the activation of the resulting char. In this stage, most of the non-carbon elements, such as hydrogen and oxygen, are removed in a gaseous form by pyrolytic decomposition, resulting in carbon with a fixed mass and a rudimentary pore structure; (ii) secondly, the activation step can be performed as follows: an appropriate amount of obtained char is mixed with an activating agent and then initiated by heating the samples in the furnace from room temperature until the desired temperature is reached. This step is employed to enlarge the diameter of fine pores and also to create new pores so that the adsorptive power of the carbonization product is enhanced [38–40]. It is well known that the activated carbon is not only used as an adsorbent for the removal of polluting molecules in the environment but is also widely used for supercapacitors' electrode materials, catalyst carriers, and reinforcement materials [41–44]. Recently, several studies have reported about the method of activated carbon production, but there have been no studies into the effect of the activating temperature on the surface area and porosity structures of the obtained activated carbon. Therefore, it is worthwhile to further examine this parameter and how it affects the properties of activated carbon.

In this research, carbonized durian husk was used as a precursor for the production of activated carbon by chemical activation with potassium hydroxide; due to its cost effectiveness, it is used as an activating agent for producing activated carbon. The effects of the stages of the activating temperature on the properties of activated carbon were investigated. Then, a comparison of the iodine absorption capacity of activated carbon was performed to study the specific surface area and the amount of porosity in adsorption. In addition, the results of Fourier transform infrared spectroscopy (FTIR), scanning electron microscopy (SEM), and Brunauer–Emmett–Teller (BET) surface area are investigated and discussed.

2. Materials and Methods

2.1. Materials

Durian husk (*Monthong varieties*) was collected from a local durian market in the Chonburi province of Thailand. The characteristics of durian husk waste are given in Table 1. Hydrochloric acid with a purity of 85% was purchased from Sigma–Aldrich (Germany). Potassium hydroxide and sodium hydroxide were purchased from Ajex chemical (Germany) and sodium thiosulfate of AR-grade was purchased from Loba Chemie (India). Other chemical agents were analytical grade purity and used as received without further purification.

Table 1. The characteristics of durian husk waste.

Contents	Result (%)
Moisture	30.34
Ash	4.25
Volatile	76.70
Carbon	21.10

2.2. Activated Carbon Preparations

As a first step, the durian husk waste was cut to 3–5 cm. Then, it was repeatedly washed with distilled water to remove dust and other inorganic impurities. After that, the clean durian husk was dried at a temperature of 100 °C for 24 h in the hot-air oven to reduce its moisture contents, and it was stored in a desiccator. In the carbonization process, the char was produced by carbonizing the samples of dried durian husk in a fixed-bed batch reactor in an atmosphere without oxygen. The reactor was filled with samples and was placed in a furnace (Model 30400, Barnstead Thermolyne, USA). The furnace was heated from room temperature to 600 °C, kept at that temperature for 3 h, and then cooled down to room temperature overnight. The obtained char was crushed in a crusher (Model ZM200, Restch, Germany) to obtain particles with sizes between 800 to 1000 μm and then stored in desiccators for further use. After the carbonization process, the mass yield of the char was calculated, and we found that the yield average was 73.58% of the total mass. The activation process of char was carried out as follows: the char was impregnated by soaking appropriate amounts of the char with different ratios of potassium hydroxide at room temperature for 24 h, as shown in Table 1. After the impregnation process was completed, we used into two processes for activations: (i) in Process I, denoted as DH1–DH3, the impregnated precursors were poured onto a porcelain disc and placed in the furnace for their first activation at a temperature of 400 °C for 1 h. Then, the activation process was carried out by adjustment to the second activation at a temperature of 600 °C for 3 h, and then the material was cooled down to room temperature. (ii) In Process II, defined as DH4–DH6 code samples, these activated carbons were heated to the second activation temperature of 800 °C for 3 h under the same conditions as in the Process I preparation. Details of the process conditions for preparing activated carbon are shown in Table 2.

Table 2. Sample identification, activation process conditions, and yield for activated carbon preparation.

Samples	Carbonization Temperature (°C)	KOH:Char Ratio	First Activation Temperature (°C)	Second Activation Temperature (°C)	Yield (%)
DH1	600	1:4	400	600	83.95
DH2	600	1:2	400	600	86.48
DH3	600	1:1	400	600	80.27
DH4	600	1:4	400	800	79.95
DH5	600	1:2	400	800	85.45
DH6	600	1:1	400	800	75.47

All activated carbons were washed sequentially with a 0.5 N HCl solution in order to eliminate excess KOH. Consecutively, activated carbon powders were repeatedly washed with distilled water until reaching a neutral pH. After that, these powders were dried in an oven at 110 °C for 24 h and stored in a desiccator before further use.

2.3. FTIR Spectroscopy

The chemical structure of the obtained activated carbon was verified by using a FTIR spectrometer (Invenio, Bruker, USA) in ATR mode equipped with a diamond crystal, which was used for the tests. All the spectra were recorded in transmittance mode with 4 cm^{-1} resolution in the wavenumber range of 4000–400 cm^{-1} under ambient conditions.

2.4. Iodine Number Analysis

The iodine number is a technique commonly employed to determine the adsorption capacity of activated carbons. In this paper, the iodine number was determined according to the standard test method ASTM D 4607–94 [45]. The procedure of the experiment can be briefly explained as follows: 1.0 g of dried activated carbon samples were placed in a 250 mL flask, and 10 mL of 5%wt. hydrochloric acid was added. The flask was swirled until the activated carbon became wet; then, the flask was boiled for 1 minute to degas the sulfur dioxide on the surface samples and cooled down to room temperature. Then, 100 mL of stock iodine solution (12.7 g of Iodine (Merck) and 19.1 g of potassium iodide (Merck) in 1 L of de-ionized water) was added into the sample flask, and the mixture was shaken for 1 minute in the magnetic stirrer (Model C-MAG HS7, IKA, Germany) and the samples were filtered. Then, 50 mL of filtrate was titrated with 0.1 N sodium thiosulphate until the solution become pale yellow. Then, approximately 1 mL of starch solution (1 wt.%), used as an indicator, was added, and titration slowly continued with sodium thiosulphate until the solution changed from blue color to colorless. The iodine number is the amount of adsorbed iodine per gram of samples (X/M), the value of which was calculated by Equations (1) and (2):

$$(X/M) = \frac{\{(N_1 \times 126.93 \times V_1) - [(V_1 + V_{HCl})/V_F] \times (N_{Na_2S_2O_3} \times 126.93) \times V_{Na_2S_2O_3}\}}{M} \quad (1)$$

$$C = (N_{Na_2S_2O_3} \times V_{Na_2S_2O_3}) \quad (2)$$

where N_1 and $N_{Na_2S_2O_3}$ are the concentration of iodine solution and sodium thiosulfate solution (N), respectively. V_1, V_{HCl}, V_F, and $V_{Na_2S_2O_3}$ refer to the volume of iodine solution, hydrochloric acid, filtrate used in titration, and the consumed volume of sodium thiosulfate solution (mL), respectively. M is the mass of activated carbon (g).

2.5. Morphological Measurement

The morphological surfaces of the obtained activated carbon were examined via SEM (JSM-6610LV, JEOL, Japan) at an accelerating voltage of 10 kV. Each sample was attached to the stubs with double-sided carbon tape. Then, the samples were sputtered with gold under vacuum before observation. The average pore sizes for each activated carbon were investigated by using the image analysis software Image J (ver. 1.45, USA).

2.6. Specific Surface Area

The specific surface area, total pore volume (V_T), micropore volume ($V_{\bullet}$), mesopore volume (V_m), and average pore size (D_p) of the obtained activated carbon were scrutinized via nitrogen adsorption Brunauer–Emmett–Teller (BET) analysis on a Micrometrics 3Flex Version 5.02 apparatus. The cell containing the sample was weighed before degassing. Approximately 0.2 g of sample powders were placed in a test tube in order to removes contaminants such as water vapor and adsorbed gases from the samples. The samples were degassed at 200 °C for 16 h before the measurements.

3. Results and Discussion

3.1. FTIR Spectroscopy

FTIR spectra were collected for the characterization of the surface functional groups of prepared activated carbon. Figure 1 depicts the FTIR spectrum of all activated carbons obtained after the chemical treatment with KOH followed by the different conditions, as shown in Table 1. The IR spectrum of the char sample was used to compare the spectra of the obtained activated carbons. We found that the weaker broad band around 3380 cm^{-1} revealed the stretching vibrations of the O-H bonds of the mixtures that were adsorbed onto the surface of activated carbon. A peak at around 2100 cm^{-1} shows the presence of a triple bond between carbon–carbon atoms (alkynes group) with stretching vibration [46]. Compared to the char spectrum, it was found that some changes could be noted in the position and intensity of some peaks in the activated carbon spectrum. A new strong peak appeared around 1030–1100 cm^{-1}, which was because of the C-O stretching vibrations of carboxyl or carbonyl groups. This probably indicates the formation of an oxidation reaction during treatment with KOH in the activation process. Furthermore, we also found that the absorption peak in the regions between 1500 and 1600 cm^{-1} could be ascribed to C = C aromatic ring stretching vibration, which is generally found in carbonaceous material [47].

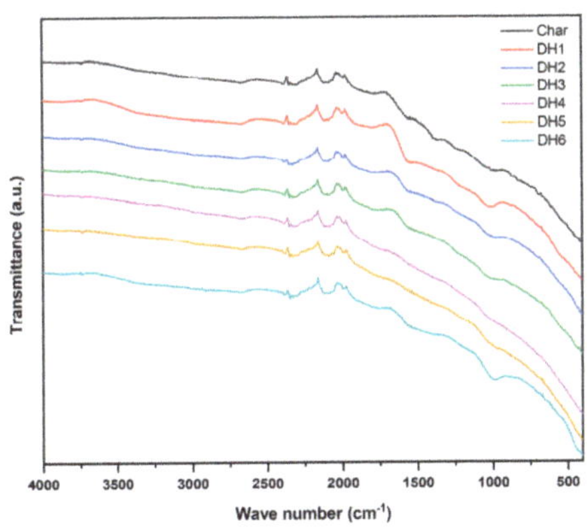

Figure 1. FT-IR analysis of the obtained char and activated carbons.

3.2. Iodine Number Analysis

It is well known that the indirect value of the iodine number also indicates the porosity of activated carbon, which is correlated with the micropore volume. A higher iodine number meant a higher porosity on the surface determined by iodine absorption. Therefore, iodine number analysis was used to analyze the porosity of DH1–DH5, and char was also used for comparison. The iodine adsorption capacity of all the activated carbon is illustrated in Figure 2. It is confirmed that the process parameters affected the porosity of the obtained activated carbon, which was found to be higher than char. Besides this, we found that the impregnation ratio between KOH and char has a profound effect on the adsorption properties of activated carbon. At the given activation temperatures, the iodine numbers of the activated carbons increased with an impregnation ratio increased from 1:4 to 1:2 and decreased with the KOH:char ratio increased to 1:1. From our experimental results, the highest iodine number was achieved at the impregnated ratio of 1:2, as shown in Figure 2. The iodine numbers of char, DH1, DH2, and DH3 are 717, 959, 971, and 933 mg/g, respectively. In addition, the effect of the activation process on the adsorption capacity was studied. It was found that the iodine numbers increased with increases in the temperature of the second activation process, which can be compared with DH2 and DH5 in Figure 2. At the given impregnation ratio, DH5 showed the highest values, at 1037 mg/g, of its iodine number compared with the other activated carbon. This is probably due to the effect of pore structure formation within the char structure during the activation process.

As mentioned above, the iodine number is commonly used in industries as a measure of adsorption capacity, reflecting the number of micropores in the carbon. This value indicates the approximate internal surface area of activated carbon. The relationship between the iodine number and the internal surface area of activated carbon has been well-established in the literature for various precursors [48–50]. It is well known that the iodine number has a direct proportional relationship with the specific surface area. Comparing recent studies, the iodine number of activated carbon from agricultural wastes has been reported to be up to 1000 mg/g, which is in accordance with the previous studies by Saka [51], Toledo et al. [52], and Anderson et al. [53]. Thus, the results of this study can suggest that the biochar from durian husk can be used as a candidate as a precursor in the manufacture of activated carbon with a highly porous structure for specialized industrial sorbents.

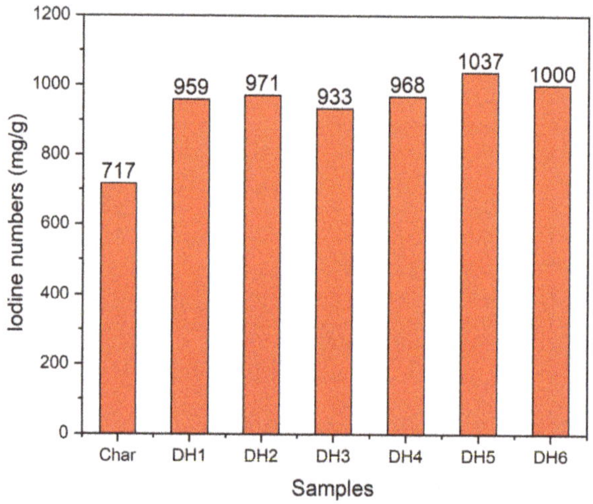

Figure 2. Effect of various preparation conditions on iodine number of activated carbon.

3.3. *Morphological Measurement*

An SEM instrument was used to observe the surface morphology of the materials obtained. Based on the iodine numbers, DH2 and DH5 were selected as desired activated carbons due to their high exhibited iodine number values. Therefore, we focused on investigating their surface morphologies, and the char was also compared in this study. Figure 3 shows the SEM micrographs of char, DH2, and DH5, respectively. From the image, it is clear that the durian husk can produce char and activated carbon composed of porous structures, with the appearance of roughly similar structures from all samples. However, the images suggest that the average pore diameters of the samples are quite different. As mentioned above, the pore sizes of activated carbon would be directly affected in terms of several of their properties, such as adsorption and mechanical properties. Therefore, it is worthwhile to further examine the pore sizes of the obtained activated carbon. Figure 4 shows the SEM images and the surface porous diameter distributions for the char and activated carbon materials. DH5 shows that the surface porosity is smaller than DH2 and char, respectively. In Figure 4, it can be seen that the average surface porous diameters of char, DH2, and DH5 are 6150 ± 2.86, 5060 ± 2.01, and 3790 ± 2.37 nm, respectively. The averages diameter of the pore structure decreased with an increasing activation temperature. This investigation qualitatively confirms in the same trends the results of the iodine number discussed earlier. Thus, we suggest that the surface structures of obtained activated carbon depend on the temperature and method of the activation process.

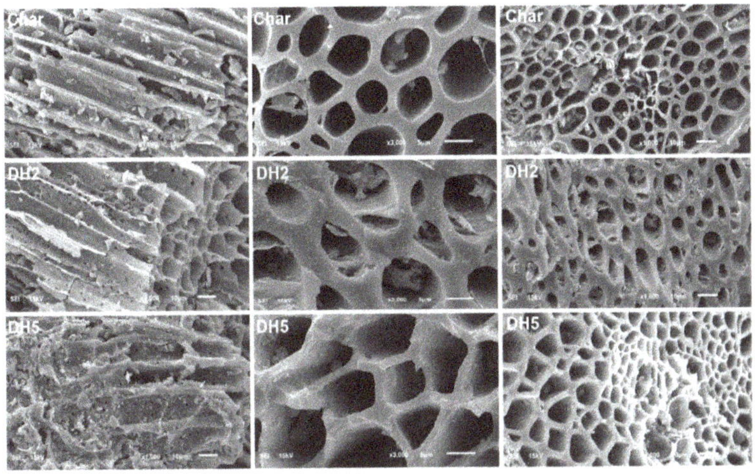

Figure 3. SEM images of the surface morphologies of char (upper), DH2 (middle), and DH5 (lower) with different magnifications.

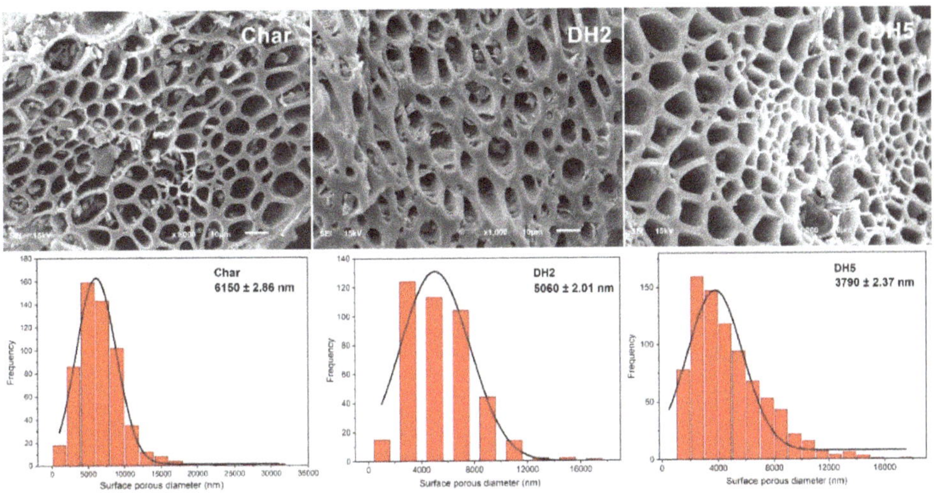

Figure 4. SEM images and the surface porous diameter distributions of char, DH2, and DH5.

3.4. Specific Surface Area

In this part, DH2 and DH5 were selected as desired activated carbons due to their high iodine number values. In addition, char was also investigated for comparison. Figure 5 illustrates the nitrogen adsorption/desorption isotherms of the char, DH2, and DH5. The adsorbed volume of DH5 is higher than DH2 and char, respectively. However, it was found that the isotherm curves for the char sample showed that N_2 adsorption/desorption exhibited a pattern of a type II isotherm, according to IUPAC classification of isotherms [54]. This type of isotherm is a general characteristic of porous carbons that have large-sized pores. On the other hand, the N_2 adsorption/desorption isotherms of DH2 and DH5 were exhibited as types I and II, indicating that there were many micropores in the materials. The rapid rise of adsorption/desorption isotherms in the low relative pressure region also indicated that there were microporous or small mesoporous structures. The textural characteristics of char and activated carbon prepared under different activation conditions are listed in Table 2. The S_{BET} and $S_{Langmuir}$ increased significantly compared to char. Additionally, it can be observed from the data that the BET and Langmuir surface area increased with increases in the temperature of the second activation process when comparing DH2 and DH5. On the one hand, the pore size of DH5 was found to be the smallest of the three materials, which corresponded with similar trends in the results for the surface area. This is probably due to the increase in the activation temperature indicating the increase in extent of the reaction between KOH and char. Thus, the increased extent of the reaction aids in the enhancement and generation of new pores in the char, resulting in a higher surface area and pore volume, as presented in Table 3. It is well known that the activation process can contribute to the carbon structure, allowing the process to produce a more ordered structural skeleton. Typically, the pore creation process would occur in four stages: (i) the opening of previously accessible pores, (ii) creation of new pores, (iii) widening of the existing pores, and (iv) combination of the existing pores due to pore wall breakage [55].

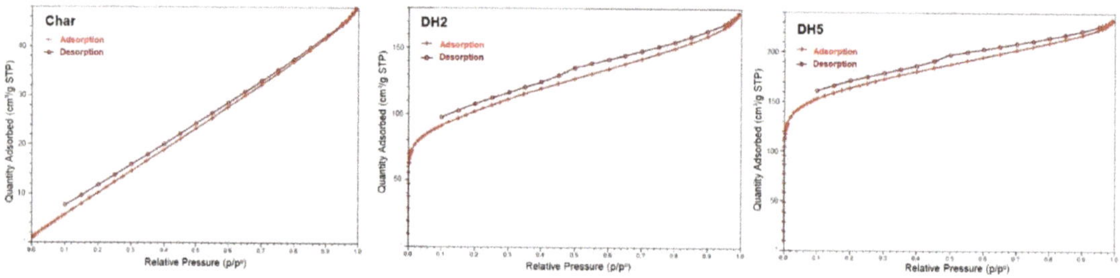

Figure 5. Nitrogen adsorption–desorption isotherms for char, DH2, and DH5, respectively.

Table 3. Microstructural characteristics of char and activated carbons.

Samples	S_{BET} (m^2/g)	$S_{Langmuir}$ (m^2/g)	V_T (cm^3/g)	$V_{\oplus}$ (cm^3/g)	V_m (cm^3/g)	D_p (nm)
Char	68.10	13.97	0.07	0.02	0.07	4.63
DH2	359.98	398.30	0.26	0.04	0.18	4.10
DH5	608.03	666.22	0.34	0.12	0.08	3.96

4. Conclusions

We successfully produced activated carbons with a high porosity and high surface area to volume structures from durian husk waste via a chemical activation method. The condition of the activation process directly affected the physical properties of the obtained activated carbons. The porosity and surface areas of activated carbons increased with the increases in temperatures during the second activation process. The average iodine number was exhibited to be up to 1320 mg/g. FTIR spectra also indicated only characteristic peaks of activated carbons without any new sign of functional groups. SEM images also revealed that all the activated carbons with various pore sizes and porosities were related to the temperatures in the second activation process. The specific surface area, pore volume, and pore size were further confirmed by BET investigations. It was found that the DH5 exhibited the highest surface area and pore volume of the three samples, which agreed with the results of iodine numbers and SEM investigations. The adsorption/desorption isotherms of activated carbons produced within the range of temperatures and impregnation ratios were a mixture of type I and II isotherms. In summary, we demonstrated the advantages of the utilization of such an inexpensive material through a simple preparation method. This work opens up a new approach to producing activated carbons for applications in the environment and industries, such as for adsorbents and battery electrodes.

Author Contributions: Conceptualization, T.J. and C.-F.H.; methodology and investigation, N.I., N.K., K.P. and W.S.; formal analysis, T.J.; resources, P.W.; writing—original draft preparation, T.J.; writing—review and editing, R.M.; K.-Y.A.L. and C.-F.H. All authors have read and agreed to the published version of the manuscript.

Funding: This research was financially supported by the Two-Institution Co-Research Scholarship provided by Kasetsart University and National Chung Hsing University.

Institutional Review Board Statement: Not applicable.

Informed Consent Statement: Not applicable.

Data Availability Statement: Not applicable.

Acknowledgments: We also acknowledge the Faculty of Science at Sriracha, Kasetsart University for providing funding and convenient laboratories.

Conflicts of Interest: The authors declare no conflict of interest.

References

1. Payus, C.M.; Refdin, M.A.; Zahari, N.Z.; Rimba, A.B.; Geetha, M.; Saroj, C.; Gasparatos, A.; Fukushi, K.; Oliver, P.A. Durian husk wastes as low-cost adsorbent for physical pollutants removal: Groundwater supply. *Mater. Today Proc.* **2021**, *42*, 80–87. [CrossRef]
2. Pansuk, J.; Junpen, A.; Garivait, S. Assessment of air pollution from household solid waste open burning in Thailand. *Sustainability* **2018**, *10*, 2553. [CrossRef]
3. Abuelnoor, N.; AlHajaj, A.; Khaleel, M.; Vega, L.F.; Abu-Zahra, M.R.M. Activated carbons from biomass-based sources for CO_2 capture applications. *Chemosphere* **2021**, *282*, 131111. [CrossRef] [PubMed]
4. Al-Maharma, A.Y.; Patil, S.P.; Markert, B. Effects of porosity on the mechanical properties of additively manufactured components: A critical review. *Mater. Res. Express* **2020**, *7*, 122001. [CrossRef]
5. Chattopadhyaya, G.; Macdonald, D.G.; Bakhshi, N.N.; Soltan, J.S.; Dalai, A.K. Preparation and characterization of chars and activated carbons from Saskatchewan lignite. *Fuel Process. Technol.* **2006**, *87*, 997–1006. [CrossRef]
6. Jibril, B.Y.; Al-Maamari, R.S.; Hegde, G.; Al-Mandhary, N.; Houache, O. Effects of feedstock pre-drying on carbonization of KOH-mixed bituminous coal in preparation of activated carbon. *J. Anal. Appl. Pyrol.* **2007**, *80*, 277–282. [CrossRef]
7. Pietrzak, R.; Wachowska, H.; Nowicki, P.; Babel, K. Preparation of modified active carbon from brown coal by ammoxidation. *Fuel Process. Technol.* **2007**, *88*, 409–415. [CrossRef]
8. Achaw, O.W.; Afrane, G. The evolution of the pore structure of coconut shells during the preparation of coconut shell-based activated carbons. *Micropor. Mesopor. Mater.* **2008**, *112*, 284–290. [CrossRef]
9. Laine, J.; Calafat, A.; Labady, M. Preparation and characterization of activated carbons from coconut shell impregnated with phosphoric acid. *Carbon* **1989**, *27*, 191–195. [CrossRef]
10. Castro, J.P.; Nobre, J.R.C.; Napoli, A.; Bianchi, M.L.; Moulin, J.C.; Chiou, B.S.; Williams, T.G.; Wood, D.F.; Avena-Bustillos, R.J.; Orts, W.J.; et al. Massaranduba sawdust: A potential source of charcoal and activated carbon. *Polymers* **2019**, *11*, 1276. [CrossRef]
11. Shaaban, A.; Se, S.M.; Ibrahim, I.M.; Ahsan, Q. Preparation of rubber wood sawdust-based activated carbon and its use as a filler of polyurethane matrix composites for microwave absorption. *New Carbon Mater.* **2015**, *30*, 167–175. [CrossRef]
12. Rahmawati, F.; Ridassepri, A.F.; Chairunnisa; Wijayanta, A.T.; Nakabayashi, K.; Miyawaki, J.; Miyazaki, T. Carbon from bagasse activated with water vapor and its adsorption performance for methylene blue. *Appl. Sci.* **2021**, *11*, 678. [CrossRef]
13. Li, Z.; Li, Y.; Zhu, J. Straw-based activated carbon: Optimization of the preparation procedure and performance of volatile organic compounds adsorption. *Materials* **2021**, *14*, 3284. [CrossRef] [PubMed]
14. Zhang, S.C.; Abdalla, M.A.S.; Luo, Z.J.; Xia, S.B. The wheat straw biochar research on the adsorption/desorption behaviour of mercury in wastewater. *Desalination Water Treat.* **2018**, *112*, 147–160. [CrossRef]
15. Uraki, Y.; Tamai, Y.; Ogawa, M.; Gaman, S.; Tokurad, S. Preparation of activated carbon from peat. *Bioresources* **2009**, *4*, 205–213.
16. Tao, H.; Zhang, H.; Li, J.; Ding, W. Biomass based activated carbon obtained from sludge and sugarcane bagasse for removing lead ion from wastewater. *Bioresour. Technol.* **2015**, *192*, 611–617. [CrossRef]
17. Kadirvelu, K.; Kavipriya, M.; Karthika, C.; Radhika, M.; Vennilamani, N.; Pattabhi, S. Utilization of various agricultural wastes for activated carbon preparation and application for the removal of dyes and metal ions from aqueous solutions. *Bioresour. Technol.* **2003**, *87*, 129–132. [CrossRef]
18. Williams, P.T.; Reed, A.R. Development of activated carbon pore structure via physical and chemical activation of biomass fiber waste. *Biomass Bioenergy* **2006**, *30*, 144–152. [CrossRef]
19. Rosson, E.; Garbo, F.; Marangoni, G.; Bertani, R.; Lavagnolo, M.C.; Moretti, E.; Talon, A.; Mozzon, M.; Sgarbossa, P. Activated carbon from spent coffee grounds: A good competitor of commercial carbons for water decontamination. *Appl. Sci.* **2020**, *10*, 5598. [CrossRef]
20. Sun, S.; Yu, Q.; Li, M.; Zhao, H.; Wu, C. Preparation of coffee-shell activated carbon and its application for water vapor adsorption. *Renew. Energy* **2019**, *142*, 11–19. [CrossRef]

21. Nham, N.T.; Tahtamouni, T.M.A.; Nguyen, T.D.; Huong, P.T.; Jitae, K.; Viet, N.M.; Noi, N.V.; Phuong, N.M.; Ahn, N.T.H. Synthesis of iron modified rice straw biochar toward arsenic from groundwater. *Mater. Res. Express* **2019**, *6*, 115528. [CrossRef]
22. Liu, H.; Feng, S.; Zhang, N.; Du, X.; Liu, Y. Removal of Cu(II) ions from aqueous solution by activated carbon impregnated with humic acid. *Front. Environ. Sci. Eng.* **2014**, *8*, 329–336. [CrossRef]
23. Eeshwarasinghe, D.; Loganathan, P.; Vigneswaran, S. Simultaneous removal of polycyclic aromatic hydrocarbons and heavy metals from water using granular activated carbon. *Chemosphere* **2019**, *223*, 616–627. [CrossRef] [PubMed]
24. Wang, Z.F.; Lin, T.; Chen, W. Occurrence and removal of microplastics in an advanced drinking water treatment plant (ADWTP). *Sci. Total Environ.* **2020**, *700*, 134520. [CrossRef]
25. Septian, A.; Kumar, A.V.N.; Sivasankar, A.; Choi, J.; Hwang, I.; Shin, W.S. Colloidal activated carbon as a highly efficient bifunctional catalyst for phenol degradation. *J. Hazard. Mat.* **2021**, *414*, 125474. [CrossRef]
26. Gil, R.R.; Ruiz, B.; Lozano, M.S.; Martín, M.J.; Fuente, E. VOCs removal by adsorption onto activated carbons from biocollagenic wastes of vegetable tanning. *Chem. Eng. J.* **2014**, *245*, 80–88. [CrossRef]
27. Ma, C.; Lu, T.; Shao, J.; Huang, J.; Hu, X.; Wang, L. Biomass derived nitrogen and sulfur co-doped porous carbons for efficient CO_2 adsorption. *Sep. Purif. Technol.* **2022**, *281*, 119899. [CrossRef]
28. Qin, T.; Song, M.; Jiang, K.; Zhou, J.; Zhuang, W.; Chen, Y.; Liu, D.; Chen, X.; Ying, H.; Wu, J. Efficient decolorization of citric acid fermentation broth using carbon materials prepared from phosphoric acid activation of hydrothermally treated corncob. *RSC Adv.* **2017**, *7*, 37112. [CrossRef]
29. Chen, J.K.; Huang, H.Y.; Tu, C.W.; Lee, L.T.; Jamnongkan, T.; Huang, C.F. SI ATRP for the surface modifications of optically transparent paper films made by TEMPO-oxidized cellulose nanofibers. *Polymers* **2022**, *14*, 946. [CrossRef]
30. Jamnongkan, T.; Wattanakornsiri, A.; Pansila, P.; Migliaresi, C.; Kaewpirom, S. Effect of poly(vinyl alcohol)/chitosan ratio on electrospun-nanofiber morphologies. *Adv. Mat. Res.* **2012**, *463–464*, 734–738.
31. Khankhuean, A.; Kuratsameethong, W.; Santibenchakul, S.; Laobuthee, A.; Sugimoto, M.; Srisawat, N.; Jamnongkan, T. Oriented ZnO nanoflowers obtained after calcination of electrospinning poly (vinyl alcohol)/zinc oxide/zinc acetate composite mats. *S. Afr. J. Chem. Eng.* **2021**, *37*, 179–185. [CrossRef]
32. Zhou, J.; Luo, A.; Zhao, Y. Preparation and characterisation of activated carbon from waste tea by physical activation using steam. *J. Air Waste Manag. Assoc.* **2018**, *68*, 1269–1277. [CrossRef] [PubMed]
33. Şahin, Ö.; Saka, C. Preparation and characterization of activated carbon from acorn shell by physical activation with H_2O-CO_2 in two-step pretreatment. *Bioresour. Technol.* **2013**, *136*, 163–168. [CrossRef] [PubMed]
34. Hayashi, J.I.; Kazehaya, A.; Muroyama, K.; Watkinson, A.P. Preparation of activated carbon from lignin by chemical activation. *Carbon* **2000**, *38*, 1873–1878. [CrossRef]
35. Diao, Y.; Walawender, W.P.; Fan, L.T. Activated carbons prepared from phosphoric acid activation of grain sorghum. *Bioresour. Technol.* **2002**, *81*, 45–52. [CrossRef]
36. Yossa, L.M.N.; Ouiminga, S.K.; Sidibe, S.S.; Ouedraogo, I.W.K. Synthesis of a cleaner potassium hydroxide-activated carbon from baobab seeds hulls and investigation of adsorption mechanisms for diuron: Chemical activation as alternative route for preparation of activated carbon from baobab seeds hulls and adsorption of diuron. *Sci. Afr.* **2020**, *9*, e00476.
37. Islam, M.A.; Ahmed, M.J.; Khanday, W.A.; Asif, M.; Hamee, B.H. Mesoporous activated carbon prepared from NaOH activation of rattan (*Lacosperma secundiflorum*) hydrochar for methylene blue removal. *Ecotoxicol. Environ. Saf.* **2017**, *138*, 279–285. [CrossRef]
38. Gul, E.; Alrawashdeh, K.A.B.; Masek, O.; Skreiberg, Ø.; Corona, A.; Zampilli, M.; Wang, L.; Samaras, P.; Yang, Q.; Zhou, H.; et al. Production and use of biochar from lignin and lignin-rich residues (such as digestate and olive stones) for wastewater treatment. *J. Anal. Appl. Pyrolysis* **2021**, *158*, 105263. [CrossRef]
39. Hu, Z.H.; Srinivasan, M.P.; Ni, Y.M. Novel activation process for preparing highly microporous and mesoporous activated carbons. *Carbon* **2001**, *39*, 877–886. [CrossRef]
40. Peng, C.; Yan, X.B.; Wang, R.T.; Lang, J.W.; Ou, Y.U.; Xue, Q.J. Promising activated carbons derived from waste tea-leaves and their application in high performance supercapacitors electrodes. *Electrochim. Acta* **2013**, *87*, 401–408. [CrossRef]
41. Guo, Y.P.; Qi, J.R.; Jiang, Y.Q.; Yang, S.F.; Wang, Z.C.; Xu, H.D. Performance of electrical double layer capacitors with porous carbons derived from rice husk. *Mater. Chem. Phys.* **2003**, *80*, 704–709. [CrossRef]
42. Kuratani, K.; Okuno, K.; Iwaki, T.; Kato, M.; Takeichi, N.; Miyuki, T.; Awazu, T.; Majima, M.; Sakai, T. Converting rice husk activated carbon into active material for capacitor using three-dimensional porous current collector. *J. Power Sources* **2011**, *196*, 10788–10790. [CrossRef]
43. Sousa, J.P.S.; Pereira, M.F.R.; Figueiredo, J.L. Catalytic oxidation of NO to NO_2 on N-doped activated carbons. *Catal. Today* **2011**, *176*, 383–387. [CrossRef]
44. Naeem, S.; Baheti, V.; Militky, J.; Ali, A. Multifunctional polylactic acid composites filled with activated carbon particles obtained from acrylic fibrous wastes. *Polym. Compos.* **2019**, *40*, 578–590. [CrossRef]
45. *ASTM D 4607-94*; Standard Test Method for Determination of Iodine Number of Activated Carbon. ASTM International: West Conshohocken, PA, USA, 2006.
46. Foo, K.Y.; Hameed, B.H. Preparation of activated carbon from date stones by microwave induced chemical activation: Application for methylene blue adsorption. *Chem. Eng. J.* **2011**, *170*, 338–341. [CrossRef]

47. Wongcharee, S.; Aravinthan, V.; Erdei, L.; Sanongraj, W. Mesoporous activated carbon prepared from macadamia nut shell waste by carbon dioxide activation: Comparative characterisation and study of methylene blue removal from aqueous solution. *Asia Pac. J. Chem. Eng.* **2018**, *13*, e2179. [CrossRef]
48. Mianowski, A.; Owczarek, M.; Marecka, A. Surface area of activated carbon determined by the iodine adsorption number. *Energy Sources A Recovery Util. Environ. Eff.* **2007**, *29*, 839–850. [CrossRef]
49. Bestani, B.; Benderdouche, N.; Benstaali, B.; Belhakem, M.; Addou, A. Methylene blue and iodine adsorption onto an activated desert plant. *Bioresour. Technol.* **2008**, *99*, 8441–8444. [CrossRef]
50. Ceyhan, A.A.; Sahin, Ö.; Baytar, O.; Saka, C. Surface and porous characterization of activated carbon prepared from pyrolysis of biomass by two-stage procedure at low activation temperature and it's the adsorption of iodine. *J. Anal. Appl. Pyrolysis* **2013**, *104*, 378–383. [CrossRef]
51. Saka, C. BET, TG–DTG, FT-IR, SEM, iodine number analysis and preparation of activated carbon from acorn shell by chemical activation with $ZnCl_2$. *J. Anal. Appl. Pyrolysis* **2012**, *95*, 21–24. [CrossRef]
52. Toledo, R.B.C.; Aragón-Tobar, C.F.; Gámez, S.; de la Torre, E. Reactivation process of activated carbons: Effect on the mechanical and adsorptive properties. *Molecules* **2020**, *25*, 1681. [CrossRef]
53. Anderson, N.; Gu, H.; Bergman, R. Comparison of novel biochars and steam activated carbon from mixed conifer mill residues. *Energies* **2021**, *14*, 8472. [CrossRef]
54. Sing, K.S. Reporting physisorption data for gas/solid systems with special reference to the determination of surface area and porosity (Recommendations 1984). *Pure Appl. Chem.* **1985**, *57*, 603–619. [CrossRef]
55. Yang, K.; Peng, J.; Srinivasakannan, C.; Zhang, L.; Xia, H.; Duan, X. Preparation of high surface area activated carbon from coconut shells using microwave heating. *Bioresour. Technol.* **2010**, *101*, 6163–6169. [CrossRef] [PubMed]

sustainability MDPI

Article

Eco-Friendly Clothing Market: A Study of Willingness to Purchase Organic Cotton Clothing in Bangladesh

Md Mehedi Hasan [1], Liling Cai [2,*], Xiaofen Ji [2] and Francisca Margarita Ocran [3]

1 School of Fashion Design and Engineering, Zhejiang Sci-Tech University, Hangzhou 310058, China; mehedi.mehedi00@gmail.com
2 Department of Fashion Design and Engineering, School of International Education, Zhejiang Sci-Tech University, Hangzhou 310058, China; xiaofenji@zstu.edu.cn
3 School of Textile Science and Engineering, Zhejiang Sci-Tech University, Hangzhou 310058, China; margariocran@gmail.com
* Correspondence: caililing@zstu.edu.cn

Abstract: This research study aims to achieve a developing country's sustainable development in the clothing industry by exploring consumer behavior to a willingness to purchase organic cotton clothing (OCC) and apparel retailers' responsibilities. Organic cotton clothing development in the fashion industry could play an important role in textile and environmental pollution and create new business opportunities for green clothes. Survey data was collected from top cities in Bangladesh, and 303 useable responses were collected (81.5% male and 18.5% female). In our survey, 60.7% of the participant was employed. This research model was inspired by the Theory of Reasoned Action (TRA) and added some new variable that influences purchase willingness under OCC fashion in developing countries like Bangladesh. The findings of this study stated that consumer environmental concerns and consumer attitudes positively impact the OCC purchase willingness of Bangladeshi consumers. Also, the authenticity and fashionable of OCC products have a significant impact on Bangladeshi consumer purchase intention. Product Performance found an indirect effect on Bangladeshi consumers' intention. Furthermore, this study will find that the Bangladeshi market is already very aware of the sustainability movement and concerned about environmental issues. Retailers should focus more on environmental awareness of OCC textile and authentic OCC items. Also, this study will update previous research findings on consumer attitudes toward OCC fashion in the Bangladesh market.

Keywords: organic cotton clothing (OCC); purchase willingness; brand responsibility; sustainable clothing

Citation: Hasan, M.M.; Cai, L.; Ji, X.; Ocran, F.M. Eco-Friendly Clothing Market: A Study of Willingness to Purchase Organic Cotton Clothing in Bangladesh. *Sustainability* **2022**, *14*, 4827. https://doi.org/10.3390/su14084827

Academic Editor: Andrea Pérez

Received: 22 February 2022
Accepted: 13 April 2022
Published: 18 April 2022

1. Introductions

Different types of unsustainable consumption patterns in today's world face several environmental problems such as pollution, greenhouse gas, global warming, etc. It's now become a global issue [1]. The textile and fashion industry is one of the environmental polluters in the industrial sectors [2]. For covering our bodies from weather and other social purposes, human beings use natural materials found in the environment. Nowadays, these natural materials are used in different products and have different types of applications. Cotton is the most widely recognized fiber as well as widely used raw material in this textiles industry. Except for a few artificial fibers, this industry is based on different types of cotton. We get cotton through our agriculture system. In the modern days, many things are involved in our agriculture system that has serious effects on the environment. Our cotton consumption increases every year, with world cotton consumption growing 2.66 percent in the 2021-22 seasons [3]. Naturally grown cotton does not harm the environment unless fertilizers, pesticides, and other harmful chemicals are used [4]. But now, the use of fast-growing cotton and the production of traditional cotton fiber using various chemicals harms the soil, water, and air, resulting in pollution of the environment [5]. Organic agricultural practices advance a more secure, eco-friendly and do not harm living creatures; organic fertilizers, natural pesticides, and insecticides are used [6]. So, organic cotton is much better for farmers, consumers, and all living things. According to the US Census Bureau, the current world population is 7.8 billion [7], and UN DESA predicted 8.5 billion by 2030 and 9.7 billion by 2050 [8]. According to current solutions, people need to have cloth and cotton consumption, and demand will increase. One of the most challenging issues is global warming and climate changes directly connected with CO2 emissions [9]. 1.22 to 2.93 billion tonnes of CO_2 are added to our environment by textile industries. The carbon footprint of cotton is a remarkable top, around 2–4 t per hectare. On the other side, compared to traditional cotton, organic cotton has 40% less "global warming potential" and suggests a 91% reduction in natural water consumption [10]. In this situation, we strongly believe Organic cotton clothing practice can be one of the great solutions for this industry, making it more sustainable and environmentally friendly. Not only in western developed countries but also in developing countries like Bangladesh. Because many recent studies on sustainable development and the green movement are based on western developed countries while less intention on developing countries such as Bangladesh [5]. Sustainable fashion practice offers a way to solve many environmental problems related to production and fashion consumption [11].Recent studies on consumer willingness to purchase behavior find that growing awareness and environmental concerns influenced consumer purchase decisions on organic clothing cotton [12]. We believe that consumers can influence the transformation of fashion companies towards sustainability through their purchasing decisions [13]. Also, the fashion industry should be aware of environmental safety, human safety, and other remarkable corporate social responsibilities. A study on European consumers believed that fashion brands should take the challenge this climate change and environmental protection [5].In many developed countries, fashion retailers focus on organic clothing products to increase their market share [14]. Many renowned brands and retailers were making a profit from organic clothing. Brands like Nike, H&M, C&A, and Wal-Mart have introduced 100% organic materials [15]. '2025 Sustainable Cotton Challenge' from May 2017 textile industry motivate clothing brands and retailers to make 100% of the cotton they use come from sustainable sources by 2025 [16]. Clothing Companies should reduce their environmental carbon footprint; in many ways, ethical sales are growing but not enough categories in the market [12]. Consumer willingness depends on many things; relevant literature from previous research summarizes that high product costs, less choice, aesthetic difficulties, the credibility of information, and uncertainty about the actual environmental benefits were the main limitation for consumers to purchase eco-friendly products, including clothing [12]. Consumer demand for green products is also connected with price and income strategies, especially green products on the market sold at a higher price than regular products (Wilson and Ilartsen, 2010) [17]. However, nowadays, environmentally concerned consumers are growing and they are shown to pay a premium price for eco-friendly products [18]. Boks And Stevels indicate that consumers are willing to buy green products when their income and budget increases [19]. A study on organically grown products (OGPs) already proved that monthly household income is statistically significant and positively influences the consumer purchase intentions of organically grown food products [20].

There are a lot of studies available on consumer behavior toward Organic personal care products and organic food products. But only a few studies are available on OCC purchasing behavior intention [14]. Therefore this study inspired by The Theory of Reasoned Action (TRA), also modified and extended other essential variables that impact willingness to purchase intention OCC. TRA is an old model, but many researchers of consumer behavior-related studies inspired their research fields and questionnaires on this model. TRA model is widely used and extended in many ways [21,22].

Bangladesh Fashion Market

Economic growth changes rapidly, and most manufacturing units are developed in underdeveloped countries. Environmental issues are not a priority in developing countries, but they are a significant threat to the world [4]. World population ranking Bangladesh is now world rank 8 [23], besides Bangladesh also 2nd largest clothing manufacture whole over the world [5]. According to the world bank, this COVID-19 limitation Bangladesh's economic growth rates pick up to 6.9 percent in the year 2021, and last decade fastest-growing economy in the world [24]. According to overall economic growth, Bangladesh's domestic market size is also increasing, and some global brands are already active in the Bangladeshi market. This includes the German leading sports brand Puma, and the popular Japanese brand Uniqlo which started their retail business in Bangladesh a few years ago [25]. Similarly, in the last decades, several local brands have had strong positions in the Bangladeshi market like 'sailor', 'yellow', 'Aarong' and other brands which are doing well, and these brands are also undertaking sustainable initiatives. For example, 'sailor' and 'Aarong' are introducing natural cotton examples: organic cotton, Bamboo fiber and Jute fiber products. Therefore, organic products and their impact on the environment has always been an important research topic, although less focus is placed on studying consumer purchase behavior in developing countries.

2. Literature Review and Hypothesis Development

Today's generations are more concerned word is sustainable fashion, sustainable fabrics, and natural fibers. Abrar, M investigated consumer purchase behavior of green organic textile products in developing countries such as Pakistan. This study finds that consumers have a positive attitude toward purchasing organic products. The study suggests that retailers create consumer attitudes towards organic clothing products and increase consumers' purchase intention for organic clothing products [14]. Hae Jin Gam also studies mothers' willingness to purchase OCC for childrens' product segment. The result found that mothers' environmental concerns significantly impact their involvement in OCC, significantly influencing mothers' willingness to purchase OCC [12]. In a case study on organic cotton clothing in Hawaiian consumers, Lin's also discussed and found that consumers are willing to pay more when consumers are more concerned about protecting the environment or involved in environmental issues [26]. Gwendolyn Hustvedt [27], has shown that consumers are motivated by their belief in the benefits of buying organic clothing. It is clear that it explains how organic cotton clothing supports organic farming; it is another way to promote the environmental benefits of buying organic cotton clothing. Previous studies on similar topics, green eco-friendly marketing, and consumer purchase behavior mentioned that intricacy of information, product advantages, functional benefits and positioning of green products, and the celerity of environmental effects benefits were the crucial barriers to purchasing eco-friendly products [28].

2.1. *Theory of Reasoned Action (TRA)*

The theory of reasoned action (TRA) used to predict and understand consumer behavior, and their behavioral intention developed by Ajzen and Fishbein 1980 [29] is a belief-attitude-behavioral intention model. In the TRA model, when consumers implement a purchasing activity, they benefit from that behavior and gain approval from others. TRA seeks to predict consumer purchasing and intent [21]. TRA theory has been extensively accepted and strongly used in multiple studies such as marketing, consumer motivations, promoting recycling behavior, the intent to engage in sustainable behaviors, and the attitude toward luxury fashion goods [29]. Researchers also modified and added the TRA model to their research plan. A recent study's effects of government subsidies on green smartphones successfully used the TRA model and added new variables [21]. We will also find some variables that also affect consumer buying behavior. This study is also an expansion of TRA. In addition, consumer environmental concerns, fashionable products in organic segments, product authenticity, performance, and consumer economic conditions are also important research values in this study. Environmental concern is assessed as a subjective norm in the TRA in this research. This study also introduces brand responsibilities like original and authentic organic cotton products and focuses on fashionable organic apparel that can strongly affect consumer purchasing behavior.

2.2. Environmental Awareness and Concerns

Environmental awareness and concerns have obtained great attention from consumers all over the world. When consumers are conscious of environmental issues, they raise concerns about the environment and human life. Previous researchers have also found that consumers more aware of the environment are more likely to be involved in ethical purchases. Another research paper also investigated environmental concerns related to ethical behaviors and consumers' willingness to pay more for green electricity [30]. In a case study on organic cotton clothing in Hawaii, resident consumers found that when people are aware of protecting the environment, they have more intention to purchase OCC items [26]. In this study, we similarly investigate in our target consumers whether environmental concerns have affected consumers' purchase intentions under OCC buying intentions.

Hypothesis H1. *Environmental concerns have a significant positive impact on willingness to purchase OCC.*

According to the TRA model, attitudes affect the intent to engage in behavior, which affects actual behavior. Attitude is the level at which a person has a favorable or unfavorable opinion of behavior [30]. Many studies have already proved it in different research fields. Kim and Chung found that attitude influenced consumer purchase behavior for organic skincare items [31]. In previous studies, Yan, Hyllegard, and Blaesi also mentioned that attitude predicts consumer purchasing intent for ethically produced fashion items [30]. Therefore, we assume that the results may be similar. Thus, we suggest the below:

Hypothesis H2. *Attitudes have a significant positive impact on willingness to purchase OCC.*

2.3. Brand Responsibilities in Eco-Friendly Fashion

Brand responsibilities and awareness allow the retailer to provide a long-term ethical plan. It is also an important marketing strategy to attract consumers, especially in the eco-friendly market. Eco-friendly brands also benefit if they are promoted with an image of nature with environmental and social responsibility [32]. In previous research works, it has been found that ethical fashion and style are related important factors for making a decision, and was also mentioned by consumers willing to boycott unethical brand leaders [33]. Some consumers have a different idea about OCC, and they still doubt its quality, finishing, and design. It is common for consumers to expect something good from green products when they make purchasing decisions [32]. This study investigates whether the fashion and trend of an organic apparel product affect consumers' purchase intention. Therefore, in this research, we proposed that:

Hypothesis H3. *Fashionable and trendy products have a significant positive impact on willingness to purchase OCC.*

Consumers who want to buy organic cotton clothing are also expected to focus on the performance of the clothing. The previous research found that consumers who used "organic" as a criterion when purchasing clothing products believed that a quality product is a result of purchasing their organic cotton clothing [34]. Product performance and quality are important variables that influence consumers' decisions to purchase. When the consumer is not satisfied with the product's effectiveness, it evaluates the purchasing experience as a time risk [35]. Thus, in this research, we proposed that:

Hypothesis H4. *Product performance has a significant positive impact on willingness to purchase OCC.*

Some clothing brands claim their product are organic, but most of the time, they do not carry any certified organic label. Previous researchers have already noted that certified environmental claims can enhance consumers' confidence in the environmental performance of a product and the quality of the organic label product [34]. In addition, some consumers have doubted the authenticity of the product. This study investigates whether products authenticity affects consumers' purchase intention of eco-friendly OCC products. We proposed that:

Hypothesis H5. *Product authenticity significantly impacts willingness to purchase eco-friendly OCC fashion.*

2.4. Purchasing Power of "Green Clothes"

Consumer purchasing power and economic growth are also important factors in their purchase decision, especially organic clothing, as many researchers mentioned that green products are usually available in the market at a higher price, and when consumers' income increases, their shopping budget also increases. Researchers point out that consumers are more interested in purchasing green products when their income and budget increase [32]. Consumers are willing to pay more for green and renewable products when they are aware of the environmental advantages [30]. Therefore, in this research, we proposed that:

Hypothesis H6. *Consumer economic growth and condition can significantly influence consumers' willingness to purchase eco-friendly OCC Fashion and willingness to pay premium purchasing.*

Based on the above literature review and theoretical analysis, this study proposes the following conceptual research model. The model is illustrated in Figure 1.

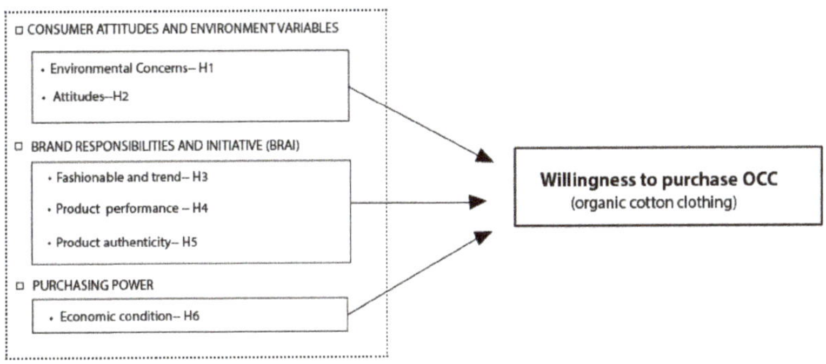

Figure 1. Conceptual model of this study.

3. Methods

The choice of research methods must be fair and appropriate to the research problem and should be selected based on the study objectives. The online survey was taken with more than 303 adult consumers. We believed that a quantitative research method using a survey is the best way to collect basic information from Bangladeshi consumers. Final data collection and all data evaluation will be IMB SPSS Statistics.,version 22, Release 22.0.0.0,64-bit edition (Hangzhou, China). Our survey data was examined through a reliability and validity test, factor loading, Cronbach's Alpha, and other reliability tests were done. The comprehension of participants' questions from a survey may differ from what the researchers anticipated. It is perhaps impossible to answer the research question evaluation. At that point, the validity of the construction and its reliability needs to be verified [36,37]. Understanding both directions, as well as the strength of the relationship between the research constructs within this study, a Pearson correlation analysis was also done. Pearson's method of correlation evaluates the linear relationship between two variables. The coefficient can be between −1.00 to 1.00, where −1.00 shows a negative relationship, 0 shows no relationship, and 1.00 is a positive relationship. After confirming that the coefficient of a reciprocal relationship is statistically significant, the strength of the relationship can be discussed and measured [37,38]. Finally, a regression analysis is done for understanding the impact of all variables because researchers utilize regression coefficients and their significance to estimate the importance of each construct. The most commonly used is multiple simultaneous regressions, where all IV regression calculations are taken at once. The regression coefficient gives an analysis of the direct impact of each IV on the DV when considering other IVs [39].

3.1. *Question Design*

This research questionnaire was adopted from previous research and a few of the author's questions according to the research direction [5,15,32,37,40–42]. Based on an online survey written in English. The first few questions are consumer demographic questions that determine their social position. The other question is to determine their involvement, knowledge, daily activities, attitude toward OCC, and willingness to purchase on OCC. Multiple-choice questions and a rating scale plan the question layout for this survey. This survey asks for the responses to each item question is placed on a 5-point Likert scale with "1 = strongly disagree, to 5 = strongly agree" [5].

3.2. *Sample*

This study was conducted quantitatively using the online survey method, Google online survey form used, and distributed survey links mainly using social networks. The survey was distributed to a random sample of more than 303 young consumers. Overall, 303 valid questionnaires have been completed. All questionnaire surveys were self-administered to respondents who lived in the top two cities in Bangladesh. Out of 303 responses, 81.5% were male consumers, and 18.5% were female. Almost 60.7% of responders were employed, and the rest of the, worked in business, were unemployed, and were students. The demographic information of the participants is presented in Table 1.

Table 1. Demographic information of the sample.

Variable	Description	Percentage (%)
Gender	Male	81.5
	Female	18.5
	Others	0
Age	<18	1.3
	19–25	28.1
	26–35	64.4
	36–45	5.9
	>46	0.3
Education	Below undergraduate	5.3
	Undergraduate	28.4
	Graduate	46.5
	postgraduate	19.8
Employment level	student	20.8
	Employed	60.7
	Unemployed	6.3
	Business holder	11.2
Income level	<14,000 BDT	21.1
	15,000–25,000 BDT	19.8
	25,000–50,000 BDT	30
	50,000–100,000 BDT	22.1
	>100,000 BDT	6.9

4. Results Analysis

Many scholars mentioned that a coefficient of 0.7 or above represents the threshold for reliability. Others believe that values of 0.6 or above are also acceptable [37]. Our Cronbach's Alpha is 0.851 and KMO is 0.830; both are reliable and suitable for our studies, presented in Table 2.

Table 2. Cronbach's Alpha Reliability Statistics.

Reliability Statistics		
Cronbach's Alpha	Cronbach's Alpha Based on Standardized Items	N of Items
0.851	0.853	33

Table 3 shows the results of this analysis. Most of the correlations were significant at the p level of <0.01. However, product performance (PP) correlates with purchase intention (PI) at the significance level of $p < 0.05$. There are some differences also; some correlations did not have statistical significance, such as FT and PI. Great correlation strength is between FT and PP with an r of 0.635. In the other EC, AT, PA, and Econ constructs, this variable significantly correlated with all other constructs.

Table 3. Inter-correlations for all constructs.

Variables		EC	AT	FT	PP	PA	ECon	PI
EC	Environmental concern	—	0.404 **	0.314 **	0.244 **	0.404 **	0.364 **	0.361 **
AT	Attitude	—	—	0.292 **	0.258 **	0.448 **	0.509 **	0.483 **
FT	Fashionable and trend	—	—	—	0.635 **	0.379 **	0.295 **	0.111
PP	Product performance	—	—	—	—	0.321 **	0.271 **	0.115 *
PA	Product authenticity	—	—	—	—	—	0.578 **	0.526 **
ECon	Economic condition	—	—	—	—	—	—	0.592 **
PI	Purchase intention	—	—	—	—	—	—	—

** Correlation is significant at the 0.01 level (2-tailed). * Correlation is significant at the 0.05 level (2-tailed).

The independent variables significantly predict consumer purchase intention (PI), $F(6,296) = 42.336$, $p < 0.001$, which indicates that our independent factors are an impact on consumer purchase intentions. Moreover, The $R^2 = 0.462$ depicts this model explaining 46.2% variance in purchase intention. Table 4 shows the summary of the findings.

Table 4. Model Summary.

Model	R	R Square	Adjusted R Square	Sig. F Change
1	0.680 [a]	0.462	0.451	0.000

[a] Predictors: (Constant), EConv, PPv, ECv, ATv, PAv, FTv.

Table 5 reveals that applying all the IVs at once creates a multiple correlation coefficient ($R^2 = 0.462$). It implies that 46.2% present of the variance in purchase intention of organic-cotton clothing can be explained by all constructs taken together. The standardized beta coefficient (ß) is considered to compare each IV's influence size directly. Table 5 shows that economic condition (Econ, ß = 0.362) has the greatest impact on PI, product authenticity (PA) (ß = 0.259), AT (ß = 0.198), FT (ß = 0.158) and EC (ß = 0.105). Econ's positive sign of ß means that a 1 standard deviation (SD) increase in Econ's generates a 0.362 SD increase in the predicted PI. The p-value facing every IV points out if that variable is important, adding to the calculation for explaining PI from the entire group of IVs. As a result, Econ ($p = 0.000$), PA ($p = 0.000$), AT ($p = 0.000$), FT ($p = 0.006$) and EC ($p = 0.034$) are the sole constructs that are importantly contributing to the explanation.

Table 5. Regression analysis overview for all constructs.

Hypotheses	Regression Weights	ß	T	*p*-Value	Hypotheses Supported
H1	EC > PI	0.105	2.130	0.034	Accepted
H2	AT > PI	0.198	3.779	0.000	Accepted
H3	FT > PI	−0.158	−2.747	0.006	Accepted
H4	PP > PI	−0.043	−0.765	0.445	Rejected
H5	PA > PI	0.259	4.626	0.000	Accepted
H6	ECon > PI	0.362	6.463	0.000	Accepted

R^2 = 0.462, F (6296) = 42.336, $p < 0.001$.

We provide a summary of this study's findings, both accepted and rejected hypotheses. This study examines the factors influencing Bangladeshi people's purchasing intention toward eco-friendly organic cotton clothing (OCC). In addition, the authors strive to find the effects of those factors, including environmental concern (EC), attitude (AT), fashionable and trend (FT), product authenticity (PA), and economic condition (Econ), as well as product performance (PP). The overall model generated from linear regression analysis is statistically significant. The findings of our studies are consistent with many previous studies.

In this study, H1 EC toward eco-friendly OCC was found to have a positive effect on PI, with a standardized regression coefficient (ß) of 0.105 (p = 0.034). Past research examined influencers of PI toward eco-friendly OCC among American mothers' willingness to purchase OCC [12]. It was revealed that the EC is significantly influenced by consumer involvement in OCC, and the effect was found to be positive (ß) of 0.33 (p = 0.04). In addition, another researcher examined factors motiving male consumers Awareness of the environmental consequences has a positive impact on male consumers' engagement in eco-friendly clothing. Its results are also positive and (ß) of 0.180 (p = 0.001) [2]. In addition, a study on purchase intention towards OCC and the result shows that EC had a insignificant positive impact on purchase intention (β) of 0.216 (p =0.168) [14]. The results of Table 5 show that there was a significant relationship between the environmental concern variable and the purchase intention in OCC (p = 0.034, R^2 = 0.462). The p-values were positive and statistically significant, indicating that participants with greater environmental concerns showed a higher intent to purchase OCC. Therefore, H1 was accepted. Furthermore, H2 is confirmed as AT significantly and positively influences the PI, with a standardized regression coefficient (ß) of 0.198 (p = 0.000). Previous research studies on purchasing green apparel products found that attitudes will positively influence the consumer purchasing intent towards green clothing products and it was found to be significant and β = 0.17 for American consumers similar to beta β value with our studies [43]. In addition, another examined factor motiving on male consumer and AT had a significantly positive impact on male consumers for engaging in eco-friendly clothing. Its results were also positive and (ß) of 0.284 (p = 0.000) [2]. A past researcher also examined influencers of PI toward eco-friendly clothing among American university students. It was revealed that the PI is significantly determined by AT, and the effect was found to be positive. Similarly, in purchasing bio-cotton apparel among female Korean households items, researchers found AT as a significant and positive influencer (ß = 0.34) [37]. There are some differences, however; recent surveys of the USA and Bangladeshi consumers' organic cotton clothing purchase behavior comparative studies found that AT is not significant in PI toward OCC purchases in the Bangladesh market. With a standardized regression coefficient for Bangladesh consumer (ß) of 0.19 (p = 0.349) [5].We can argue this point because this result variation may be due to the number of participants. The studies were based on only 51 participants from the Bangladeshi market. Moreover, it has been shown that fashionable and trendy products (FT) have a significant impact on PI (ß = −0.158) (p = 0.006), which supports H3. In addition, the negative sign indicates that more perception of limited availability regarding eco-friendly bio cotton clothing is associated with lower degrees of PI. A previous study already investigated consumer fashion consciousness having a positive impact on purchase intention of fast fashion products with (ß = 0.375) (p = 0.000) [44]. Previous researchers also found that fashion and style are related factors in making decisions to purchase clothes [33]. Fashionable products are among the trendiest items, and consumers share and find the latest trend in real-time [45]. In our studies, we also find that fashionable products have a significant impact on PI. Therefore, a lot of large retailers may have to rethink their design practices according to eco consumers' requirements and introduce more organic cotton clothing.

Next, product authenticity (PA) was found to have a significant influence on PI (ß = 0.259, p = 0.000). H5 is supported. In addition, the positive sign indicates that more positive PA is related to greater degrees of PI. Previous studies also mentioned that the effect of authenticity on purchase intention is significant [46]. Recent studies also found that product authenticity had a positive impact on PI (β = 0.221) for Iraqi Facebook users. It is recommended that marketers should provide practical and honest information in their promotion messages [47].

In this study, economic condition (Econ) toward purchasing organic cotton clothing was found to be the highest of all constructs and has a positive effect on PI, with a standardized regression coefficient (ß) of 0.362 (p = 0.000). The results in Table 5 show that consumer purchasing power and consumer economic growth positively influence consumer purchase intention. Therefore H6 was accepted. Consumer economic growth and income always influence purchase intentions. Previous studies have shown that high-income respondents have a strong desire to purchase through social media websites [48]. However, a survey of Pakistani consumers found that there was no relationship between income and purchase intentions [49].

Though another hypothesis in our model, product performance (PP) had a relation in our correlation model, H4 was not accepted in consumers' PI toward green OCC products in the Bangladesh market. Product performance (PP) with a standardized regression coefficient (ß) of −0.043 (p = 0.445). In our studies, both the lowest Beta (ß) values and p-value are more than our acceptable level. When we surveyed organic products that are not durable, 29.4% of participants replied neutral and 21.1% disagreed. The same participants were asked another survey question about organic cotton styles and fashion, 32.3% agreed, and almost 15% strongly agreed that organic cotton did not fully fit their style and fashion. Therefore, in that survey, we can point out that consumers in the Bangladesh market are more concerned about fashion and style than product lifetime. Finally, we determine the effect of environmental concerns, attitudes, fashionable products on organic clothing segments, product performance, and economic growth on the purchase intention of consumers of OCC products. This study will find that the Bangladeshi market is already very aware of the sustainability movement and concerns about environmental issues. Bangladeshi consumers show a positive attitude toward OCC items. Bangladeshi markets have a good potential for all kinds of organic cotton clothing. Retailers should be concerned and take the initiative in their eco-friendly clothing design practice, also take the initiative and emphasis the values of organic cotton in our environment and human life. Fashion retailers should be concerned promote honest and transparent information to consumers because product authentication is one of the important factors in consumer purchase decisions, especially when consumers are aware and willing to purchase eco-friendly OCC items. This research can be conducted in different sectors and countries using the qualitative method through focus groups. This study will be beneficial for the environmental improvement and economic growth efforts among policymakers, our academy, and all over the textile industry. This study enables them to design strategies that ensure interaction with consumers to promote desirable behavior towards organic cotton clothing products. Figure 2 illustrates the effect of IV on dependent variables (e.g., PI). Arrows with a solid line represent a statistically significant effect when accompanied by them; the dashed line shows a non-significant effect.

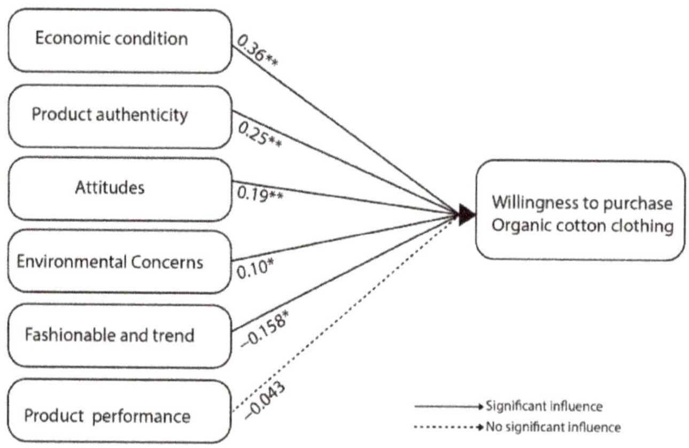

Figure 2. Summary of revised model. All values are standardised regression coefficients. ** $p < 0.00$, * $p < 0.05$.

5. Limitations and Future Research

This study was conducted in the top two busiest metropolises in Bangladesh, Dhaka, and Gazipur. Bangladesh is a country of population density. Future research can examine both urban and rural contexts such as income by accurately presenting different population profiles such as age, level of education, and population. Furthermore, this research work can be replicated in other developing countries, and future research should include other theories or methods. Future researchers could verify the proposed research model on other products such as home textiles and luxurious items that are made from organic cotton or other sustainable textiles.

Research survey items on "Willingness to Purchase Organic Cotton Clothing in Bangladesh" added below in Appendix A Table A1.

Author Contributions: Conceptualization, M.M.H. and L.C.; methodology, M.M.H. and L.C.; software (SPSS), M.M.H. and F.M.O.; formal analysis, M.M.H.; resources, M.M.H.; data curation, M.M.H. and F.M.O.; writing—original draft preparation, M.M.H.; supervision, L.C.; project administration, X.J.; funding acquisition, X.J. All authors have read and agreed to the published version of the manuscript.

Funding: The paper is supported by Zhejiang Provincial Philosophy and Social Science Planning Project (21NDJC062YB), National Social Science Foundation of China art program (20BG134).

Institutional Review Board Statement: Not applicable.

Informed Consent Statement: Informed consent was obtained from all subjects involved in the study.

Data Availability Statement: The information presented in this study is available in the manuscript. Additional data is available at the request of the author concerned.

Conflicts of Interest: The author declares no conflict of interest.

Appendix A

Table A1. Research Survey item and sources.

Variable	Item	Source
Environmental concern	It is time for environmental groups to get more radical. I am extremely worried about the state of the environment. We are in a serious negative impact of environmental issues. I feel personally helpless to have much impact on the environment.	[40,42]
Attitudes	Buying organic cotton clothing instead of conventional clothing would feel like the morally right thing to do. Buying organic cotton clothing instead of conventional apparel would make me feel like a better person. Buying organic cotton clothing instead of traditional clothing will feel more beneficial to the environment.	[15]
Fashionable and trend	In my opinion, organic cotton clothing is only popular at the moment. Organic cotton clothing does not come in different styles that fulfill my need. In my opinion, organic cotton clothing is not trendy and fashionable.	[37,41]
Product Performance	organic cotton clothing is of good quality and has a better lifetime. Organic cotton clothing is not as high-quality then conventional cotton clothing. Organic cotton clothing will not be durable when cleaned (e.g., color fades, form changes)	[37]
Product authenticity	I suspect that I can't distinguish real organic cotton clothing. I only believe in organic cotton garments that have been authoritatively certified. I will be more likely to purchase eco-friendly OCC fashion if their apparel Brand makes sure their product is Authentic. Not all businesses claiming to sell organic cotton clothing are credible Those businesses that often advertise environmental friendliness do not all sell real organic cotton clothing I sometimes don't believe that some businesses claim that their clothes are organic cotton clothing.	[15,42]
Economic condition	The cost of organic apparel was the same as nonorganic apparel. I would be more likely to buy more organic apparel. I am willing to pay more for eco-friendly products (i.e., green, organic). Is the current financial development and growth affecting your shopping behavior? According to financial development, did you concern that clothing and shopping expenses are increased?	[15,32,41]
Purchase intention	I would gladly buy more organic cotton apparel if I could find it easily. I desire to buy organic cotton clothing if it's health and environment friendly. I am willing to pay more for organic cotton clothing if it's fashionable and fulfills my needs. The next time you go shopping, I will likely purchase organic apparel products.	[5,15]

References

1. Mahmoud, T.O. Impact of green marketing mix on purchase intention. *Int. J. Adv. Appl. Sci.* **2018**, *5*, 127–135. [CrossRef]
2. Taljaard, H.; Sonnenberg, N.C.; Jacobs, B.M. Factors motivating male consumers' eco-friendly apparel acquisition in the South African emerging market. *Int. J. Consum. Stud.* **2018**, *42*, 461–468. [CrossRef]
3. Global Cotton Production & Consumption to Improve: TexPro. Available online: http://www.fibre2fashion.com/news/textile-news/global-cotton-production-consumption-to-improve-texpro-277436-newsdetails.htm (accessed on 16 April 2022).
4. Ali, M.A.; Sarwar, M. Sustainable and Environmental Freindly Fibers in Textile Fashion (A Study of Organic Cotton and Bamboo Fibers). Master's Thesis, University of Borås, Borås, Swiden, 2010. Available online: https://www.semanticscholar.org/paper/Sustainable-and-Environmental-freindly-fibers-in-(A-Ali-Sarwar/6d263e1621c8ef3814a8e8d7f5ce3ab4f5bbae89 (accessed on 20 February 2022).

5. Hasan, M.N.U.; Liu, C.; Ahmed, B. Organic Cotton Clothing Purchase Behavior: A Comparative Study of Consumers in the United States and Bangladesh. *Textiles* **2021**, *1*, 376–386. [CrossRef]
6. Kumar, P.S.; Yaashikaa, P.R. Organic Cotton and Its Environmental Impacts. In *Organic Cotton*; Gardetti, M.A., Muthu, S.S., Eds.; Textile Science and Clothing Technology; Springer: Singapore, 2019; pp. 127–139, ISBN 978-981-10-8781-3.
7. World Population Surged 74 mn in 2021, Expected to be 7.8 bn: Report. Business Standard India. 2021. Available online: https://www.business-standard.com/article/international/world-population-surged-74-mn-in-2021-expected-to-be-7-8-bn-report-121123101389_1.html (accessed on 17 January 2022).
8. The World Population Prospects: 2015 Revision. 2015. Available online: https://www.un.org/en/development/desa/publications/world-population-prospects-2015-revision.html (accessed on 17 January 2022).
9. Ali, K.A.; Ahmad, M.I.; Yusup, Y. Issues, Impacts, and Mitigations of Carbon Dioxide Emissions in the Building Sector. *Sustainability* **2020**, *12*, 7427.
10. Clothes and Climate: Is Cotton Best? Available online: https://ejfoundation.org/news-media/clothes-and-climate-is-cotton-best (accessed on 17 January 2022).
11. Hasanspahic, M. Sustainable Fashion-Practices, Strategies, and Meanings. 2016. Available online: http://lup.lub.lu.se/student-papers/record/8879292 (accessed on 20 February 2022).
12. Gam, H.J.; Cao, H.; Farr, C.; Kang, M. Quest for the ecoapparel market: A study of mothers' willingness to purchase organic cotton clothing for their children. *Int. J. Consum. Stud.* **2010**, *34*, 648–656. [CrossRef]
13. Blasi, S.; Brigato, L.; Sedita, S.R. Eco-friendliness and fashion perceptual attributes of fashion brands: An analysis of consumers' perceptions based on twitter data mining. *J. Clean. Prod.* **2020**, *244*, 118701. [CrossRef]
14. Abrar, M.; Baig, S.A.; Bashir, M.; Shabbir, R.; Ayub, M. Consumer attitude and purchase intention towards organic textile products. *Amazon. Investig.* **2018**, *7*, 472–485.
15. Maloney, J.; Lee, M.-Y.; Jackson, V.; Miller-Spillman, K.A. Consumer willingness to purchase organic products: Application of the theory of planned behavior. *J. Glob. Fash. Mark.* **2014**, *5*, 308–321. [CrossRef]
16. 2025 Sustainable Cotton Challenge-Textile Exchange. Available online: https://textileexchange.org/2025-sustainable-cotton-challenge/ (accessed on 19 January 2022).
17. Wilson, J.W.; Eilertsen, S. How did strategic planning help during the economic crisis? *Strategy Leadersh.* **2010**, *38*, 5–14. [CrossRef]
18. Chi, T.; Gerard, J.; Dephillips, A.; Liu, H.; Sun, J. Why U.S. Consumers Buy Sustainable Cotton Made Collegiate Apparel? A Study of the Key Determinants. *Sustainability* **2019**, *11*, 3126. [CrossRef]
19. Boks, C.; Stevels, A. Theory and practice of environmental benchmarking in a major consumer electronics company. *Benchmarking Int. J.* **2003**, *10*, 120–135. [CrossRef]
20. Wekeza, S.; Sibanda, M. Factors Influencing Consumer Purchase Intentions of Organically Grown Products in Shelly Centre, Port Shepstone, South Africa. *Int. J. Environ. Res. Public. Health* **2019**, *16*, 956. [CrossRef]
21. Liu, H.T.; Tsaur, R.-C. The Theory of Reasoned Action Applied to Green Smartphones: Moderating Effect of Government Subsidies. *Sustainability* **2020**, *12*, 5979. [CrossRef]
22. Haris, A.; Kefeli, Z.; Ahmad, N.; Daud, S.N.M.; Muhamed, A.; Shukor, S.A.; Kamarubahrin, A.F. Consumers' Intention To Purchase Dates: Application of Theory of Reasoned Action (Tra). 16. Available online: https://www.researchgate.net/publication/332015191_Consumers\T1\textquoteright_Intention_to_Purchase_Dates_Application_of_Theory_of_Reasoned_Action_TRA (accessed on 10 February 2022).
23. Bangladesh Population (2022)–Worldometer. 2022. Available online: https://www.worldometers.info/world-population/bangladesh-population/ (accessed on 18 January 2022).
24. Overview. Available online: https://www.worldbank.org/en/country/bangladesh/overview (accessed on 20 January 2022).
25. Local and Foreign Fashion Brands in a Race to Grab Bangladesh Market. Textile News, Apparel News, RMG News, Fashion Trends. 2019. Available online: https://www.textiletoday.com.bd/local-and-foreign-fashion-brands-in-a-race-to-grab-bangladesh-market/ (accessed on 20 January 2022).
26. Lin, S.-H. A case study in Hawaii: Who will pay more for organic cotton?: Organic cotton. *Int. J. Consum. Stud.* **2010**, *34*, 481–489. [CrossRef]
27. Hustvedt, G.; Dickson, M.A. Consumer likelihood of purchasing organic cotton apparel: Influence of attitudes and self-identity. *J. Fash. Mark. Manag. Int. J.* **2009**, *13*, 49–65. [CrossRef]
28. Meyer, A. What's in it for the customers? Successfully marketing green clothes. *Bus. Strategy Environ.* **2001**, *10*, 317–330. [CrossRef]
29. Neffling, C.M.M. Can CSR and Slow Fashion Influence Purchase Intent? A Quantitative Study. 2002. Available online: https://www.diva-portal.org/smash/get/diva2:1436309/ATTACHMENT01.pdf (accessed on 15 February 2022).
30. Weiner, H.E. Wearing your Ethics: Investigating Consumer Purchase Intention of Ethically Produced Fashion Products. Master's Thesis, University of South Carolina, Columbia, SC, USA, 2017.
31. Yeon Kim, H.; Chung, J. Consumer purchase intention for organic personal care products. *J. Consum. Mark.* **2011**, *28*, 40–47. [CrossRef]
32. Wang, C.C.; Tsai, S. Consumer Attitudes towards Sustainable and Environmental Strategies in Fashion Clothing. Master's Thesis, University of Manchester, Manchester, UK, 2010.
33. Shaw, D.; Hogg, G.; Wilson, E.; Shiu, E.; Hassan, L. Fashion victim: The impact of fair trade concerns on clothing choice. *J. Strateg. Mark.* **2006**, *14*, 427–440. [CrossRef]

34. Tong, X.; Su, J. Exploring young consumers' trust and purchase intention of organic cotton apparel. *J. Consum. Mark.* **2018**, *35*, 522–532. [CrossRef]
35. Nepomuceno, M.V.; Laroche, M.; Richard, M.-O. How to reduce perceived risk when buying online: The interactions between intangibility, product knowledge, brand familiarity, privacy and security concerns. *J. Retail. Consum. Serv.* **2014**, *21*, 619–629. [CrossRef]
36. Saunders, M.; Lewis, P.; Thornhill, A. Research Methods for Business Students, 7th Edition. Available online: https://www.pearson.com/uk/educators/higher-education-educators/program/Saunders-Research-Methods-for-Business-Students-7th-Edition/PGM1089011.html (accessed on 30 March 2022).
37. Liu, S.; Tiger, B. Antecedents of buying intention towards bio-cotton clothing: A quantitative study among young business students at USBE. 2017. Available online: http://www.diva-portal.org/smash/record.jsf?pid=diva2%3A1111532&dswid=-4083 (accessed on 7 February 2022).
38. Pearson Correlation Coefficient–An overview | ScienceDirect Topics. Available online: https://www.sciencedirect.com/topics/social-sciences/pearson-correlation-coefficient (accessed on 30 March 2022).
39. Milford, T.M.; Keith, T.Z. *Multiple Regression and Beyond: An Introduction to Multiple Regression and Structural Equation Modeling*, 2nd ed.; Taylor & Francis: New York, NY, USA, 2015.
40. Bianchi, C.; Birtwistle, G. Consumer clothing disposal behaviour: A comparative study: Consumer clothing disposal behaviour. *Int. J. Consum. Stud.* **2012**, *36*, 335–341. [CrossRef]
41. Montero, C. Consumer Behavior and Eco-Friendly Fashion Apparel. Master's Thesis, Old Dominion University, Norfolk, VA, USA, 2009.
42. Hasbullah, N.N.; Sulaiman, Z.; Mas'od, A.; Sugiran, H.S.A. Drivers of Sustainable Apparel Purchase Intention: An Empirical Study of Malaysian Millennial Consumers. *Sustainability* **2022**, *14*, 1945. [CrossRef]
43. Ko, S.B.; Jim, B. Predictors of Purchase Intention toward Green Apparel Products in the U.S. and China. *J. Fash. Mark. Manag.* **2017**, *21*, 70–87.
44. Gia Vuong, H.; Tan Nguyen, M. Factors Influencing Millennials' Purchase Intention towards Fast Fashion Products: A Case Study in Vietnam. *Int. J. Soc. Sci. Humanit.* **2018**, *8*, 235–240. [CrossRef]
45. Won, J.; Kim, B.-Y. The Effect of Consumer Motivations on Purchase Intention of Online Fashion–Sharing Platform. *J. Asian Financ. Econ. Bus.* **2020**, *7*, 197–207. [CrossRef]
46. Bergamelli, E. The role of Brand Orientation on Consumers Perceived Authenticity. Available online: https://run.unl.pt/bitstream/10362/39713/1/Bergamelli_2018.pdf (accessed on 10 January 2022).
47. Mosa, R.A. The Impact of Advertising Credibility on Purchase Intentions: An Empirical Study among Iraqi Facebook Users. *Eur. J. Bus. Manag. Res.* **2021**, *6*, 228–234. [CrossRef]
48. Kian, T.P.; Boon, G.H.; Fong, S.W.L.; Ai, Y.J. Factors That Influence the Consumer Purchase Intention in Social Media Websites. *Int. J. Supply Chain. Manag.* **2017**, *6*, 8.
49. Arshada, M.S.; Aslam, T. The Impact of Advertisement on Consumer's Purchase Intentions. *SSRN Electron. J.* **2015**. [CrossRef]

sustainability MDPI

Article

Application of Best Available Techniques to Remove Air and Water Pollutants from Textile Dyeing and Finishing in South Korea

Gahee Kim [1], Phil-Goo Kang [1,*], Eunseok Kim [1] and Kyungae Seo [2]

1 Integrated Pollution Prevention and Control Research Team, Natural Environment Research Division, National Institute of Environmental Research (NIER), Gyeongseo-dong, Seo-gu, Incheon 22689, Korea; gahkim@korea.kr (G.K.); jugaru@korea.kr (E.K.)
2 Environmental Inspection and Investigation Team, Hanriver Basin Environmental Office, Misagangbyeonhangang-ro, Hanam-si 12902, Gyeonggi-do, Korea; nnke02@korea.kr
* Correspondence: philgkang@korea.kr

Abstract: The textile industry in South Korea is characterized by excessive water consumption, high concentrations of wastewater, hazardous chemicals, and high energy consumption. This study aimed to analyze Best Available Techniques Reference Documents (BREFs) based on best available techniques (BATs) and BAT-associated emission levels (BAT-AELs) and identify potential solutions for tackling environmental pressure from the South Korean textile industry. Therefore, the existing practices of the textile dyeing and finishing industry in South Korea were compared with those from the BREFs of the European Union. Many existing BATs in South Korea are related to reducing water consumption. There is also a strong focus on BATs for reducing wastewater discharge and achieving energy-saving during treatment rather than after treatment, which differs from other industries. Moreover, BAT-AELs were derived for chemical oxygen demand, suspended solids, and total nitrogen for treating non-biodegradable, highly polluted wastewater. Furthermore, BREFs related to atmospheric pollution included dust generated from the heated fabrics in the finishing process that contained cadmium and phenolic hydrogen chloride from dyes and raw materials in the fabrics. Notably, the European Union has not specified BAT-AELs for the textile industry, whereas South Korea has tailored BAT-AELs for toxic and hazardous chemicals. Thus, numerous green techniques to reduce emissions and energy consumption are being implemented in South Korea.

Keywords: BAT; BAT-AEL; K-BREF; environmental integrated permit; textile industry; South Korea

Citation: Kim, G.; Kang, P.-G.; Kim, E.; Seo, K. Application of Best Available Techniques to Remove Air and Water Pollutants from Textile Dyeing and Finishing in South Korea. *Sustainability* **2022**, *14*, 2209. https://doi.org/10.3390/su14042209

Academic Editors: Hafeezullah Memon, Xiaoke Jin, Tian Wei and Chengyan Zhu

Received: 23 December 2021
Accepted: 11 February 2022
Published: 15 February 2022

1. Introduction

The Integrated Environmental Permit System was introduced in the European Union (EU) in 1996 as an integrated environmental management system for pollution prevention and control in different industries. Subsequently, the EU published Best Available Techniques Reference Documents (BREFs), which specify techniques for controlling pollutants, pollution emission levels, and other information, to support this system [1,2]. South Korea enacted the Act on the Integrated Control of Pollutant-Discharging Facilities on 1 January 2017, which applies to water and air quality for Type 1 and Type 2 business sites across 19 industries. Type 1 sites include workplaces that generate >80 tons of air pollutants per year or emit >2000 m^3 of water pollutants per day. In contrast, Type 2 sites include workplaces that generate 20–80 tons of air pollutants per year and emit 700–2000 m^3 of water pollutants per day [3,4]. The Integrated Environmental Permission System of South Korea applies to each business site (unlike in the EU) because it is focused on large-scale business sites. This permit system aims to establish a method that can control discharge facilities in an integrated manner and facilitate the implementation of optimal environmental management techniques. These techniques can be applied to each business site in line with their features to effectively reduce pollutants generated from their corresponding economic activities [5]. Therefore, a customized license/permission system can be introduced with one integrated permit instead of approximately 10 licenses and permits [6,7]. In particular, the Integrated Environmental Permission System is composed of BREFs, which are required for integrated environmental control planning, emission impact analysis, and an integrated environmental control system [8,9]. From this perspective, BREFs include the current status and overall processes of representative industries, including the primary industrial knowledge regarding various processes and techniques, ranging from raw material input to pollutant emissions and treatment techniques.

BREFs are normally compiled via a comprehensive examination of business sites included in the Integrated Environmental Permission System, which considers the current status of each industry, emitted pollutants, and the best available techniques (BATs) that are economically achievable. BREFs play the role of providing business sites in the industries subject to integrated environmental control with BATs to reduce the generation and emission of pollutants and support them to efficiently enforce environmental control. In practice, BREFs are used as the reference tool to elucidate the processes when preparing an integrated environmental control plan and evaluate existing techniques and emissions. Furthermore, BAT-associated emission level (BAT-AEL) can be introduced and used as the primary data for quantifying the maximum emission criteria and emission impact analysis. Although BATs are not mandatory, they are becoming practical permission standards for setting the maximum emission criteria. In general, BREFs are published by the BREF Secretariat at the National Institute of Environmental Research (Incheon, South Korea), which provides practical support, such as for the preparation and publication of BREFs. Practical support implies the determination of details, such as the procedure and composition of BREFs, composition, and operation of technical working groups (TWGs), and operation of BREF Deliberation Committee under the Central Environmental Policy Committee [10]. Generally, BREFs are created to establish the standards for an integrated permit, and 17 BREFs have been published for 19 industries. The BREF for the textile dyeing and finishing industry was published in 2020. The domestic textile industry imports, dyes, and processes textiles produced internationally. Unlike other industries, it features multiple materials and fibrous assembly processes. Additionally, the domestic textile industry is mainly composed of small businesses, which use a larger amount of water and discharge high volumes of wastewater in relation to the size of the business. From an environmental perspective, chemicals such as dyes result in high concentrations of hazardous chemicals in discharged water.

The textile industry involves excessive water consumption, high concentrations of wastewater discharge, hazardous chemicals [11], and consumption of large amounts of energy, thereby requiring efficient BREFs to overcome these challenges. The present study aimed to assess the improvement of standards by analyzing the BREFs of South Korea and comparing them with those of the EU. The present study introduced BREFs based on BATs and BAT-AELs for South Korea. The specific objectives of the study were to (1) analyze BREFs of the textile dyeing and finishing industry in South Korea and (2) propose an efficient national management route by comparing the existing practices in South Korea with the BREFs of the EU textile industry.

2. Materials and Methods

2.1. Processes of the Textile Dyeing and Finishing Industry

The textile industry processes include (1) spinning, which produces yarns; (2) fabric manufacturing, which produces fabrics; (3) dyeing and finishing of the fabricated fabrics, which produce the final clothing. The textile industry is composed of processes with high value-added organic correlations that lead to finished products for clothing, manufacturing, and exports. Notably, the textile industry in South Korea is only composed of the dyeing and finishing industry because all the raw materials are imported from other countries. The domestic textile industry in South Korea is diminishing as it advances towards international markets.

The textile dyeing and finishing industries focus on processing; they differ considerably from other domestic industries in terms of integrated permits. The textile industry features multiple raw materials, processes, and emitted pollutants, which are changed by the combination and use of materials, fibrous assemblies, and manufacturing processes, as well as by many overlapping and similar processes. Although other industries compose classification systems by-products and processes in the BREF, the textile industry has overlapping processes that are minimized by distinguishing contents by material, fibrous assembly, and process. Thus, all contents of the textile industry can be included when analyzing BATs (Table 1). The BREFs of the textile industry in the EU comprise a wider range of processes than those in South Korea because the former industries grow cotton, produce yarns, and manufacture and process fabrics by themselves.

The textile dyeing and finishing industry inevitably consumes large volumes of water to dye fabrics, thereby generating large amounts of wastewater. In particular, dyeing 1 kg of fabrics requires approximately 200 kg of water. Moreover, the water consumption is large relative to the size of the business [12]. The business size and daily water consumption were compared using the 2019 Water Emission Management System (WEMS) data in this study. To this end, the average daily water consumption of businesses with Type 1 water quality, which are subject to integrated permits, was used.

The comparison of the textile dyeing and finishing industry processes between South Korea and the EU is indicated in Figure 1. Figure 2 shows that the textile industry uses large amounts of water relative to the business size; the textile industry is the third-largest consumer of water, whereas the business size is the smallest, among 17 industries (businesses with 50 employees or less are small according to the Ministry of SMEs and Startups [13]).

Table 1. Classification system of the textile dyeing and finishing industry in the South Korean Best Available Techniques Reference Documents (BREFs) [14].

Classification	Content
Materials	Cotton and cellulose Polyester Nylon Wool Silk, etc.
Processes	Pre-treatment (singeing, desizing, scouring, mercerizing, bleaching, etc.) Dyeing/Printing Post-treatment (dehydration, drying, coating, finishing)
Fibrous assembly types	Raw materials Yarns Weaving Knitting

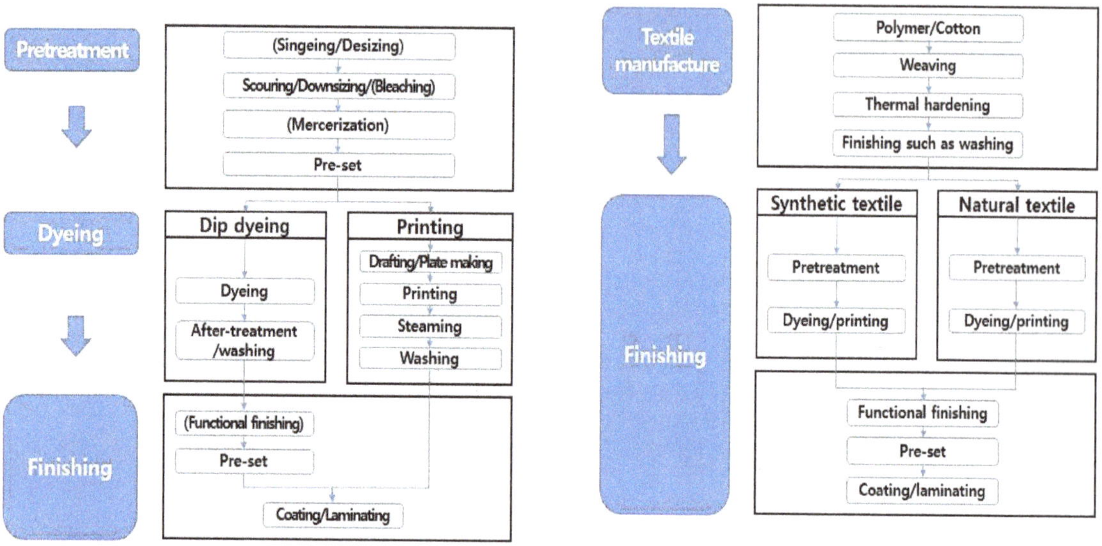

(a) (b)

Figure 1. Comparison of the textile dyeing and finishing industry processes between (**a**) South Korea and (**b**) the European Union (EU) [14,15].

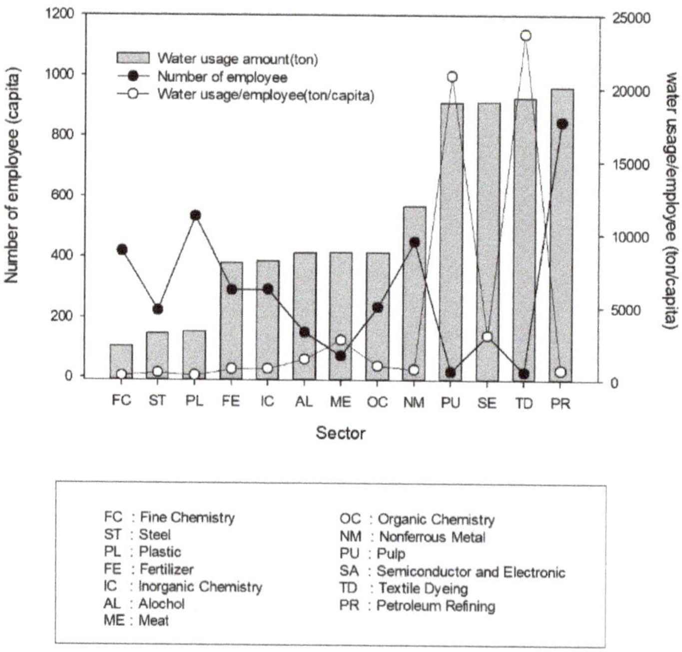

Figure 2. Comparison of water consumption by the size of the manufacturing business [16].

2.2. Environmental Problems in the Textile Dyeing and Finishing Industry

Owing to the inherent characteristics of the industry, the textile dyeing and finishing industry consumes a large amount of water in the dyeing process and generates high-concentration wastewater because of the use of chemicals such as dyes. Thus, water pollution has emerged as a major environmental issue associated with the textile industry [17]. Wastewater contains high concentrations of hazardous pollutants from dyes and chemicals that are challenging to treat [18]. According to the Ministry of Environment Guidebook for Licensing and Permission of Wastewater Discharge Facilities, 32 types of water pollutants have been identified in the textile dyeing and finishing industry. This list includes 18 types of water pollutants that can exist, such as biochemical oxygen demand(BOD), chemical oxygen demand (COD), and total nitrogen (T-N), and 14 types of specific substances harmful to water quality, such as chrome, cadmium, and formaldehyde. The discharge list is shown in Table 2 [18]. Moreover, large amounts of energy are required to treat industrial wastewater, which is challenging, owing to its high temperature in the industrial treatment systems [19].

The major environmental pollutants discharged from the textile dyeing and finishing industry can be classified into air and water pollutants (see the process-related classification in Tables 2 and 3). All the processes of this industry generate wastewater, as they all use water. Moreover, because chemicals represent the input in the first process, and the following processes are continuously conducted, heavy metals and unreactive dyes are discharged from all processes due to the chemicals in wastewater. Additionally, the COD of the emissions is high. From the perspective of air pollution, dust aerosol (i.e., harmful particulate matter) is mainly generated in the drying process. Moreover, gaseous pollution can be generated by producing formaldehyde from the use of chemicals as raw materials. According to the BREFs of the textile dyeing and finishing industry in the EU, the use of water and chemicals for dyeing and the generation of wastewater are reported as the emerging environmental problems in South Korea. The use of pesticides is also an emerging environmental problem because pesticides are involved in the processes of manufacturing yarns and textiles, such as cotton farming.

Table 2. List of water pollutants discharged from the textile dyeing and finishing industry [19].

Classification		Pollutants
Water pollutants discharge list (32 types)	Specific substances harmful to water quality (14 types)	Copper and copper compounds, lead and lead compounds, mercury and mercury compounds, cyanide compounds, hexavalent chrome compounds, cadmium and calcium compounds, dichloromethane, chloroform, 1,4-dioxane, diethyl hexyl phthalate (DEHP), acrylonitrile, naphthalene, formaldehyde, and epichlorohydrin
	Water pollutants (18 types, excluding specific substances)	Organic matter (biochemical oxygen demand (BOD) and chemical oxygen demand (COD)), suspended solids (SS), total nitrogen (T-N), total phosphorus (T-P), oils (mineral oil), oils (animal and vegetable oils), nickel and nickel compounds, manganese and manganese compounds, barium compounds, fluorine compounds, detergents, zinc and zinc compounds, iron and iron compounds, chrome and chrome compounds, perchlorate, phenols, and acids and alkalis (pH)

Table 3. Raw materials, process descriptions, and major environmental problems associated with major manufacturing processes in the domestic textile dyeing and finishing industry.

Process	Raw Materials	Process Description	Major Environmental Problems
Desizing	Weak acid, oxidative acid, water, alkali (Na_2CO_3)	A process of adding a chemical (thickener) to impart strength	Air → Volatile organic compounds (VOCs) Water → High BOD load, discharges many non-biodegradable matters, high concentration of solids
Scouring	Alkali (NaOH), surfactant, water	A process of removing impurities for uniform dyeing	Water → High pH, BOD, and COD; high alkalinity; and toxic substances
Bleaching	Bleach (e.g., H_2O_2), water	A process of removing coloring matters after scouring	Water → Bleach residues, high pH, SS, and BOD
Mercerizing	Alkali (NaOH)	A process of improving the gloss and strength of textiles	Water → Strong alkalinity
Dyeing	NaOH, dye, surfactant, water, additives (e.g., oxidant)	A process of coloring textiles using chemicals	Air → VOCs, energy Water → High COD and BOD, odor, alternative oxidase (AOX), heavy metals, additives, discharge of undyed dyes
Printing	Pigment, dye, additives (e.g., oxidant)	A process of applying specific patterns or designs to textiles	Air → VOCs, methanol, formaldehyde, etc. Water → SS, AOX, COD is higher than BOD
Finishing	Heat, resin, softener	A process of imparting desired properties to textiles	Air → dust, sulfur oxides, formaldehyde, energy, etc. Water → SS, energy

2.3. BATs and BAT-AELs of the Textile Dyeing and Finishing Industry

BATs are formulated as a result of the experts' meetings in the corresponding research/industry domain [1,5]. Such expert meetings include TWGs, whose members are primarily the experts at workplaces, industries, and academia. According to Article 24, Paragraph 5 of the Act on the Integrated Control of Pollutant-Discharging Facilities, BREFs are created based on the opinions and decisions of TWG members. They are published after being edited under the supervision of the National Institute of Environmental Research through the deliberation of the Central Environmental Policy Committee. BATs are delivered through the investigation of current status, in-field investigation, and license/permit review for 236 business sites in the textile dyeing and finishing industry subject to the integrated environmental control, primary data survey for three years for Korea, and 20 TWG meetings and small group meetings [14].

Unlike those of other industries, BATs for the textile dyeing and finishing industry have been formulated according to the classification systems that rely mostly on material, fibrous assembly, and process. Thus, the overlapped contents were minimized while describing all three characteristics (material, fibrous assembly, and process). They were classified according to the material within each process, and the shapes of fibrous assemblies were described in the detailed process description of each material. Thus, BATs were largely divided into general BATs, which comprised commonly applied and process BATs. Then, the process BATs were divided into pre-treatment, dyeing, and finishing processes. Additionally, BATs were separately derived according to the material within each process. Moreover, raw materials for each process, pollutants emitted, and BATs for treating pollutants were presented in integrated process diagrams.

BAT-AELs are mainly the lower and upper limits established for a range of pollutant discharge levels during normal operation by applying the derived BATs. Thus, these are the maximum permissible discharge levels that serve as a reference point for the permitted emission standards [20]. The maximum emission criteria are selected based on the pollutant items and upper limits derived from BATs. They are explicitly described in the Act on the Integrated Control of Pollutant-Discharging Facilities and are directly applied to the business sites. Additionally, the national certified data are used for setting the BAT-AEL values. Therefore, we applied the 2015 data for Stack Emission Management System (SEMS), Tele-Monitoring System (TMS), Water Emission Management System (WEMS), and Water Tele-Monitoring System (WTMS) of the business sites subject to the integrated environmental control of the textile dyeing and finishing industry [14]. The air pollutants were classified by the type of facilities according to the Clean Air Conservation Act. In contrast, the water pollutants were classified by region according to the Water Environment Conservation Act. Subsequently, the classification system for calculating linked emission levels was established, and BAT-AELs were calculated according to Seo et al. [21] by considering the classification system of BREFs.

During this process, COD, SS, and T-N were derived; they are all essential for treatment because wastewater is highly concentrated and contains mostly non-biodegradable components. Furthermore, the total suspended particles generated from heating the fabrics in the finishing process and the cadmium and phenolic hydrogen chloride generated from the dye residues in fabrics were included in the considered air pollutants. In particular, pollutants such as cadmium and phenol are classified as hazardous chemicals, and only trace amounts are discharged from the textile dyeing and finishing industry. It should also be noted that they were additionally selected for BAT-AEL through TWG meetings because they were considered as the subject of priority control, due to being hazardous chemicals that can harm the human body.

3. Results and Discussion

3.1. BATs

The integrated process diagrams related to the textile dyeing and finishing processes that apply BATs are presented in Figures 3 and 4.

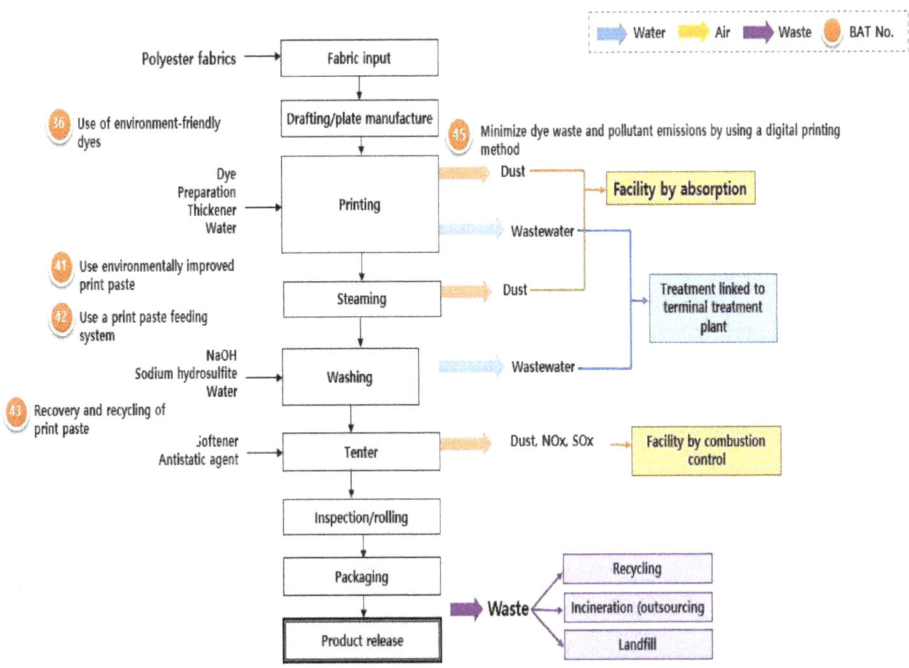

Figure 3. Example of integrated process diagram and BATs for the printing process (BAT No. is the number of BREF as shown in Table 4).

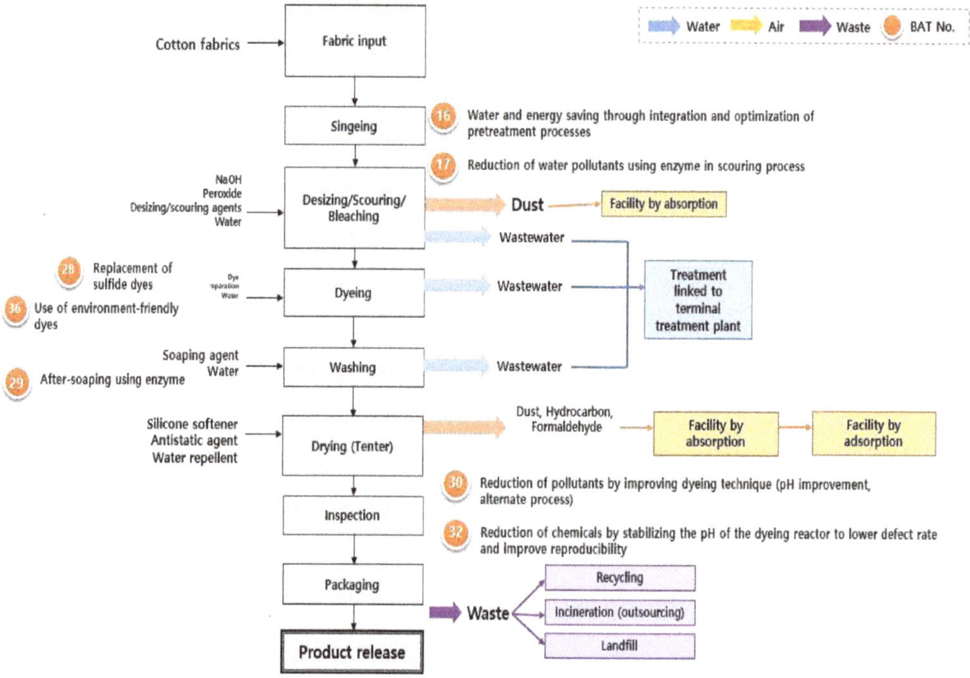

Figure 4. Example of integrated process diagram and best available techniques (BATs) for the cotton and cellulose dip dyeing process (BAT No. is the number of BREF as shown in Table 4).

The textile dyeing and finishing industry features numerous BATs for processes to reduce water consumption, considering the discharge of wastewater-centered pollutants and excessive water consumption. For the final versions of BATs, discharge and energy-saving techniques were particularly high in the manufacturing and finishing processes, rather than after treatment, compared with the currently published BATs of other industries (Table 5). In other industries, most BATs are often associated with post-treatment techniques, that is, treatment after discharge. However, the ratio of process BATs in the textile dyeing and finishing industry was found to be 84%, which is nearly twofold higher than that of discharge reduction techniques in processes (47.5%) among the 13 currently published BREFs. Additionally, the BATs for saving water by reusing water or integrating processes were derived and shown in Table 5. This indicates that many green techniques to reduce discharges and energy consumption are being developed and used.

Table 4. Examples for pollutant reduction BATs in the production processes of the textile dyeing and finishing industry.

Classification	BAT No *	Description
Dip dyeing	16	A technique to save water and energy consumption by integrating and optimizing individual pre-treatment processes applied to cotton materials. a. Application of integrated pre-treatment process for cotton textiles b. Optimization of cotton warp pre-treatment
	32	A technique to decrease the defect rate by stabilizing the pH of the bath in which dyeing reaction occurs and to reduce the use of chemicals by improving reproducibility
Printing	43	A technique to reduce pollutant discharge through the recovery and recycling of print paste a. Recovery of print paste from the supply system of a rotary screen printing machine b. Recycling of residual print pastes
	45	A technique to minimize the waste of dyes and the discharge of pollutants using a digital printing method a. Use of a digital jet printing method for bulky textiles b. Use of inkjet digital printing method for flat textiles

* BAT number of BREF for textile industry [14].

Table 5. Comparison of pollutant reduction BAT ratios in production processes.

Classification	Number of BATs	
	In Manufacturing Facilities	In Prevention Facilities
BATs in textile manufacturing BREF, Korea	42 (84%)	8 (12%)
BATs in BREFs, Korea	590 (48%)	651 (52%)

3.2. Comparison with EU-BREF

Similar to South Korea, the EU has formulated general BATs, which can be applied to the entire textile industry, and process BATs, which are applicable only to certain processes. Similar techniques are included in the general and process BATs. This implies that the businesses of the textile industries in South Korea and the EU are currently using similar techniques. The number of BATs in the EU is 113, which is nearly twofold, compared with BATs in South Korea (Table 6). This difference is driven by the existence of yarn- and fabric-manufacturing processes. Furthermore, in the EU, the processes and BATs of the carpet industry are also specified. Moreover, separate BATs for effluents and wastes are derived among BATs in the EU; for example, BAT for monitoring effluents, wherein the discharged cooling water is reused as process water, and the solid waste is collected and disposed of separately [15]. Unlike South Korea, the EU does not use many process techniques, while the wastewater of the textile industry is generally non-biodegradable. Notably, wastewater treatment is important because of the high concentration of organic matter in the wastewater. In particular, the dedicated BATs for wastewater must be highlighted for this purpose. Among the conventional BATs, those associated with improving the water and energy consumption efficiencies by integrating and optimizing individual processes in pre-treatment can be used for this purpose. There were common BATs for using automated systems in dye injection and supply and recycling water used in the dyeing process. In the printing process, common BATs included the techniques for minimizing dye wastes and pollutant discharges by utilizing a digital printing method and advanced techniques such as environmentally friendly printing thickener. In the finishing process, as in the printing process, many common BATs for saving energy have been formulated, such as heat recovery or smart inverter installation when operating tenter (dryer) facilities.

For example, in South Korea, there are wastewater-related BATs to reduce the discharge of water pollutants using the enzymes in the scouring process for pre-treatment and reduce pollutant discharge, such as minimizing the use of chlorine-based bleach. In the EU, there are BATs to facilitate wastewater treatment using either biodegradable lubricant or an easily degradable thickener. From the dyeing perspective, there are BATs in both the EU and South Korea for reducing chemical use by stabilizing pH in dyeing reaction and for reducing chemical discharge by removing non-permeated dyes on the textile surface. In addition, techniques to apply a chrome-free dyeing method for wool dyeing have been derived as a BAT. Specifically, the chrome-free dyeing method has been listed as an environmental control technique in the EU BREF, but it has not been considered as a BAT in South Korea. However, BATs have been proposed for decreasing discharge by the reduced use of chemicals in the dyeing process. The EU has also formulated techniques to reduce chemical use, but they are characterized by many BATs pertaining to the dyeing process for carpet and wool. Notably, these techniques are frequently used in Europe. From the printing perspective, BATs for minimizing dyes have been derived using a digital printing method, thereby reducing pollutant discharge. This is achieved using environmentally friendly print pastes in South Korea and the EU. However, the technique for improving the economy using pigments with a lower production cost than dyes has been proposed as a BAT only in South Korea.

BATs in the EU are focused on wastewater, whereas South Korean BATs are focused on various pollution sources, including air pollution, odor, and noise. The EU BATs have generally adopted primary industrial techniques by investigating and registering the techniques of many countries that belong to the EU. In contrast, BATs in South Korea have adopted the methods directly used in workplaces through a complete survey of workplaces, resulting in more reasonable results because the individual permit system is based on workplaces rather than on pollutants. Therefore, the BATs of South Korea include environmental techniques in general, such as those pertaining to air pollutants, odor, and wastewater. Furthermore, South Korea could derive more subdivided BATs of the textile industry (based on materials) than the EU. This is recommended because South Korea, in contrast to the EU, seemingly only deals with dyeing and finishing processes.

Table 6. Comparison of BATs between Korea BREF (K-BREF) and European Union BREF (EU-BREF).

Classification	K-BREF	EU-BREF
BATs	50 BATs in total	113 BATs in total
	- General (15) - Process (35)	- General (39) - Process (63) - Effluent and wastes (11)
Characteristics	BATs about the reduction in general pollutant discharges from business sites as well as about water quality	Separate BATs about the carpet industry and about the treatment of effluent and wastes (sludge)
	- Reduction of air pollutants - Reduction of noise and vibrations - Reduction of odor	- Sludge treatment of the carpet industry - Effluent and waste treatment

3.3. BAT-AEL

Table 7 shows the classification system for each textile dyeing and processing industry deriving BAT-AEL. The BAT-AEL derived from analysis based on the classification system is shown in Tables 8 and 9.

Table 7. Classification system for pollutant emission facilities in textile dyeing and processing industry for BAT-AEL derivation.

Main Category	Middle Category	Small Category	Detailed Category
Textile dyeing and finishing industry	Process facilities	Pre-treatment facilities	
		Dyeing facilities	
		Finishing facilities	
	Common facilities	Solid particle storage facilities	
	Incineration facilities	Waste gas Incineration facilities	200 kg·h^{-1}
			Less than 2 t·h^{-1}
		Wastewater/waste incineration facilities	200 kg·h^{-1}
			Less than 2 t·h^{-1}

Table 8. BAT-AELs for air pollutants.

Classification		BAT-AELs for Air Pollutants		
		Pollutants	**Unit**	**BAT-AEL**
Pretreatment		Dust	mg·Sm$^{3^{-1}}$	4–24
		HCl	ppm	1–4
		Cadmium	mg·Sm$^{3^{-1}}$	0–0.1
Dyeing/finishing		HCl	ppm	1–2
		Phenol	ppm	0–2
		Hydrocarbon	ppm	7–34
Incineration facilities	Waste gas incineration	Sulfur oxides	mg·Sm$^{3^{-1}}$	4–27
		Carbon monoxide	ppm	21–154
		HCl	ppm	2–13
	Wastewater/waste incineration	Sulfur oxides	mg·Sm$^{3^{-1}}$	5–27

Table 9. BAT-AELs for water pollutants.

Classification	BAT-AELs for Water Pollutants		
	Pollutants	**Unit**	**BAT-AEL**
≥2000 t/day—clean area	COD	mg·L^{-1}	18–30
≥2000 t/day—clean area	SS	mg·L^{-1}	2–17
Clean area	T-N	mg·L^{-1}	12–30

3.4. Limitations and Further Studies

3.4.1. BATs

In South Korea, all environmental management techniques applied in business sites have been considered BATs. Compared with the EU BATs, the South Korean BATs are advantageous because they are site-specific and directly applied. This is because these BATs had been developed through field surveys of all business sites, thereby providing and supporting tailored measures for business sites.

Moreover, although the EU BATs are focused on techniques related to wastewater in the textile industry, the BATs of South Korea have been derived for environmental management techniques. They are tailored to deal with air pollution, odor, and waste managed by business sites, although techniques related to wastewater also exist. However, BATs in South Korea have considerable limitations. As some techniques used by business sites were derived as BATs, the evaluation of economic feasibility and efficiency was insufficient. Moreover, it is not obligatory for business sites to measure economic feasibility and efficiency; therefore, only the design values can be applied. This considerably compromises the economic feasibility analysis of BATs. In contrast, the EU has considered the economic feasibility and environmental benefits of BATs. Thus, South Korean BATs should also be evaluated for economic feasibility and environmental benefits in the future. This can be achieved by analyzing and reviewing the actual data post-management, whereas the annual reports of business sites should be published to reflect the business reality. Additionally, new technologies can be proposed to be used as BATs when BREFs are revised, using more suitable techniques.

3.4.2. BAT-AELs

Thus far, BAT-AELs have been derived by the industry only for basic BAT environmental threats: aerosols (dust) and gaseous pollutants (sulfur oxides) in the air and water pollutants such as COD and SS. Additionally, the data used to derive BAT-AELs currently include national certified data such as SEMS, WEMS, TMS, and WTMS. However, they have long measurement and input cycles, and the number of analyzed items remains limited. Therefore, some items cannot be derived due to insufficient sample size in statistical analysis. This problem can be alleviated by collecting the data through annual reports and post-management, as with BATs. From the perspective of water pollution, the analysis of specific substances harmful to water quality is presently mandatory. Therefore, BAT-AELs can be derived further for a broader range of pollutants by utilizing these data.

Author Contributions: Conceptualization, G.K., P.-G.K., E.K. and K.S.; investigation, G.K.; resources, G.K. and P.-G.K.; data curation, E.K.; writing—original draft preparation, G.K.; writing—review and editing, G.K. and P.-G.K. All authors have read and agreed to the published version of the manuscript.

Funding: This research was funded by the Ministry of Environment, National Institute of Environment Research, Grant Number NIER-2017-01-02-061.

Acknowledgments: We would like to express our gratitude to Jaehong Park from the National Institute of Environmental Research, who reviewed this research paper comprehensively, and Joonseok Koh from Konkuk University, who provided support during the field inspection of the fiber dyeing and finishing industries in South Korea.

Conflicts of Interest: The authors declare no conflict of interest.

References

1. Integrated Pollution Prevention and Control (IPPC). Laying down Rules Concerning Guidance on the Collection of Data and on the Drawing up of BAT Reference Documents and on Their Quality Assurance Referred to in Directive 2010/75/EU of the European Parliament and of the Council on Industrial Emissions, EU. 2012. Available online: https://eur-lex.europa.eu/eli/dec_impl/2012/119/oj (accessed on 24 November 2010).
2. Seo, K.A.; Kim, G.H.; Kim, E.S.; Seok, H.J.; Shin, S.J.; Kim, Y.L.; Kang, P.G. Comparative Analysis of Best Available Techniques Reference Documents on the Fertilizer Manufacture Between Korea and European Union. *J. Environ. Sci. Int.* **2020**, *29*, 307–318. [CrossRef]
3. Ministry of the Environment (MOE). Act on Clean Air Conservation Act. 2021. Available online: https://law.go.kr/lsSc.do?section=&menuId=1&subMenuId=15&tabMenuId=81&eventGubun=060101&query=%EC%88%98%EC%A7%88%ED%99%98%EA%B2%BD#undefined (accessed on 30 December 2021).
4. Ministry of the Environment (MOE). Act on Water Environment Conservation Act. 2022. Available online: https://law.go.kr/lsSc.do?section=&menuId=1&subMenuId=15&tabMenuId=81&eventGubun=060101&query=%EC%88%98%EC%A7%88%ED%99%98%EA%B2%BD#undefined (accessed on 13 January 2022).

5. Ministry of the Environment (MOE). Act on Integrated Management of Environmental Pollution Facilities. 2021. Available online: https://law.go.kr/lsSc.do?section=&menuId=1&subMenuId=15&tabMenuId=81&eventGubun=060101&query=%ED%86%B5%ED%95%A9%ED%99%98%EA%B2%BD%EA%B4%80%EB%A6%AC%EB%B2%95#undefined (accessed on 21 July 2021).
6. Shin, S.; Park, J.; Park, S.; Lee, D.; Kim, D. Understanding and Improvement of Best Available Techniques for Waste Incineration Facility. *J. Korean Soc. Atmos. Environ.* **2017**, *33*, 533–543. [CrossRef]
7. Kim, H.K. The Appraisal and the Tasks of the Act on Integrated Pollution Prevention and Control. *J. Kor. Sox. Environ. Law* **2016**, *38*, 327–361.
8. Kim, G.H.; Kim, E.S.; Seo, K.A.; Kang, K.H.; Khan, J.B.; Seok, H.J.; Kang, P.G. Application of Best Available Techniques Reference Document for the Paint Manufacturing Industry. *Korean J. Hazard. Mater.* **2019**, *7*, 36–41.
9. Kim, E.S.; Kim, G.H.; Seo, K.A.; Khan, J.B.; Seok, H.J.; Kim, Y.L.; Kang, P.G. Introduction on Best Available Techniques Reference Document for the Display Manufacturing Industry in Korea. *Korean J. Hazard. Mater.* **2019**, *7*, 42–49. [CrossRef]
10. National Institute of Environmental Research (NIER) SOP. Operation Rules for the Preparation of Best Available Techniques Reference Documents. 2020. Available online: https://www.nier.go.kr/NIER/cop/bbs/selectNoginBoardArticle.do?menuNo=13001&bbsId=BBSMSTR_000000000031&nttId=27117&Command=READ/?pMENUMST_ID=95 (accessed on 6 January 2020).
11. Sana, K.; Abdul, M. *Environmental and Health Effects of Textile Industry Wastewater, Environmental Deterioration and Human Health*; Springer: Berlin/Heidelberg, Germany, 2014; pp. 55–71.
12. Kim, T.W.; Seo, B.J.; Park, G.H.; Lee, Y.W. Predicting Diffusion Behavior of Disperse Dyes in Polyester Fibers by a Method Based on Extraction. 2020. *J. Supercritical. Fluids.* **2020**, *157*, 104685. [CrossRef]
13. Ministry of SMEs and Startups. Framework Act on Small and Medium Enterprises. 2021. Available online: https://law.go.kr/lsSc.do?section=&menuId=1&subMenuId=15&tabMenuId=81&eveGubun=060101&query=%ED%86%B5%ED%95%A9%EA%B4%80%EB%A6%AC#undefined (accessed on 30 December 2021).
14. National Institute of Environmental Research (NIER). Reference Document on Best Available Techniques in the Textile Dyeing Industry. 2019. Available online: https://ieps.nier.go.kr/web/board/5/666/?pMENUMST_ID=95 (accessed on 6 January 2020).
15. Integrated Pollution Prevention and Control (IPPC). Integrated Pollution Prevention and Control Reference Document on Best Available Techniques for the Textiles Industries, EU. 2003. Available online: eippcb.jrc.ec.europa.eu/sites/default/files/2019-11/txt_bref_0703.pdf (accessed on 6 July 2003).
16. National Institute of Environmental Research (NIER). Water Emission Management System. 2018–2019. Available online: htpps://wems.nier.go.kr/board/viewMpReferenceList.do (accessed on 9 September 2020).
17. Dos Santos, R.F.; Ramlow, H.; Dolzan, N.; Machado, R.A.F.; de Aguiar, C.R.L.; Marangoni, C. Influence of Different Textile Fibers on Characterization of Dyeing Wastewater and Final Effluent. *Environ. Monit. Assess.* **2018**, *190*, 693. [CrossRef] [PubMed]
18. Wang, Z.; Xue, M.; Huang, K.; Liu, Z. Textile Dyeing Wastewater Treatment. In *Advances in Treating Textile Effluent*; Hauser, P., Ed.; Intech Open Sci: London, UK, 2011; pp. 91–116.
19. National Institute of Environmental Research (NIER). Guidebook for Licensing and Permission of Wastewater Discharge Facilities. 2017. Available online: me.go.kr/home/web/policy_data/read.do?pagerOffset=0&maxPageItems=10&maxIndexPages=10&searchKey=&searchValue=&menuId=10259&orgCd=&condition.deleteYn=N&seq=7396 (accessed on 24 October 2019).
20. Hong, S.I.; Nam, C.W.; Lee, W.S. Analysis of the Dye Absorption Behavior in Accordance with the Dye Structure in the Cold Pad Batch Dyeing of Cotton Knit. *Text. Color. Finish.* **2016**, *28*, 175–182. [CrossRef]
21. Seo, K.A.; Bae, Y.J.; Park, J.H.; Shin, D.S.; Rhew, D.H. Determination of Best Available Techniques Associated Emission Level (BAT-AEL). *J. Environ. Sci. Int.* **2019**, *28*, 455–464. [CrossRef]

sustainability

MDPI

Article

Evolution of Overall Cotton Production and Its Determinants: Implications for Developing Countries Using Pakistan Case

Muhammad Umer Arshad [1,*], Yuanfeng Zhao [2,*], Omer Hanif [3] and Faiza Fatima [4]

1 Pakistan Research Center, Inner Mongolia Honder College of Arts and Sciences, Hohhot 010010, China
2 College of Economics and Management, Inner Mongolia Agriculture University, Hohhot 010010, China
3 School of Business and Tourism, Southern Cross University, East Lismore, NSW 2480, Australia; omer.hanif@scu.edu.au
4 Department of Botany, University of Agriculture, Faisalabad 38000, Punjab, Pakistan; student.pak@yahoo.com
* Correspondence: umer.pk@yahoo.com (M.U.A.); zhaoyf@263.net (Y.Z.); Tel.: +86-13134717195 (M.U.A.)

Abstract: Managing the declining yield of non-food crops has opened new strategic challenges amidst global uncertainties. The COVID-19 scenario has increased awareness of natural lifestyle and eco-friendly products, largely dependent on non-food crop material. This strategic shift requires moving beyond traditional farm practices to improve agricultural production efficiency, and developing countries in particular have shown a consistent loss in their self-sufficiency of industrial crops despite being major exporters of non-food crop materials. However, existing studies analyze production efficiencies of non-food crops from general or theoretical aspects often by virtual estimates from breaking down the multiple factors of crop productivity. This study examined multiple factors of crop production to identify "which crop inputs have been inefficiently used overtime" by tracking efficiency changes and various input issues in overall cotton production from practical aspects, i.e., scaling non-constant returns of those multiple factors would allow for the violation of various situations. Accordingly, a stochastic frontier approach was employed to measure the production frontier and efficiency relationship using time-series data of Pakistan's cotton production from 1971–2018—a specific non-food crop perspective from a top-ranked cotton-producing country that has recently been shifted towards being a non-exporter of cotton due to low yield. The coefficient of area, seed, and labor indicates the positive relationship with cotton production, while fertilizer, irrigation, electricity, and machinery are statistically negative. This implies that policymakers need priority-based strategies for the judicial use of synthetic fertilizers, irrigation, a subsidy policy, and technology adoption, which could significantly improve the efficiencies of cotton productivity from the same land resources. Being adaptable to other developing economies, the analysis would strategically facilitate designing and developing affordable technology-driven solutions and their customized extensions towards sustainable non-food crop production practices and Agri-Resources efficiencies.

Keywords: crop production inefficiencies; stochastic frontier approach; total factor productivity; non-food crop; developing countries

Citation: Arshad, M.U.; Zhao, Y.; Hanif, O.; Fatima, F. Evolution of Overall Cotton Production and Its Determinants: Implications for Developing Countries Using Pakistan Case. *Sustainability* **2022**, *14*, 840. https://doi.org/10.3390/su14020840

Academic Editors: Hafeezullah Memon, Xiaoke Jin, Tian Wei, Chengyan Zhu and Marc A. Rosen

Received: 19 October 2021
Accepted: 9 January 2022
Published: 12 January 2022

1. Introduction

Recent global uncertainties and life-threatening concerns have increased awareness of a healthier and natural lifestyle. Subsequently, it has increased the consumption of eco-friendly products and natural fibers, which demand the management of the declining cotton yield in developing countries to look beyond traditional farming practices [1,2]. Managing the constant declining productivity of non-food crops remains a challenge for developing countries [1]. To handle this challenge, producers in these countries mainly follow traditional practices either by bringing new lands into production (extensification) or increasing the use of crop inputs (intensification) to increase the yield [3]. This way of raising agricultural productivity in traditional economies makes up a shortage of resources due to a large population base with relatively little arable land and water, as well as the low acceleration of agricultural technological modernization [4]. Accordingly, the traditional farming model has become more and more unsustainable because of its reliance on the increased use of agricultural factor inputs and land resources [3], forcing producers in those countries to prefer growing food crops or shifting to other business practices to worsen the situation further. Consequently, some developing countries, which also top producers of non-food crops, have started importing these crops (USDA cotton outlook 2019). Since a significant global portion of global industrial crop produce relies on farmers' technological expertise from developing countries [5], this trend would gradually increase the demand for non-food crops at the global level. In addition, these practices, coupled with global crises-driven uncertainties, hint at an upcoming alarming increasing demand for agricultural raw materials such as cotton, natural fiber, and jatropha in coming years, which could shift a typical global crisis into more intense and prolonged consequences [6]. For instance, since COVID-19, global consumption patterns have shifted towards natural lifestyle and eco-friendly products that significantly depend on industrial crops like cotton and its byproducts [7]. In this context, most researchers agree that developing countries would have to anticipate the upcoming industrial crop demand within the scarce financial and technological resources for sustainable non-food crop production practices [8], which is impractical without introducing affordable technology-driven solutions and their extensions [9].

Debates over the affordable technology-driven solutions continue to heavily focus on doing "more with less" in developing economies [3,10]. Subsequently, most researchers believe that a sustainable rise in agricultural output is possible with judicial use of agricultural inputs, i.e., an overall crop yield could proportionally be expanded without altering the input quantities used, or in other words, the crop input quantities could be reduced without changing the total crop production [11]. This assumption draws on a socio-ecological trade-off along the pathways of extensification or intensification in achieving sustainable crop growth, which demands us to identify favorable combinations of various factors of productivity growth, such as labor, land, capital, and inputs—the total factor productivity (TFP) in agriculture [3]. Accordingly, the functional scope of such a hypothesis is well exhibited through farming practices of modern economies that have remained the main engine of production growth and the increase in income as a way to modernize their economies and, rather than depending upon intensification or extensification, these systems concentrate on identifying the positive synergies between agriculture inputs [3]. Therefore, the critical element of crop productivity growth is estimating the technical inefficiency changes of various factors of crop production, which could facilitate the understanding of those factors of production involved in increasing production cost or low yield of non-food crops over the years [12]. However, understanding crop inputs inefficiencies over time from an empirical perspective is sensitive against agricultural-crop diversity and regions [13–15]; this hints to a need for a much closer production growth perspective to track technical efficiency change and technological factors of non-food crop production.

Given the proximity of perspective, cotton production becomes a typical case of the non-food crop for four reasons. First, cotton, also known as white gold, is the world's most demanding non-food crop as a source of natural fiber for eco-friendly products, byproducts, lint, and edible oil [1]. It is the most wide-spreading profitable industrial crop for its role with more than 20% of the global share in all fiber use and is the global income to more than 250 million people worldwide [16]. Even with such a broad global reach, the current cotton production methods are environmentally unsustainable—ultimately undermining the industry's ability to maintain future demands [17]. Moreover, various coalitions of international partners promote the sustainable production and use of cotton in multiple ways [18]. Second, developing countries are among world's largest cotton producers with a 7% contribution to overall labor employment opportunities [16]; accordingly, their priorities in bringing cotton production in line with even minimally acceptable environmental standards are significantly challenged despite owning a huge tract of fertile agriculture land [1].

Third, the main challenge of these developing countries is lower cotton productivity than the rest of the world [19]. Even the irrigated area of these countries are underperformers for a given yield per unit from the rain-fed cotton-growing regions of the world (World textile information report 2019, USDA annual gain report 2017). Subsequently, farmers in these areas get comparatively less profit when harvesting cotton crops than other food crops, leading to a decrease in cotton crop area. Unfortunately, this situation has been worsening since 2010 in some developing countries and continues to decrease in more than 10 percent of the cropping area. Its reflection is observable because some of the world's top cotton-producing developing countries have started to import cotton in recent years—an alarming sign of the upcoming increase in demand for raw cotton. Fourth, the low crop productivity of developing countries has a range of contributing socio-ecological factors, and knowing "which factors or their combinations are the main contributors" becomes more critical in identifying the positive integration of these factors for improving crop productivity [19]. In contrast, very few studies have examined the detailed resources of the productivity and technical efficiency of cotton over time [1,19,20], besides some mentioned limitations. For instance, this feature accommodates the accurate estimation of productivity measures breakdowns when multiple inputs lead to a single output [19], rather than concentrating on multiple-output and multiple-input technology [20] and failing to account for multiple input leads to single output technology when analyzing the productivity trend [12]. Some researchers have attempted a more specific perspective, yet productivity remains a secondary objective in those studies [1,21], or they did not account for multifactor analysis [20]. More importantly, these studies take the theoretical assumption of the accurate estimation of cotton productivity measures' breakdowns even when accounting for multiple input technology [21]. This theoretical objective believes in obtaining the maximum output from inputs [1], which contrasts to the reality where not all inputs can achieve it; consequently, inefficiency may arise. From an implication perspective, scaling the non-constant returns is necessary when considering the overall impact on productivity, and violation of the various situations is one of the required conditions for long-run competitive equilibrium [22]. To summarize the points mentioned earlier, the situation reveals a significant knowledge gap that demands an implication-oriented multifactor analysis to track which factors of production drive the low crop productivity of cotton, particularly for developing countries. In the long term, such empirical analysis is critical in knowing whether a developing country's farmers are progressive concerning technical efficiency change, technological change, and speed of knowledge-based learning skills to counter with the existing raw cottons' supply and demand gap effectively.

Given this background, the study adopts a more specific case using a top-ranked cotton-producing country that has recently started importing raw cotton and struggling to manage the declining cotton yield, i.e., Pakistan's cotton production represents the non-food crop perspective for developing countries. Accordingly, the main objective of this study is to estimate the production frontier and efficiency relationship of various factors from the implication aspect to identify those factors of production that influence overall cotton production over the years. Accordingly, a stochastic frontier approach was employed to measure the efficiency of the changes by using Pakistan's cotton production time-series data from a much larger period, 1971–2018. This framework contributed to the existing literature in the following three ways. First, it estimates the indices of crop-wise technical efficiency changes, covering a longer time period using SFA, a parametric procedure for non-food crops. As the main benefit, the parametric approach is capable of accounting for composed error, thereby separating the noise from inefficiencies. This way, it tracks efficiency changes and technological changes in the overall productivity from practical aspects, i.e., scaling non-constant returns of those multiple factors that allow violation of various situations. Second, the SFA allows the functional specific need to accommodate (a) single output from multiple inputs, (b) it can test the hypothesis, and (c) it presents the availability of the maximum likelihood econometric estimates. This crop-specific imperial finding would identify the main drivers of inefficient crop production that might be influencing Pakistan's self-sufficiency in cotton. Moreover, the study would also estimate the hypothesis of whether subsidies affect cotton crop production. Third, the study puts forward corresponding policy recommendations, specifically for improving cotton production efficiencies. Furthermore, in general, applying such analysis to other developing countries could realize the potential of affordable technology-driven solutions and Agri-Resource efficiencies for sustainable non-food crop production.

Cotton Production and Pakistan Case

The study uses Pakistan's cotton production as a typical case to analyze production factors from a much closer input issues and efficiency relationship perspective to represent the non-food crop in a developing country perspective for the following reasons. First, Pakistan is ranked among the top five largest cotton producers (Figure 1), is the 7th largest cloth producer worldwide, and cotton contributes 10% to the national GDP compared to the overall agriculture sector GDP share of 18.9% [22]. This sector also contributes to 42.3% of the labor force with employment and provides the raw materials for many value-added sectors [23], with 55% of the foreign earnings contribution, a share majorly dominated by cotton-based finished products [24]. Since Pakistan is a developing country, these states apparently favor the export; in contrast, Pakistan has been importing raw cotton for many years. Actually, Pakistan has not been exporting raw cotton since 2010, and it is the fourth country that ends up ending stalk (USDA cotton outlook 2019). Second, the main reason for increasing the import of raw cotton is that Pakistan's cotton yield per hectare has had a gradually decreasing trend over the years, and the numbers are among the lowest in the world; even the countries that have a much smaller geographic area have a higher yield per hectare than Pakistan [25]. On average, the yield of cotton in Pakistan is 730 kg/ha with 10,671 million bales, which is 1.5–2% lower than to rest of the world, and even irrigated areas of the countries are lagging in terms of lint per hectare from the rain-fed cotton-growing areas of the world (Figure 2). As a result, cotton cultivation has become less attractive than growing other crops, which leads to the minimizing the area of cotton crops (World textile information report 2017, USDA economic survey of Pakistan (2019–2020)). It is true that the area of cotton crops has been declining in Pakistan since 2004–2005; unfortunately, this situation has been aggravated since 2013–2014, and by the ongoing year, 14.2 percent sowing area of the crop had decreased [26]. Before 2014–2015, it was 2.902 million hectares, while currently 2.489 million hectares of cotton are being cultivated in Pakistan. Third, this low productivity of Pakistan's cotton crop mainly results from inefficient use of multiple factors related to irrigation water, plant population, disease protection, plant nutrition, resource management skills, and insufficient technology, resulting in a gap between potential production and actual production (agriculture statistics of Pakistan (2019)). However, as a general trend, Pakistan farmers seek to remedy this low yield per hectare through traditional practices, such as increasing the use of crop inputs (intensification) or bringing new lands into production (extensification), which leads to constantly deteriorating the agricultural ecological environment. Such practices reduce the economic effect of agriculture and destroy the resources and environment on which agriculture depends on for survival and development [27]. Besides, the scarcity of resources coupled with a large population and lagging agricultural infrastructure is keeping Pakistan's cotton input and output in a bad state; hence, affordable agricultural technology and its extension become an urgent need to identify the positive balance between the cotton input and output. Fourth, most studies use Pakistan's agriculture sector to analyze TFP growth in the sector at the national or provincial level. Accordingly, very few studies have analyzed the Pakistan cotton productivity, and efficiency [1], particularly when "multiple inputs leads to a single output" are involved. Such studies either measure the total factor productivity for multi-crops [28] or the sub-sector of Pakistan [26] to analyze the relationship between productivity and agricultural research expenditures or conduct research on the basis of comparative analysis [1]. As agriculture productivity is expected to be influenced by different factors unique for each crop, a crop-wise analysis at the national level would be more useful [19]. Moreover, these studies take the theoretical assumption of the accurate estimation of cotton productivity measures' breakdowns even when accounting for multiple input/output technology [1]; however, violation of the various crop productivity factors is one of the required conditions for long-run competitive equilibrium. Therefore, this study is being conducted because no prior multifactor analysis is available from an implication perspective that not only scales the non-constant returns when considering the overall impact on cotton production, but also takes a much closer perspective on non-food crops of developing countries like Pakistan concerning the various input and efficiency relationships.

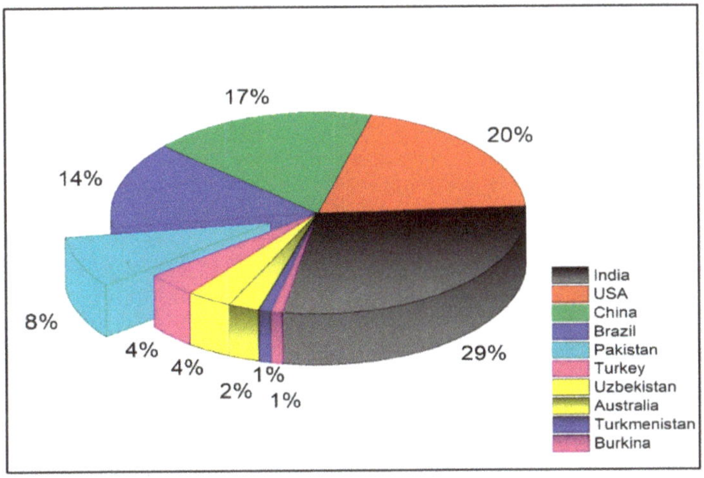

Figure 1. Cotton Production by country Worldwide in 2018/2019 (Source: world bank, Graph 2).

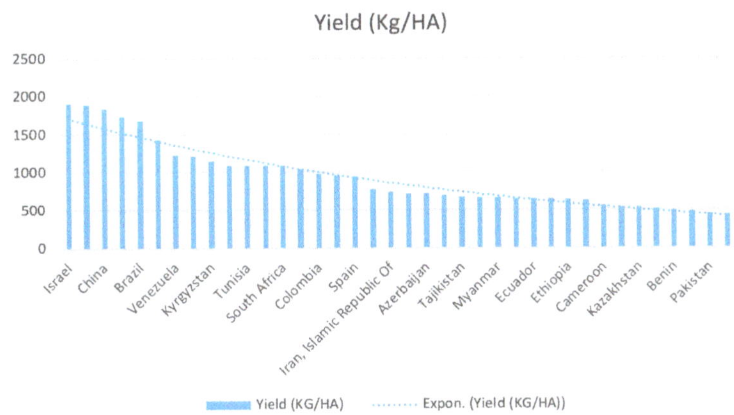

Figure 2. Cotton yield by country in 2019 (Source: Author own computation).

The rest of the paper is arranged as follows: Section 2 describes the research methods, and Section 3 represents the results and discussion step-by-step in support of the methodology. Finally, Section 4 concludes the paper with some recommendations.

2. Research Methods

2.1. Theoretical Concept

This study draws a great deal from a stochastic frontier production function, proposed by Battese and Coelli, which is assumed to be distributed as truncated standard random variables that permit the data to vary systematically with time [29]. The SFA can obtain the maximum likelihood estimates of a subset of the production, the cost function. Besides, this study applies time-series data. The time series data offer to remove a few rigidities because each observation is observed at several different points. In time series data, introducing the ai (time-invariant and individual-specific) enables us to take some heterogenicity that otherwise would be unable to control in cross-section data. Moreover, time-series data also allow us to examine whether the inefficiency is varying or persistent over time. Likewise, the time series data can be simply treated as cross-section data, which implies that the data of the same units, observed at different points of time, are treated as a separate unit; however, the inefficiency and noise at this component point should be heteroscedastic. One of the main advantages of SFA is the possibility of offering the decomposition of productivity change into parts that have straightforward interpretation [30]. As the most important faction, the stochastic frontier approach (SFA) can separate data noise from variations inefficiency. Given the inherent crop diversity in agricultural production, it is assumed that all deviations from the frontier are associated with inefficiency (as assumed in the data envelopment analysis approach), which is challenging to accept in this sector. Based on the following previous literature in the agriculture field [12,31,32], this study narrows down the focus to cotton output and digs into the possibility of realizing similar production frontier and efficiency relationships with the SFA method. Following Battese and Coelli, this paper applies the one stage modeling approach to more comprehensive data representing various factor included in the production frontier and efficiency analysis.

2.2. Measurment of Technical Inefficieny

We employed the stochastic production function, which was proposed by Ainger, Lovell, and Schmidt in 1977 [33], as well as Meeusen and Broeck [34]. The model can be expressed in the format, as shown in Equation (1).

$$Y_i = X_i\beta + (V_i - U_i) \tag{1}$$

where Y_i is total crop output from agriculture input X_i ($i = 1, \ldots N$, and X_i is vector of N inputs). The β is the vector of the technology parameters and V_i represent the two-sided random error accounting for input measurement and statistical errors. The inefficiencies variable is $U_i > 0$; it is measured as the difference between actual crop yield Y_i and the maximum agricultural output $X_i\beta$. Later on, in 1995 Battese and Coelli formulated a model with the exception that the allocative efficiency is imposed and this model permits panel data [35]; it was equivalent to the model proposed by Kumbhakar, Ghosh, and McGukin [36]. In this model, the issue of the two-stage estimation procedure, which was unlikely to provide estimates, was addressed as expressed in Equation (2).

$$Y_{it} = X_{it}\,\beta + (V_{it} - U_{it}) \tag{2}$$

$$i = 1 \ldots . N, 1 \ldots T$$

The V_{it} is random variables and represents the total crop output in the t-th time period from agriculture input X_{it} and it is assumed to be independent and identically distributed distribution $\sim N(0, \sigma v^2)$, as well as being independent of the U_{it}, which is also a random variable, assumed to account for technical inefficiency in production, and assumed to be distributed as truncations at 0 of the $N(m_{it}, \sigma u^2)$ distribution, as shown in Equation (3).

$$m_{it} = Z_{it}\delta \tag{3}$$

where m_{it} represents technical inefficiency effects and Z_{it} is a p × 1 vector of a variable, which may influence the efficiency in the time period *t*-th, while δ is a 1 × p vector of the parameters to be estimated.

This study used the stochastic frontier approach with time-series data considering a stochastic production frontier model specified as follows:

$$\ln(Y_{it}) = \beta_0 + \beta_1 \ln(X_1) + \beta_2 \ln(X_2) + \beta_3 \ln(X_3) + \beta_4 \ln(X_4) + \beta_5 \ln(X_5) + \beta_6 \ln(X_6) + \beta_7 \ln(X_7) + V_{it} + U_{it} \quad (4)$$

where

Y_{it}	is the logarithm of the production in the *t*-th time period.
X_{it}	is a $k \times 1$ vector of input quantities of the *t*-th time period.
β	is a vector of unknown parameters.
η	is a parameter to be estimated.
V_{it}	is a non-negative random variable that is assumed to be *iid* $(N(0, \sigma v^2)$ and it is independent and identical to the U_{it}, distributed normal random errors, with a mean of zero and variance σv^2, and distributed independently of $U_i N(\mu, \sigma v^2)$.
U_{it}	is a non-negative technical inefficiency effect representing management factors, and it is assumed to be independently distributed with mean U_{it} and variance σv^2 [32].

The *t*-th time exploits the full technological production potential when the value of U_{it} comes out to be equal to zero, and at that time, the production is produced further than the production frontier, beyond which it cannot be produced [33]. The greater the magnitude of U_{it}, and the farther away the product from the production frontier, the more inefficiently it will be operating [9]. In this study, we utilized the parameterization and replaced σv^2 and σu^2 with the $\sigma^2 = \sigma v^2 + \sigma u^2$ and $\gamma = \sigma u^2 / (\sigma u^2 + \sigma u^2)$ [37]. The above model specification also encompasses several other model specifications; if T = 1 and zit contains the value 1 and only a constant term, then the model can reduce to the truncated normal specification, where $\delta 0$ would have the same interpretation as the μ parameter. Besides, specifications mentioned in Equations (2) and (3) are non-nested and no set of restrictions are defined to permit a test of one specification versus the other.

2.3. Data and Source

This section of the paper presents the Stochastic Frontier approach corresponding index numbers useful in efficiency analysis. Accordingly, seven observations were taken for the Stochastic production analysis, and the time series data were taken from 1971 to 2018. The data were collected from agriculture statistics, Pakistan Bureau of Statistics, Ministry of Labor and Employment, Fertilizer use by Crop-FAO [38]. Seven different production variables, defined to encompass the function's variables, were used in cotton production: $[Y_1]$ represents the total production of cotton in one thousand bales; $[X_1]$ cultivated area for cotton crop in one thousand hectares; $[X_2]$ human labor in the agriculture sector, which is taken as the proxy of cotton labor; $[X_3]$ the available number of tractors in the country and available agriculture machinery, which is taken as the proxy variable of infrastructure, such as roads and the improvement in marketing facilities; $[X_4]$ consumption of cotton seed; $[X_5]$ electricity consumption in the agriculture sector taken as an explanatory variable; $[X_6]$ fertilizer consumption for the cotton crop taken as a major determinate of the production; and $[X_7]$ availability of irrigated water for monsoon crops, which is taken as an essential variable. The next section briefly explains the model specification.

2.4. Method

2.4.1. Statement of Hypothesis

The Maximum likelihood estimates of the parameters of the inefficiency model and stochastic production frontier. For the estimation parameters of the production frontier and the influencing factor of the cotton productivity, this study investigates the model validation for the analysis.

$$H_0 = \gamma = \delta 0 = \delta 1 = \delta 2 = 02$$
$$H_1 = \gamma = \delta 0 = \delta 1 = \delta 2 = 02 = 0$$

H_0: Null hypothesis to be tested that there is no inefficiency in the model.

H_1 = Technical Inefficiency variables are not affected by the independent variables included in the model.

For the estimation parameters of the production frontier and the influencing factor of the cotton productivity, this study investigates the model validation for the analysis. Because the parametric SFA requires a particular functional form, the hypotheses considered for investigation are (A): H_0: and H_1, where H_0 is restricted model and H_1 is an unrestricted model; with the assumption that the unrestricted model is better than a restricted model, as estimated by likelihood ratio LR, defined as, $LR = 2[L(H_0)/L(H_1)$; and L1 is the value of the log likelihood function ration under the specification of the null and alternate hypotheses. The likelihood test statistics have an asymptotic Chi-square.

2.4.2. Dependent Variable

$\ln(Y_{it})$	is the natural log of cotton production output in bales.
Y_{it}	is the only variable that represents the total production during the time period.

2.4.3. Independent Variable

$\ln(X_1)$	is the natural log of the area under cotton in acres.
$\ln(X_2)$	is the natural log of the consumption of fertilizer nutrients for the cotton crop (i.e., nitrogen (N), phosphorus (P) and potash).
$\ln(X_3)$	natural log of cotton seed consumption.
$\ln(X_4)$	natural log of the source of irrigation for the monsoon crops: canal, tube, well, in million-acre feet.
$\ln(X_5)$	natural log of the consumption of electricity access to the farm/rural area taken as the dummy variable of electricity consumption of cotton crops because there is unavailability of electricity consumption specific to cotton crops.
$\ln(X_6)$	natural log of the labor force employed in the agriculture sector.
$\ln(X_7)$	natural log of no of available tractors and agriculture machinery.

2.4.4. Method Determination the Return Scale with a Trans Log Function Diagnostic Tests on Data

The current study adopted the computer program Frontier 4.1 to estimate the inefficiency model and maximum likelihood ratio of the stochastic production frontier parameters. Since checking the model's validity is necessary before evaluating the production function parameter and the factor affecting the cotton crop inefficiency, the model assumed two hypotheses. The hypotheses were tested using the generalized likelihood-ratio LR, which is defined as $LR = -2ln[L(H_0)/L(H_1)$. The results of these estimation's likelihood function values are under the specification of the null and alternate hypotheses, respectively. The Chi-square value is asymptotically distributed with a degree of freedom equal to the difference between the restricted and unrestricted model parameters.

3. Results and Discussion

3.1. *Descriptive Statistic of Production Factors*

The production function model is used to determine production efficiency. At the first stage, we run a descriptive test to measure the sample's tendency and dispersion to decide whether it is normally distributed or not and to find the data's outlier. The summary of the descriptive statistics for the input and output variables included in the stochastic frontier model are presented in Table 1. The result showed that the mean cotton production in the study area was 8264 tons (in thousands) bales during the period. The mean number of cultivated areas to producers' cotton was 2564 (in thousand) hectares, with a minimum value of 1733.3 and a maximum of 3192.6. The available data show that, from small-scale farms to large scale farms, the minimum and maximum used fertilizer were 46 tons and 1259.75 tons during the period, respectively. The farmers' average labor force (wage and family labor) to produce all types of crops in the agriculture sector, including cotton, were 37,693 labors (in thousands) for the production cycle. The mean of the machinery use was 376,628 Nos., while the mean of irrigation water was 71 MAF. Similarly, the average seed cost was 4541 tons (in thousands).

Table 1. Descriptive statistics of the variable vectors used in the SFA under seven different combinations of distributional assumptions.

	Variable	Unit	Mean	SD	Minimum	Maximum	Skewness	Kurtosis
(Y_i)	Production (000)	Bales	8264.448	3770.031	2557.8	14,265.2	−0.1428	1.617904
(X_1)	Area (000)	Hectares	2564.706	453.4895	1733.3	3192.6	−0.45123	1.799832
(X_2)	Fertilizer	Tones	500.1513	359.958	46	1259.75	0.346148	1.768292
(X_3)	Seed (000)	Tons	4302.503	1801.136	1304.1	7779.4	−0.26666	1.848038
(X_4)	Irrigation	MAF	71.26875	9.01249	56.33	86.93	0.029849	1.737987
(X_5)	Electricity	GWH	5463.545	3311.117	876.688	12,525.94	0.156374	1.901286
(X_6)	Labor force (000)	Nos.	37,693.65	15,976.42	17,281.51	70,602.09	0.531648	2.094445
(X_7)	Machinery	Nos.	376,628.1	331,078.2	20,000	1,211,898	1.020978	3.006516

Note: summary statistics of input and output variables.

This descriptive analysis shows that, in sample data, Y_1, X_1, X_3, are negatively skewed, which means the tail lies on the left side on the central value, while X_2, X_4, X_5, X_6 has the long right tail. All the variables in the model are in platykurtic form except Machinery (X7), which is in a leptokurtic form that means that it has positive kurtosis (peak curve). Jarque–Bera statistic results also show that the data are normally distributed. Probability statistics show that the Jarque–Bera statistic exceeds the observed value under the null hypotheses, while the small values lead to the acceptance of the null-hypotheses, which means the curve is a normal distribution. These results of the descriptive analyses do not permit us to evaluate the contribution of each factor of production to the achievement of the production objectives nor the productive performance of the cotton production during the study period. Hence, only an econometric estimation of the production frontier was used to assess efficiency status of the cotton production inputs and to isolate the probable sources of its growth.

3.2. Unit Root Test

It is a standard procedure to check the stationarity of the time series data, as time series data usually demonstrate the unit root problem. Therefore, before performing the proper analysis, the selected variables were subjected to a standard unit root test to examine the stationarity level of the variables, and results were verified by two tests: Kwiatkowski-Phillips-Schmidt-Shinand and Augmented Dickey-Fuller (ADF). The estimated results of the unit root test are presented in the Table 2 and show that labor is not stationary at the first difference, and it becomes stationary at the trend and intercept. The result of the Augmented Dickey-Fuller (ADF) Test and its significance level at 1 and 5 percent level indicated that all the variables are stationary at I(0) and I(I). However, none of the variables are stationary at I(2). The result of the Kwiatkowski-Phillips-Schmidt-Shin test also validates the results revealed by the ADF test.

Table 2. Result of Unit root analysis.

	Augmented Dickey-Fuller (ADF) Test				Kwiatkowski-Phillips-Schmidt-Shin Test			
	I(0)		I(I)		I(0)		I(I)	
	C	C & T	C	C & T	C	C & T	C	C & T
Production	−1.75	−3.13 *	−8.87 *	−8.83 *	0.79 *	0.14 *	0.18	0.13
Area	−1.95	0.08	−9.17 *	−6.44 *	0.74 *	0.21 *	0.23	0.09 **
Seed	−1.84	−3.49 **	−9.37 *	−5.13 *	0.79 *	0.16 *	0.50	0.50
Irrigation	−1.52	−3.14 **	−7.58 *	−7.51 *	0.75 *	0.10 *	0.12	0.11
Electricity	0.100	−3.61 *	−3.84 *	−3.82 *	0.88 *	0.05	0.20 *	0.09
Labour	8.44	0.26	−1.29	−7.63 *	0.88 *	0.23 *	0.80 *	0.08 **
Machinery	8.43	2.87	−2.52 **	4.27 *	0.83 *	0.21 *	0.68 *	0.17

Present significance: at * 1% and ** 5% level.

3.3. Testing of Hypothesis

The first null hypothesis that we tested was $H_0 = \gamma = \delta 0 = \delta 1 = \delta 2 = 02$, presented in Table 3, and it indicated that the technical inefficiency effects were not present in the selected model. It also suggests that the stochastic frontier production function was the same as the traditional average production function estimated using the OLS procedure. This null hypothesis was rejected. The second null hypothesis was tested as H_0: $\delta 1 = \delta 2 = 0$, indicating that the government's subsidies on cotton and fertilizer were not affected by the input variables included in the model; thus, this hypothesis was again rejected. The second LR result revealed that the variables included in the inefficiency showed technical inefficiency effects for cotton production. Subsequently, it was appropriate to have those input variables in the model.

Table 3. Hypothesis test for model specification.

Null Hypothesis (H_0)	LogLH_1 (Unrestricted Model)	LogLH_0 (Restricted Model)	Statistiques LR	Critical Value	Decision
H_1: OLS model is adequate	35.76	20.06	31.38	5.991465	Reject
H_0: Lack of inefficiency	22.99954	35.76	25.52	15.50731	Reject

3.4. Analysis of Econometric Model

The stochastic frontier production model results that estimate the cotton production are presented in Table 4, where 11 parameters were evaluated, including 7 in a stochastic frontier model, 2 in the efficiency model, and the remaining 2 parameters relate to the variance of the random variables. In the frontier function, out of 11 parameters, 9 parameters were significant. The test results of the likelihood ratio showed that the model was significant at the 1% level. Most of the coefficients of the variables were significant at the 1% level and 5% levels.

The random term $\sigma v2$ perceived the variables that incorporated technical efficiency in the analysis of the production frontier model. The estimation result showed that $\sigma v2$ was 0.21, and it was significant at the 1% level, indicating goodness of fit, which means the presence of technical inefficiency related to the technical errors in the cotton production. Likewise, the estimation result of γ was 0.999, indicating 99% of the deviation from the production frontier. It was statistically significant with positive elasticity, which indicated that cotton productivity differentials predominantly related to the variance in management. Thus, the specification was much more appropriate in terms of the stochastic production frontier in order to represent cotton production technology.

Table 4. Estimated parameters of variables used in the inefficiency model.

Variables	Parameters	OLS		Frontier Function	
		Coefficient	t-Ratio	Coefficient	t-Ratio
Constant	β_0	−13.0221	−2.9189	−10.6543 ***	−9.7047
Ln (Area)	β_1	1.8955 ***	4.2232	1.7678 ***	15.6427
Ln (Fertilizer)	β_2	−0.0930	−0.3652	−0.1312 **	−2.3619
Ln (Seed)	β_3	0.2764	1.5562	0.5225 ***	9.10470
Ln(irrigation)	β_4	−0.4926	−0.9554	−1.9720 **	−18.0470
Ln (Electricity)	β_5	0.0214	0.0773	−0.1923 ***	−3.0839
Ln (Labor force)	β_6	0.9268 ***	2.8568	1.3046 ***	25.8351
Ln (Machinery)	β_7	−0.1974	−1.1191	−0.1061 **	−2.4034
Inefficiency Effect					
Constant				−0.8003	−1.8158
Cotton Subsidy	$\delta 1$			0.1898	0.7696
Seed Subsidy	$\delta 2$			−0.34625	−1.0515
Variance Parameters					
	$\sigma v2$			0.2155 ***	4.1027
	Γ			0.9999 ***	8604.7
Likelihood log					35.7622
LR test					0.31
Number of observations					48

*** 1% significant; ** 5% significant.

3.5. Elasticity of Production

The contribution of each factor of production to the productivity of cotton production is demonstrated with production elasticity. The estimation results are shown in Table 4. The result of the estimation shows that the coefficient of all inputs is statistically significant, and the results consist of previous findings [1,39,40]. The output of cotton production is statistically significant, but elasticity is negative. The coefficient of the cultivated area under the cotton crop is statistically significant. It shows a positive relationship at the 1% level, revealing that the cultivated area under the cotton crop remains as an important contributor to improving the technical efficiency in cotton production. This implies that an increase in the amount of one percent induces an increase of 1.76% technical efficiency in the cotton production, indicating that the area is the first factor that would cause the rise in cotton production. The fertilizer coefficient shows a negative relation at a 5% level, which implies that one percent increase in fertilizer would decrease the 0.13% technical efficiency in cotton production. The probable cause of the negative relation between cotton crop and fertilizer could be the farmer's access to the low quality of fertilizer [41,42]. Another reason could be that the farmers are applying higher doses of chemical fertilizer on the cotton crop, due to a lack of technical knowledge of fertilizer combination [43]. Seed consumption has a positive significant relation with cotton production, which implies that a 1% increase of seed consumption would increase technical efficiency of the cotton crop by 0.52%. Irrigation shows an inverse relationship with cotton production. Empirical results reveal that a 1% increase in irrigation could decrease the technical efficiency of cotton production by −1.97%. The possible cause of this negative relation could be the outdated and least efficient irrigation and water storage system, which does not correspond to the water storage requirement as it is operated on historic canal diversion patterns. Therefore, combined with the high seasonal pattern of the river flow and insufficient reservoir capacity—which provides 85% of water during monsoon crop season (cotton) [44]—the increase in water supply will affect the growth of the cotton plant, which would reduce the cotton yield [45]. The cotton crop is resilient to the drought due to its vertical root, and much more sensitive to the availability of water at flowering and boll formation stage, which need adequate quantities of water [46,47]. The availability of a high quantity of water in canals could be a reason for the inverse relationship between cotton production and irrigation. Furthermore, access to poor irrigation water also affects crop yield, quantity, and production. In addition, the excessive accumulation of salt affects the root zone of plant [42], which could be another reason for the negative relationship between crop production and irrigation variable. The coefficients of electricity and agriculture machinery are also statistically negative with cotton production, which implies that the 1% increase in these inputs would reduce cotton production by 0.19 and 0.10, respectively. A possible cause for these inverse relationships could be due to outdated and unmaintained machinery, the lack of the adoption new agriculture technologies [48], unwillingness to use due to lack of credit facilities [49], shortage of electricity [50], and high prices of electricity and agriculture machinery. The labor force was the second affecting factor that increased the cotton production. The coefficient of labor was positively associated with the cotton production, implying that a 1% increase in labor force would improve the technical efficiency by 1.3% in cotton production. The results are similar to the previous studies, indicating that the cultivated area, labor, and seed variables under the cotton crop remain significant contributors, leading to improved technical efficiency in cotton production in Pakistan [1,28,39,40,51].

The estimated parameters of variables used in the inefficiency model are provided in Table 4. Both variables assessed in the inefficiency effect are statistically significant; however, the cotton crop's subsidy shows a positive relationship with cotton production. Additionally, in the inefficiency model (U), the seed subsidy has negative elasticity, but it is not significant. The subsidy provided by the government for cotton is statistically not substantial, but its elasticity is positive, indicating an essential factor in decision making to have a positive effect on cotton production. A prospective reason for such a relationship may be linked with government organizations' slow processing in providing those subsidies after cultivating crops. The seed subsidy's elasticity carries a negative sign, demonstrating that the overall per hectare yield would be lower after the policy is implemented, indicating a similar trend to the previous research [21].

3.6. Technical Efficiencies of Cotton Production

The presence of technical inefficiency, tested using LR test, was 0.31, which is lesser than the critical chi-square value of 21.66 [52]. Therefore, the null hypothesis of the production inefficiency is rejected.

The production efficiency score ranges between 0.36 and 0.99, with the distribution seeming to be skewed toward the frontier. The production efficiency index is in the 99 percentage. The average efficiency score is 85% for the data taken from the last 48 years where 43% of the year production scores are in the range of 0.36–0.85, while 56% of year scores range are above average. These findings of the production efficiency score shows in Table 5, indicate the need to reduce 13% input to attain the status of most economical efficiency.

Table 5. Frequency Distribution of production efficiency estimates.

Efficiency Level	Cotton Production Efficiency in Pakistan No.	Parentage
0.36–0.85	21	43
0.85–0.99	27	56
Total	48	100
Mean Efficiency	0.85	
Minimum	0.36	
Maximum	0.99	

4. Implications and Concluding Remarks

The study adopts a more specific case of cotton production to represent the non-food crop perspective for developing countries like Pakistan, a top-ranked cotton-producing country that has recently started importing raw cotton and struggling to manage the declining cotton yield. This study examined the evolution of the overall cotton production and its determinants that have been inefficiently used over time in developing countries based on the stochastic frontier approach that was employed to measure the production frontier and efficiency relationship by tracking the efficiency changes and technological factors concerning those multiple input factors. Accordingly, the analysis is exampled on Pakistan's cotton production data from 1971–2018 in view of the specific non-food crop perspective of a developing country that has been a top-ranked cotton-producing country and has recently been shifting towards a non-exporter of cotton due to low yield. Accordingly, the study offers the following contribution with concluding remarks.

4.1. Theoretical Contribution

The study offers some theoretical contribution to the few studies related to cotton technical efficiency literature. First, the study estimates the indices of crop-specific technical efficiency changes and total factor of production, covering a longer time period using SFA for the non-food crops perspective. As the main benefit, since this approach can account for composed-error by separating the noise from inefficiencies, it tracks efficiency changes and technological changes in overall productivity from practical aspects [53]. Hence, scaling non-constant returns of those multiple factors is possible to allow for the violation of various situations, which is one of the necessary conditions for the long-run completive equilibrium in productivity measurement [12]. The other benefit of using such an approach is the ability to authorize systematic variation with time [54]. Moreover, using time-series data enable us to account for some heterogenicity that we cannot control in cross-section data. It is the way the current research prefers for SFA to estimate efficiency and cotton growth data considering its efficiency over other approaches in the agricultural economic literature [54]. Second, parametric procedures also allow the functional specific need to accommodate (a) calculating single output from multiple inputs, (b) testing the hypothesis, and (c) the availability of maximum likelihood econometric estimates. These features could well accommodate a more specific crop perspective of a developing country [16], as in the case of cotton production in Pakistan, a top-ranked cotton-producing country that has stopped exporting cotton since 2010 [25]. For instance, such features accommodate accurate estimation of productivity measures breakdowns when multiple inputs lead to a single output [19], rather than concentrating more on multiple-output and multiple-input technology that many previous studies have assumed [20,55]. Moreover, the study would also estimate the hypothesis "whether subsidies affect cotton crop production." Third, the study puts forward corresponding policy recommendations specifically for improving cotton production efficiencies under practical scenarios, unlike previous studies that assume cotton production under ideal/efficient condition and treat it as a residual, using index numbers such as Törnqvist [1]. Thus, this narrow focus of cotton output could dig into the possibility of realizing similar productivity and productivity growth analysis for other developing countries

4.2. Implications

The study revealed various findings. First, the study finds a positive relationship between cotton productivity and the using cultivated area, particularly cotton growth, which was significantly higher in those years when the government provided subsidy on the cotton crop as a farmer had more to spend on other productions factors. Second, those findings also indicate that cotton productivity has a strong inverse relationship with the fertilizer to the cotton crop. Given the importance of balance fertilizer management practices [56], an inverse relationship may be associated with low quality of fertilizer or farmers applying higher doses of chemical fertilizer on cotton crop, less knowledge of fertilizer combinations [57]. Third, the technical efficiency is negatively associated with irrigation; as a result of efficiency analysis, average technical efficiency is about 85 percent over the period. But it was at 36 percent level during the period. This negative trend in cotton productivity could be due to the prevalence of leaf curl virus disease in the cotton crop [58], which appears in the 1990s, and symptoms were seen in 2002 or ineffective use of irrigation strategy. Moreover, the apparent reason for this could be the waterlogging, salinity, and high usages of the brackish underground; results endorse observations of the previous study [59]. Forth, the subsidy remains a vital policy carried on crop production; however, most literature shows that subsidy could reduce the per hectare yield, our results are parallel to the other research results [21]. In contrast, the seed subsidy policy has failed to increase Pakistan production of cotton crop effectively.

The findings mentioned above suggest few implications in Pakistan's perspective that could provide initial guidelines for policy makers in improve cotton productivity due to its contribution to the country's economy and poverty alleviation [60]. First, the findings suggest a negative impact of technical progress on cotton production growth. It seems like either implementation of new technological solutions on production input (better seed quality, better fertilizer, better infrastructure, and land leveling) or generation of new technological solutions [60], has some issue that needs to invest in Agriculture research and development. Accordingly, government, agriculture union, NGOs should develop policies to improve technical efficiency by providing advisory support to production factors and better use of inputs. Moreover, considering the strong combined impact of fertilizer and irrigation [59], the government should increase awareness regarding the irrigation system's efficiency, fertilizer use, and strict regulation regarding low-grade fertilizers to fully exploit the existing economies of scale in the sector. Such policies should be included within the agriculture development framework for small and big-level farmers that help increase efficiency, transfer, and generation of technology, implement the best practices, and provide access to credit, market opportunities, and production inputs. These policies must be accompanied by strategies to limit imports of clothes and expand the export of cotton and yarn to encourage the cotton crop's local production and improve agriculture farmers' living conditions.

Moreover, few practical implications considering developing countries' perspective could also be drawn. First, this study contributes to the literature of crop-wise efficiency or technical efficiency, an understudied research area (considering the agriculture crop diversity [60]) having great potential of relevant multifactor information that might facilitate detailed options to navigate between various factors of production for a single crop. Second, the study also offers policy guidelines for cotton crop multi-source productivity, particularly from the developing country perspective, a predominant exporter of raw cotton [1]. Such multifactor productivity information can also be applied to farm practices indirectly associated with cotton to choose between socio-ecological factors [3]. Third, the study analyzed the focus of cropping preference regarding intensification or extensification in developing countries like Pakistan. It would facilitate forming strategic guidelines for policy making institutions and other stakeholders associated with sustainable non-food crop production practices.

4.3. Conclusions

Recent global uncertainties have increased the awareness of affordable technology-driven agricultural solutions, which are contemporary demands of developing economies, mainly supported by agriculture practices and scarce resources. In seeking those solutions, the TFP remains a significant indicator. Accordingly, the crop-specific technology is viewed for the cotton production inefficiencies and its role in developing economies taking Pakistan's case study.

This study analyzes the efficiency changes and technological factors of various sources of cotton growth from practical aspects, i.e., scaling non-constant returns of those multiple factors that allow violation of various situations. The study adopted Pakistan's cotton production case to represent a typical non-food crop perspective of developing countries, i.e., a country that has recently not been involved in exporting the cotton crop despite being a top-ranked cotton-producing country. Therefore, the study's findings revealed various insights in connection with cotton production inefficiencies. Accordingly, the study finds a positive relationship between cotton productivity and the using cultivated area; however, a strong inverse relationship was observed with the chemical fertilizer to the cotton crop productivity. Moreover, technical efficiency was observed to be negatively associated with irrigation, while subsidy could reduce the over yield per hectare, particularly the seed subsidy policy. These findings need to be prioritized by policymakers in Pakistan. As the model adopts Pakistan's context, the overall statistical analysis is generalizable to other developing economies, particularly concerning the issue of the constant declining yield of non-food crops. Given the scarce financial and technological resources of developing countries, these results would facilitate their strategy workers in designing and developing affordable technology-driven solutions and Agri-resource efficiencies for sustainable non-food crop production.

This study also offers some limitations. Since the present study evaluated efficiency analysis based on tracking efficiency and technological change in overall productivity over the past five decades, this does not imply that the crop management practices were ideal in those years. There was almost certainly room for improvement for considering other factors of production that are not being considered due to limited data availability. Further studies are encouraged for management reviews.

Author Contributions: Conceptualization, M.U.A. and Y.Z.; methodology, M.U.A.; software, M.U.A.; validation, M.U.A. and Y.Z.; formal analysis, M.U.A. and F.F.; investigation, O.H.; resources, M.U.A.; writing—original draft preparation, M.U.A.; writing—review and editing, O.H.; supervision, O.H., Y.Z. All authors have read and agreed to the published version of the manuscript.

Funding: This research work was supported by the Major social science projects of Inner Mongolia Autonomous Region China (Grant No.: 2018ZGH006 and 2018ZDA004).

Institutional Review Board Statement: Not applicable.

Informed Consent Statement: Not applicable.

Data Availability Statement: The data presented in this study are available on request from the corresponding author.

Acknowledgments: The first author is very grateful to the Chinese Government Scholarship Council (C.S.C.) for their support during this research tenure. We also acknowledge the support of, Pakistan Meteorology Department, Pakistan Bureau of Statistics.

Conflicts of Interest: The authors declare no conflict of interest.

References

1. Shabbir, M.S.; Yaqoob, N. The impact of technological advancement on total factor productivity of cotton: A comparative analysis between Pakistan and India. *J. Econ. Struct.* **2019**, *8*, 27. [CrossRef]
2. An, J.; Mikhaylov, A.; Richter, U.H. Trade war effects: Evidence from sectors of energy and resources in Africa. *Heliyon* **2020**, *6*, e05693. [CrossRef] [PubMed]
3. Coomes, O.T.; Barham, B.L.; MacDonald, G.K.; Ramankutty, N.; Chavas, J.-P. Leveraging total factor productivity growth for sustainable and resilient farming. *Nat. Sustain.* **2019**, *2*, 22–28. [CrossRef]
4. Liu, J.; Dong, C.; Liu, S.; Rahman, S.; Sriboonchitta, S. Sources of Total-Factor Productivity and Efficiency Changes in China's Agriculture. *Agriculture* **2020**, *10*, 279. [CrossRef]
5. Statista. Leading Cotton Producing Countries Worldwide in 2019/2020. 2021. Available online: https://worldpopulationreview.com/country-rankings/cotton-production-by-country (accessed on 21 August 2020).
6. Siche, R. What is the impact of COVID-19 disease on agriculture? *Sci. Agropecu.* **2020**, *11*, 3–6. [CrossRef]
7. Pan, D.; Yang, J.; Zhou, G.; Kong, F. The influence of COVID-19 on agricultural economy and emergency mitigation measures in China: A text mining analysis. *PLoS ONE* **2020**, *15*, e0241167. [CrossRef]
8. Calicioglu, O.; Flammini, A.; Bracco, S.; Bellù, L.; Sims, R. The future challenges of food and agriculture: An integrated analysis of trends and solutions. *Sustainability* **2019**, *11*, 222. [CrossRef]
9. An, J.; Mikhaylov, A.; Jung, S.-U. A Linear Programming approach for robust network revenue management in the airline industry. *J. Air Transp. Manag.* **2021**, *91*, 101979. [CrossRef]
10. Morkovkin, D.; Gibadullin, A.; Kolosova, E.; Semkina, N.; Fasehzoda, I. Modern transformation of the production base in the conditions of Industry 4.0: Problems and prospects. *J. Phys. Conf. Ser.* **2020**, *1515*, 032014. [CrossRef]
11. Zhang, J.; Chen, Y.; Li, Z. Assessment of efficiency and potentiality of agricultural resources in Central Asia. *J. Geogr. Sci.* **2018**, *28*, 1329–1340. [CrossRef]

12. Benedetti, I.; Branca, G.; Zucaro, R. Evaluating input use efficiency in agriculture through a stochastic frontier production: An application on a case study in Apulia (Italy). *J. Clean. Prod.* **2019**, *236*, 117609. [CrossRef]
13. Saliou, I.O.; Zannou, A.; Aoudji, A.K.N.; Honlonkou, A.N. Drivers of Mechanization in Cotton Production in Benin, West Africa. *Agriculture* **2020**, *10*, 549. [CrossRef]
14. Arshad, A.; Raza, M.A.; Zhang, Y.; Zhang, L.; Wang, X.; Ahmed, M.; Habib-ur-Rehman, M. Impact of Climate Warming on Cotton Growth and Yields in China and Pakistan: A Regional Perspective. *Agriculture* **2021**, *11*, 97. [CrossRef]
15. Giang, M.H.; Xuan, T.D.; Trung, B.H.; Que, M.T. Total factor productivity of agricultural firms in Vietnam and its relevant determinants. *Economies* **2019**, *7*, 4. [CrossRef]
16. Statista. Distribution of Fiber Consumption Worldwide in 2019, by Type of Fiber. 2021. Available online: https://www.statista.com/statistics/741296/world-fiber-consumption-distribution-by-fiber-type/ (accessed on 6 June 2021).
17. Ali, M.A.; Farooq, J.; Batool, A.; Zahoor, A.; Azeem, F.; Mahmood, A.; Jabran, K. Cotton production in Pakistan. In *Cotton Production*; John Wiley & Sons Ltd.: Hoboken, NJ, USA, 2019; Volume 249.
18. Ting, W. Bridge Interview of Karin Malmstrom, Director of Cotton Council International China. *China Text.* **2014**, *2*, 38–40.
19. Rodríguez, X.A.; Elasraag, Y.H. Assessing the total factor productivity of cotton production in Egypt. *PLoS ONE* **2015**, *10*, e0116085. [CrossRef]
20. Mitchell, C.; Traxler, G.; Novak, J. Measuring sustainable cotton production using total factor productivity. *J. Prod. Agric.* **1996**, *9*, 289–297. [CrossRef]
21. Tan, Y.; Guan, J.; Karimi, H.R. The Impact of the subsidy policy on total factor productivity: An empirical analysis of China's cotton production. *Math. Probl. Eng.* **2013**, *2013*, 248537. [CrossRef]
22. Azumah, S.B.; Donkoh, S.A.; Awuni, J.A. Correcting for sample selection in stochastic frontier analysis: Insights from rice farmers in Northern Ghana. *Agric. Food Econ.* **2019**, *7*, 9. [CrossRef]
23. Sohaib, M.; Jamil, F. An insight of meat industry in Pakistan with special reference to halal meat: A comprehensive review. *Korean J. Food Sci. Anim. Resour.* **2017**, *37*, 329. [CrossRef]
24. Arshad, M.U.; Zhao, Y.; Gong, Y.; Guo, X.; Hanif, S.; Ge, Y.; Jun, T. The effect of climate change on cotton productivity—An empirical investigation in Pakistan. *Pak. J. Agric. Sci.* **2021**, *58*, 8. [CrossRef]
25. Pakistan, G.O. *Economic Surrvey of Pakistan 2019–2020*; Government of Pakistan, Finance Division: Islamabad, Pakistan, 2021.
26. Nadeem, A.H.; Nazim, M.; Hashim, M.; Javed, M.K. Factors which affect the sustainable production of cotton in Pakistan: A detailed case study from Bahawalpur district. In Proceedings of the Seventh International Conference on Management Science and Engineering Management, Philadelphia, PA, USA, 7–9 November 2013; Xu, J., Fry, J., Lev, B., Hajiyev, A., Eds.; Springer: Berlin/Heidelberg, Germany, 2014; pp. 745–753.
27. Chen, G.; Breedlove, J. The effect of innovation-driven policy on innovation efficiency: Based on the listed sports firms on Chinese new Third Board. *Int. J. Sports Mark. Spons.* **2020**, *21*, 735–755. [CrossRef]
28. Zulfiqar, F.; Shang, J.; Nasrullah, M.; Rizwanullah, M. Allocative efficiency analysis of wheat and cotton in district Khanewal, Punjab, Pakistan. *GeoJournal* **2020**, *86*, 2777–2786. [CrossRef]
29. Banker, R.; Natarajan, R.; Zhang, D. Two-stage estimation of the impact of contextual variables in stochastic frontier production function models using data envelopment anlysis: Second stage OLS versus bootstrap approaches. *Eur. J. Oper. Res.* **2019**, *278*, 368–384. [CrossRef]
30. Yang, Z.; Roth, J.; Jain, R.K. DUE-B: Data-driven urban energy benchmarking of buildings using recursive partitioning and stochastic frontier analysis. *Energy Build.* **2018**, *163*, 58–69. [CrossRef]
31. Zewdie, M.C.; Moretti, M.; Tenessa, D.B.; Ayele, Z.A.; Nyssen, J.; Tsegaye, E.A.; Minale, A.S.; Van Passel, S. Agricultural Technical Efficiency of Smallholder Farmers in Ethiopia: A Stochastic Frontier Approach. *Land* **2021**, *10*, 246. [CrossRef]
32. Yin, Z.; Wu, J. Spatial Dependence Evaluation of Agricultural Technical Efficiency—Based on the Stochastic Frontier and Spatial Econometric Model. *Sustainability* **2021**, *13*, 2708. [CrossRef]
33. Aigner, D.; Lovell, C.K.; Schmidt, P. Formulation and estimation of stochastic frontier production function models. *J. Econom.* **1977**, *6*, 21–37. [CrossRef]
34. Meeusen, W.; van Den Broeck, J. Efficiency estimation from Cobb-Douglas production functions with composed error. *Int. Econ. Rev.* **1977**, 435–444. [CrossRef]
35. Battese, G.E.; Coelli, T.J. A model for technical inefficiency effects in a stochastic frontier production function for panel data. *Empir. Econ.* **1995**, *20*, 325–332. [CrossRef]
36. Kumbhakar, S.C.; Ghosh, S.; McGuckin, J.T. A generalized production frontier approach for estimating determinants of inefficiency in US dairy farms. *J. Bus. Econ. Stat.* **1991**, *9*, 279–286.
37. Battese, G.E.; Corra, G.S. Estimation of a production frontier model: With application to the pastoral zone of Eastern Australia. *Aust. J. Agric. Econ.* **1977**, *21*, 169–179. [CrossRef]
38. Chang, S.-H. A pilot study on the connection between scientific fields and patent classification systems. *Scientometrics* **2018**, *114*, 951–970. [CrossRef]
39. Raza, A.; Ahmad, M. *Analysing the Impact of Climate Change on Cotton Productivity in Punjab and Sindh, Pakistan*; Pakistan Institute of Development Economics (PIDE): Islamabad, Pakistan, 2015.
40. Bakhsh, K.; Hassan, I.; Maqbool, A. Factors affecting cotton yield: A case study of Sargodha (Pakistan). *J. Agric. Soc. Sci.* **2005**, *1*, 332–334.

41. Khaliq, A.; Abbasi, M.K.; Hussain, T. Effects of integrated use of organic and inorganic nutrient sources with effective microorganisms (EM) on seed cotton yield in Pakistan. *Bioresour. Technol.* **2006**, *97*, 967–972. [CrossRef] [PubMed]
42. Khan, M.; Mahmood, H.Z.; Damalas, C.A. Pesticide use and risk perceptions among farmers in the cotton belt of Punjab, Pakistan. *Crop Prot.* **2015**, *67*, 184–190. [CrossRef]
43. Han, H.-Y.; Zhao, L.-G. Farmers' character and behavior of fertilizer application-evidence from a survey of Xinxiang County, Henan Province, China. *Agric. Sci. China* **2009**, *8*, 1238–1245. [CrossRef]
44. Faruqee, R.; Hussain, Z. Future of Irrigation and Drainage in Pakistan [with Comments]. *Pak. Dev. Rev.* **1997**, 565–591. [CrossRef]
45. Karl, T.R.; Melillo, J.M.; Peterson, T.C.; Hassol, S.J. *Global Climate Change Impacts in the United States*; Cambridge University Press: New York, NY, USA, 2009.
46. Ton, P. Cotton and climate change: Impacts and options to mitigate and adapt. *ITC* **2011**. Available online: https://www.intracen.org/Cotton-and-Climate-Change-Impacts-and-options-to-mitigate-and-adapt/ (accessed on 6 June 2021).
47. Meyer, W.S.; Ritchie, J.T. Water Status of Cotton as Related to Taproot Length 1. *Agron. J.* **1980**, *72*, 577–580. [CrossRef]
48. Mottaleb, K.A.; Krupnik, T.J.; Erenstein, O. Factors associated with small-scale agricultural machinery adoption in Bangladesh: Census findings. *J. Rural. Stud.* **2016**, *46*, 155–168. [CrossRef] [PubMed]
49. Cavallo, E.; Ferrari, E.; Bollani, L.; Coccia, M. Attitudes and behaviour of adopters of technological innovations in agricultural tractors: A case study in Italian agricultural system. *Agric. Syst.* **2014**, *130*, 44–54. [CrossRef]
50. Shahbaz, M. Measuring Economic Cost of Electricity Shortage: Current Challenges and Future Prospects in Pakistan. MPRA Paper. 2015. Available online: https://mpra.ub.uni-muenchen.de/67164/1/MPRA_paper_67164.pdf (accessed on 6 June 2021).
51. Anwar, M.; Chaudhry, I.S.; Khan, M.B. Factors affecting cotton production in Pakistan: Empirical evidence from Multan district. *J. Qual. Technol. Manag.* **2009**, *5*, 91–100.
52. Kodde, D.A.; Palm, F.C. Wald criteria for jointly testing equality and inequality restrictions. *Econometric Soc.* **1986**, *54*, 1243–1248. [CrossRef]
53. Liu, J.; Li, H.; Sriboonchitta, S.; Rahman, S. Technical Efficiency Analysis of Top Agriculture Producing Countries in Asia: Zero Inefficiency Meta-Frontier Approach. In Proceedings of the International Conference of the Thailand Econometrics Society, Chiang Mai, Thailand, 9–11 January 2019; Volume 808, pp. 702–723.
54. Lai, H.-p.; Kumbhakar, S.C. Endogeneity in panel data stochastic frontier model with determinants of persistent and transient inefficiency. *Econ. Lett.* **2018**, *162*, 5–9. [CrossRef]
55. Işgın, T.; Özel, R.; Bilgiç, A.; Florkowski, W.J.; Sevinç, M.R. DEA Performance Measurements in Cotton Production of Harran Plain, Turkey: A Single and Double Bootstrap Truncated Regression Approaches. *Agriculture* **2020**, *10*, 108. [CrossRef]
56. Pan, X.; Lv, J.; Dyck, M.; He, H. Bibliometric Analysis of Soil Nutrient Research between 1992 and 2020. *Agriculture* **2021**, *11*, 223. [CrossRef]
57. Omer, M.; Idowu, O.J.; Ulery, A.L.; VanLeeuwen, D.; Guldan, S.J. Seasonal changes of soil quality indicators in selected arid cropping systems. *Agriculture* **2018**, *8*, 124. [CrossRef]
58. Zubair, M.; Zaidi, S.S.-e.-A.; Shakir, S.; Farooq, M.; Amin, I.; Scheffler, J.A.; Scheffler, B.E.; Mansoor, S. Multiple begomoviruses found associated with cotton leaf curl disease in Pakistan in early 1990 are back in cultivated cotton. *Sci. Rep.* **2017**, *7*, 680. [CrossRef] [PubMed]
59. Masasi, B.; Taghvaeian, S.; Boman, R.; Datta, S. Impacts of irrigation termination date on cotton yield and irrigation requirement. *Agriculture* **2019**, *9*, 39. [CrossRef]
60. Nicolay, G.L. Understanding and Changing Farming, Food & Fiber Systems. The Organic Cotton Case in Mali and West Africa. *Open Agric.* **2019**, *4*, 86–97.

sustainability

Article

Relationship Analysis among Apparel Brand Image, Self-Congruity, and Consumers' Purchase Intention

Lihong Chen [1], Habiba Halepoto [2], Chunhong Liu [1,*], Naveeta Kumari [3], Xinfeng Yan [4], Qinying Du [1] and Hafeezullah Memon [5]

[1] Shanghai International Fashion Science and Innovation Center, Donghua University, Shanghai 200051, China; lhckxyy@dhu.edu.cn (L.C.); duqinying@126.com (Q.D.)
[2] Engineering Research Center of Digitized Textile and Fashion Technology, Ministry of Education, Donghua University, Shanghai 201620, China; 317111@mail.dhu.edu.cn
[3] Brunswick Campus, RMIT University, Melbourne, VIC 3001, Australia; naveeta.kumari@rmit.edu.au
[4] International Cultural Exchange School, Donghua University, Shanghai 200051, China; yanxf@dhu.edu.cn
[5] International Institute of Silk, College of Textile Science and Engineering, Zhejiang Sci-Tech University, Hangzhou 310018, China; hm@zstu.edu.cn
* Correspondence: chliu@dhu.edu.cn

Abstract: Brand image has been a crucial clue to making subjective judgment for consumers to determine the brand, which is critical to making a purchase decision. The influence mechanism from apparel brand images on consumers' purchase intention was explored for theoretical and positive analysis based on the self-congruity theory. This research first constructed a hypothetical model of apparel brand images influencing consumers' purchase intention with self-congruity and perceived quality as mediators, in which a questionnaire was designed and conducted to test the theoretical model. The research shows that apparel brand image and perceived quality can significantly influence consumers' purchase intention. The consumer purchase intention is directly related to clothing brand image provided self-consistency and perceived quality play an intermediary with the regulatory effect of self-motivation and brand familiarity. A positive attitude toward apparel brand image will stimulate consumers to build cognitive clues and associations between the consumers and the brand and strengthen cognitive consistency with the apparel brand's spirit. The research results are beneficial to textile fashion and clothing enterprises to improve brand building and marketing.

Keywords: fashion brand image; self-congruity; perceived quality; self-motivation; brand familiarity

Citation: Chen, L.; Halepoto, H.; Liu, C.; Kumari, N.; Yan, X.; Du, Q.; Memon, H. Relationship Analysis among Apparel Brand Image, Self-Congruity, and Consumers' Purchase Intention. *Sustainability* **2021**, *13*, 12770. https://doi.org/10.3390/su132212770

Academic Editor: Xiaoke Jin

Received: 19 October 2021
Accepted: 15 November 2021
Published: 18 November 2021

1. Introduction

Building substantial brand equity is a critical challenge in today's fast-paced, growing business world [1]. It is essential to continuously create a robust distribution network and establish brand extensions for consumers to compete. Brand image is the crucial element that drives brand equity and is the primary source of brand equity [2]. This is also an important marketing strategy for enterprises to obtain a sustainable economic lead [3]. However, in Chinese apparel companies' shaping of brand image, there are typical problems such as ambiguous brand positioning, similar image shaping, and low accuracy, resulting in a lack of brand recognition and poor brand image communication.

In clothing consumption, brand image has become the main clue for consumers' subjective evaluation of brands, and it plays a vital role in people's purchasing decisions [4]. Simultaneously, as a critical value asset for apparel companies' differentiated competitive advantage, brand image plays an increasingly important strategic role in business management [5]. Determining how to accurately shape a clothing brand's image, optimize its communication effect, and stimulate clothing consumption have become urgent problems for Chinese clothing brands [6]. Thus, this research was conducted to analyze the relationship among apparel brand image, self-congruity, and consumers' purchase intention to overcome these questions. This study aimed to answer the following research questions.

Do apparel brand images impact consumers' purchase intention, self-consistency, and perceived quality?
Do self-consistency and perceived quality impact consumers' purchase intention?
What role do self-consistency and perceived quality play for bridging clothing brand image and consumer purchase intention?
How does self-motivation, including self-improvement and self-confirmation, relate to self-consistency and purchase intention?
How are brand familiarity, perceived quality, self-consistency, and purchase intention connected?

This research would be beneficial to understand the problems related to the textile fashion and clothing industry of China, as well as showing the basis for understanding for other regions as well.

2. Literature Review and Hypothesis Deduction

2.1. Clothing Brand Image

One of the very critical aspects of branding that has received significant attention is the brand image [7]. Researchers generally believe that brand image is an essential factor affecting consumers' purchase decisions. A reputable brand image is a crucial aspect that helps customers make purchasing decisions efficiently and satisfactorily, which eventually gives favorable outcomes and associations in consumers' minds in the long run [8]. Brand equity is directly dependent on the brand image; however, the current literature does not contain clear evidence of how this relationship exists [9–11].

The current literature has not determined the intervention mechanism, which defines the impact of brand image background on brand equity. Therefore, it is still uncertain whether brand equity drivers play an essential role in connecting brand image and equity. Therefore, the controlling factors of brand equity may be one of the main mechanisms of brand image affecting brand equity. From the above discussion, it is essential to establish potential brand equity drivers, such as brand attitude, brand awareness, brand attachment, and brand value, which will shape the relationship between brand image and brand equity. These drivers will be considered more carefully when making purchase decisions and expanding consumer-based brand assets.

Brand image and brand equity relationships are also significantly influenced by the nature of products, and there is a big gap in understanding of how it is impacted as no research shows this critical aspect. The influence of the type of brand on consumer-based equity also differs depending on the type of products, such as search goods or experience goods, a concept developed by Philip Nelson [12]. For example, Adidas (clothing category) and Nike brands (shoe category) are classified as search goods (can be verified before purchase). In contrast, many personal services, such as personal care products, fall into the experience goods category, verified after purchase and use only. Therefore, the influence on consumer-based brand equity differs between these brands. Brand equity is measured using two approaches: financial-based and consumer-based [13]. Verhoef et al. [14] mentioned that competition becomes deeper when retailers create and market their brands and products by looking at ongoing styles and market demands. The cost of product is a highly significant component, and is being set by the reputation of certain brands in the appropriate manner and thus involves awareness, quality, uniqueness, social image, and country of origin [15]. These are definitive brand images; this study explains their impact on customers' willingness to pay price premiums. Moreover, from the factors mentioned above, it is essential to find out which one is highly crucial or worthy for customers to finalize and purchase branded clothes, and they can concentrate on certain parameters more than others to produce higher profits and lower marketing costs. There is a view that a brand can create a differentiated position, and this advantage brought by a brand relative to other brands will encourage consumers to pay more.

Tsai pointed out that consumers usually make consumption decisions based on the brand's evaluation results, significantly affecting brand image recognition [16]. Bird pointed out that consumers have different perceptions of brand image with different product usage levels and verified through empirical research that brand image affects consumers' purchase intentions [17]. Jiang revealed that the brand image's customer perception would affect their consumption decision-making and purchasing behavior tendencies [18]. Jin's research results show that company and product brand images positively affect consumer attitudes and purchasing tendencies [19]. Online shopping research shows that after the online clothing brand image forms a value perception in customers' minds, it impacts pre-purchase intentions and purchase decisions [20]. Accordingly, hypothesis H1 is proposed:

Hypothesis 1 (H1). *Apparel brand image has a significant positive impact on consumers' purchase intention.*

In brand image research, Jacoby research pointed out that the brand image strongly influences quality perception and found that the more positive the brand image, the more significant the impact [21]. Jiang and Lu divided the brand image into functional and non-functional factors and researched the impact of these two aspects on perceived quality, perceived value, satisfaction, and loyalty. The results showed that perceived quality is affected by functions, gender, and non-functional factors [18]. Wang and Zhang used empirical research to prove that store images, including product images, environmental images, and service images, significantly impact brand perception quality [22]. In research on Chinese and Western brand names' influence on perceived quality, Yulistiana proved that brand names positively affect the perceived quality [23]. Accordingly, hypothesis H2 is proposed:

Hypothesis 2 (H2). *Apparel brand image has a significant positive impact on perceived quality.*

2.2. Self-Consistency

Self-consistency is a research perspective of the theory of cognitive coherence. The core issues concerned include two aspects: When the individual's self-concept participates in the process of recognizing things, if the cognitive subject is inconsistent with the individual's self-concept, the individual Cognitive disharmony will occur; when an individual's behavior or its consequences conflict with self-awareness, it will cause a cognitive inconsistency [24,25]. Self-consistency is the matching process resulting from the consumer's self-concept and cognition of things, including product images, brand images, store images, destination images, products, brands, or services with symbolic attributes and symbolic meanings. Self-concept (such as actual self, ideal self, and social self) is the sum of people's perceptions and emotions about themselves. It includes how an individual views oneself objectively and what kind of person he thinks he is and what kind of person he wants to be [26]. In the field of consumer behavior, self-consistency is mainly used to study the relationship between symbolic consumption trends and self-concept, focusing on two aspects: paying attention to the relationship between self-consistency and consumers' pre-purchase behavior, intentions, and decision-making; exploring self the effect of consistency on various post-consumer variables, such as satisfaction, loyalty, perceived quality, and post-purchase attitude.

This article uses self-consistency theory to explore the characteristics of consumers' selective perception of brand image information and the relationship between cognitive things and self-concept fit. Scholars' studies have emphasized the association between brand symbolic attributes and self-concept; the higher the degree of matching the brand's symbolic attributes and the consumer's self-concept, the more likely consumers are to think they can satisfy their needs. Sirgy et al. showed that when the symbolic attribute is associated with self-concept (that is, when there is a higher self-consistency), consumers will have a higher preference for products and brands [27]. Rogers evaluated self-concept to explain product symbolism and believes that products, brands, or things most like self-concept are more attractive to consumers [28]. Ibrahim's research shows that when a hotel's overall brand image is better, it can stimulate the connection between consumers' self-concept and brand image and make it more self-consistent with the brand image [29]. Accordingly, hypothesis H3 is proposed:

Hypothesis 3 (H3). *Apparel brand image has a significant positive impact on self-consistency.*

Self-consistency affects consumer attitudes and behaviors. If people have a higher self-consistency when dealing with brand symbolic information, it will promote their good feelings about the brand. Otherwise, they will form an unfavorable attitude towards the brand, which will affect their evaluation of the brand. Mukherjee confirmed that self-consistency has a significant impact on consumer attitudes and behaviors and proposed that when consumers' self-concepts match their views on a sure thing, they will have a more positive attitude [30]. Therefore, people are more inclined to choose and buy brands with higher self-consistency [31]. Ericksen proposed the applicability of self-consistency in clothing consumption research and constructed a relationship between professional women's clothing preferences and self-consistency [32]. Tan Qing explored the relationship between clothing styles, colors, clothing brands, shopping environment, and the self-concept of female consumers and demonstrated empirical evidence that women's s clothing choice is related to self-concept [33]. Research by Liu et al. showed that self-consistency positively affects consumers' brand attitude and loyalty [34]. Lee confirmed that consumer self-concept and consistency of brand personality directly impact brand clothing purchase intention [35]. Accordingly, hypotheses H4 and H5 are proposed:

Hypothesis 4 (H4). *Self-consistency has a significant positive impact on consumers' purchase intention.*

Hypothesis 5 (H5). *Self-consistency plays an intermediary role in the relationship between clothing brand image and consumer purchase intention.*

2.3. Perceived Quality

Research by scholars has shown that there is a meaningful relationship between perceived quality and purchase intention. Most use perceived quality as an intermediary variable to construct a theoretical model of the influence of brand information cues on purchase intention. Aaker defines perceived quality as an overall intangible perception of consumers' brand and found that perceived quality significantly impacts purchase intention [36]. Boulding established a perceived quality model on consumer decision-making, proposing that perceived quality is a consumer's judgment on brand quality and the expected result of actual service quality and has an essential predictive effect on consumers' purchase decision-making process [37]. Research by Lin et al. found that when customers' perceived quality of a particular brand is higher, they will have a more positive attitude towards the brand, which will produce a higher willingness to buy [38]. Consumers' perception of store quality is the primary measurement index for predicting store sales, and it is pointed out that the improvement of perceived quality is conducive to promoting consumers' buying behavior and willingness [39]. Dodds proposed a conceptual model of the influence of external cues (price, brand name, and store name) on consumers' perceived quality and purchase intention [40]. Recently, Kim proved that showing prices and positive brand information perception positively impact perceived quality [41]. Based on this, hypotheses H6 are proposed:

Hypothesis 6 (H6). *Perceived quality plays an intermediary role in the relationship between clothing brand image and consumer purchase intention.*

2.4. The Moderating Role of Self-Motivation in Self-Consistency

In the research on self-consistency and consumer purchasing decision, willingness and behavior, scholars found certain boundary conditions in the influence relationship between the two, and consumer motivation and consumer personality affect the two to a certain extent [42]. Among them, consumption motivation includes self-confirmation motivation and self-improvement motivation, collectively referred to as self-motivation. Self-motivation reflects two different purposes of consumers taking self-consistency as a clue of judgment. When consumers are motivated by self-confirmation, people tend to explain and accept the same information or like self-concept [43]. This motivation is related to the self-concept's actual self and social reality and is the self-intrinsic standard and social standard for the individual to make informed judgments. Under the influence of self-confirmation motivation, consumers are more inclined to choose brands that maintain their authentic self or social self-image, have high consistency, and give positive evaluations [44]. When people aim to self-improve, they prefer brands with higher or better grades than self-image to enhance their self-image and display a higher social image or status [45]. Accordingly, hypotheses H7, H7a, and H7b are proposed.

Hypothesis 7 (H7). *Self-motivation regulates the relationship between self-consistency and purchase intention.*

Hypothesis 7a (H7a). *Self-improvement motivation regulates the relationship between self-consistency and purchase intention.*

Hypothesis 7b (H7b). *Self-confirmation motivation regulates the relationship between self-consistency and purchase intention.*

2.5. The Moderating Role of Brand Familiarity in Perceived Quality

Brand familiarity is an important indicator to measuring and evaluating customers' understanding of the target brand. It is possible to intuitively understand the degree of accumulation of consumer experience and knowledge of a particular brand [46]. Szybillo believes that consumer perception quality will be affected by internal product cues (i.e., physical product attributes) and product external cues (i.e., product's non-physical attributes) [47]. Karangi's research shows that consumers with a different product or brand familiarity have different judgment criteria for evaluating the same product or brand [48]. People with high product or brand familiarity tend to pay attention to the product and brand's actual attributes, while familiarity is low. People pay attention to external clues related to products or brands. Labeaga believes that low perceived risk caused by high brand familiarity can drive customers to have a higher brand consumption tendency without positive quality perception [49]. Accordingly, hypothesis H8 is proposed:

Hypothesis 8 (H8). *Brand familiarity regulates the relationship between perceived quality and purchase intention.*

2.6. Chain Mediation between Perceived Quality and Self-Consistency

When the symbolic utility of external clues makes people have a higher quality perception of products and brands, it can, to a certain extent, enhance the predictive value and confidence value of products and brands to consumers, thereby generating higher expectations and attention. The consistency between the concept and the brand image has a positive impact [50,51]. Quester et al. pointed out that product quality perception positively influences the cognitive fit relationship (i.e., self-consistency) between consumers and brands [52]. Graeff also confirmed that people's subjective evaluation of products positively correlates with consumer self-image and retail brand image [53]. Simultaneously, transmitting a clothing brand's image to consumers is a process of information dissemination, which has hierarchical transmission characteristics. The role of information disseminated in individual cognition formation ranges from shallow to deep, from shallow intuitive perception to the formation of thoughts and emotions, and then to the matching of deep self-personality and characteristics.

When consumers are unfamiliar with products or brands, self-concept clues cannot be obtained immediately, but a relatively complicated confirmation and cognitive processing process is required. Self-consistency is related to the fit between consumers' self-image and brand image. Consumers' self-concept and brand image comparison require more cognitive elaboration and more complex information processing. Therefore, this article speculates that in consumer brand image recognition, the perceived quality is first formed through direct perception of image information, and then this information evaluation clue is used to associate self-image with the brand image. Accordingly, hypothesis H9 is proposed:

Hypothesis 9 (H9). *Perceived quality and self-consistency play a chain-like intermediary role in the process of clothing brand image influencing purchase intention.*

Given the above literature review and theoretical analysis, this paper proposes the following research model (Figure 1).

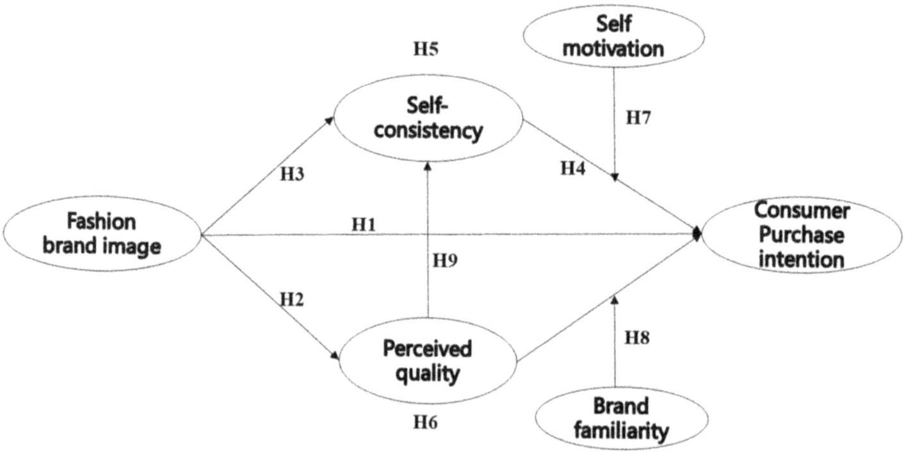

Figure 1. The theoretical model of this research.

3. Research Methods

3.1. Data Collection

This study used questionnaire surveys to collect data. The survey samples select people who had experience in online and offline clothing consumption in shops and distributed offline questionnaires and online questionnaires to investigate consumers fully. The offline questionnaires were distributed in the brand agglomeration area in Shanghai. Online questionnaires were distributed through the "Questionnaire Star" software. There were 134 offline questionnaires and 116 valid questionnaires, with an effective rate of 86.5%. Moreover, there are 412 online questionnaires and 352 valid questionnaires, with an effective rate of 85.4%. A total of 546 questionnaires were collected in this survey, of which 468 were valid questionnaires, the effective recovery rate was 85.71%, and the number of valid samples met the requirement of 5 times the measurement items.

The sampling control characteristics include consumers' age, gender, education level, income, and city of residence. The results of the descriptive statistical analysis of the sample are as follows: women accounted for 53.19%, men accounted for 46.91%; 18–25 year olds accounted for 47.01%, 25–35 years olds accounted for 27.29%, 36–50 years olds accounted for 21.32%, under 18 and over 50 years olds accounted for 4.58%; Junior high school and below accounted for 2.19%, college degree accounted for 4.38%, bachelor degree accounted for 51%, master degree or above accounted for 42.23%; income below 2000 yuan accounted for 44.62%, 2000–4999 yuan accounted for 21.71%, 5000–9999 yuan accounted for 22.31%, 10,000–14,999 yuan accounted for 7.17%, 15,000–19,999 yuan accounted for 3.17%, and 20,000 yuan or more accounted for 1.02%; first-tier cities accounted for 59.96%, second-tier cities accounted for 21.71%, third-tier cities accounted for 6.37%, and fourth-tier and below 11.95%. Overall, the survey subjects are mostly school students and newcomers, and their monthly income is below 10,000 yuan. Such a sample structure is closer to the research object's consumption characteristics, conducive to obtaining research results and general research conclusions that conform to objective facts.

3.2. Variable Measurement

This research involves seven constructs, namely clothing brand image (BI), self-consistency (SC), perceived quality (QP), brand familiarity (BF), self-motivation (SM), and purchase intention (PUI). All the constructs are derived from relevant mature scales at home and abroad, and experts in apparel and senior practitioners are invited to evaluate and correct the scale's semantic accuracy and revise it into a formal questionnaire. All items involved in this research use the Likert 6-point scale. Among them, "1" means "strongly disagree" and "6" means "strongly agree", and the scales are all derived from representative documents at home and abroad. The clothing brand image measurement draws on the brand image scale used in the research by Biel [54], Aaker [36], Keller [55], and Fan Xiucheng [56] to measure consumers' perception of clothing brands image. Self-consistency draws on the scale developed by Sirgy [27] and Escalas [57] to measure the degree of conformity between clothing brand image and consumer image. Perceived quality draws on the scale developed by Parasuraman [58] and Dodds [40] to measure consumers' overall brand quality, quality stability, and quality dependence. Self-motivation refers to the scale used in the research by Alexandrov [59] and Napper [60]. Brand familiarity draws on the scale used in research by Dursun [61] and Nepomuceno [62]. Refer to the scales used in the research by Diallo [63] and Fishbein [64] for purchase intention.

4. Empirical Analysis and Hypothesis Test Results

4.1. Reliability and Validity Test

The reliability test is judged by Cronbach's α coefficient value and the correction term's correlation coefficient (CITC). As shown in Table 1, the Cronbach's α coefficient values of clothing brand image, self-consistency, perceived quality, brand familiarity, self-motivation, and purchase intention are between 0.821 and 0.951, which are all greater than the critical value of 0.7, and the CITC value is 0.553~0.927, indicating that the scale has good reliability and a high level of reliability.

The validity test uses the AMOS 21.0 tool to perform confirmatory factor analysis on each variable to verify its convergent validity. As shown in the results in Table 1, the standardized factor load coefficients of each variable are all greater than 0.5 and reach a significant level ($p < 0.05$), and the combined reliability (CR) of each construct is more significant than 0.8, indicating that the scale has a good convergence effect degree. The average extraction amount (AVE) value of each construct reaches 0.5 judgment standard, and its square root is greater than the Pearson correlation coefficient between the two constructs (as shown in Table 2), indicating that the variable has a good discrimination validity.

4.2. Model Fitting and Hypothesis Testing

This study used the AMOS 21.0 tool to fit the collected questionnaire data and the hypothetical model that brand image affects consumers' purchase intention to fit the structural equation model. The statistical values of the fitness test indicators are shown in Table 3. The fit test index CMIN/DF value of this study is 2.778 < 3, the values of GFI, NFI, CFI, TLI, and AGFI are all greater than 0.9, the RMSEA value is 0.062 < 0.1, and the RMR is 0.037 < 0.5, all of which are relatively good Level of fitness. The results show that the model has outstanding fit test results, and the fit effect is good.

Table 1. Reliability and validity test results of measurement items.

Construct	Item	Factor Loading	CITC	Cronbach's α	CR	AVE
Clothing Brand Image	NI	0.64	0.624	0.879	0.877	0.507
	PPI	0.60	0.553			
	SE	0.66	0.633			
	BC	0.72	0.646			
	PI	0.85	0.777			
	BI	0.83	0.751			
	SI	0.70	0.665			
Self-consistency	SC-1	0.813	0.734	0.869	0.870	0.627
	SC-2	0.836	0.758			
	SC-3	0.790	0.733			
	SC-4	0.724	0.660			
Perceived Quality	QP-1	0.945	0.898	0.951	0.960	0.889
	QP-2	0.96	0.927			
	QP-3	0.923	0.918			
Purchase Intention	PUI-1	0.718	0.640	0.821	0.867	0.567
	PUI-2	0.743	0.555			
	PUI-3	0.742	0.698			
Brand Familiarity	BF-1	0.687	0.665	0.843	0.930	0.767
	BF-2	0.712	0.665			
	BF-3	0.855	0.647			
	BF-4	0.855	0.623			
	BF-5	0.911	0.646			
Self-Motivation	SM-1	0.907	0.831	0.934	0.734	0.508
	SM-2	0.819	0.834			
	SM-3	0.858	0.860			
	SM-4	0.642	0.854			

Table 2. Questionnaire discriminative validity test results.

	BI	QP	SC	PUI
BI	0.507			
QP	0.046 ***	0.889		
SC	0.037 ***	0.043 ***	0.627	
PUI	0.030 ***	0.039 ***	0.051 ***	0.508
AVE square root	0.712	0.943	0.792	0.713

Note: *** means significant at 0.05 level.

Table 3. Main indicators of model fit test.

Index	X^2/df		GFI	NFI	CFI	TLI	AGFI	RMR	RMSEA
Value	CMIN 311.171	CMIN/DF 2.778	0.932	0.941	0.961	0.953	0.921	0.037	0.062

4.3. Test of the Scale

Data testing was carried out on the measurement scale of value creation of clothing brand image, mainly including reliability and validity test, factor test, and goodness of fit test of the model. The specific results are shown in Table 3. In the exploratory factor molecule, the cumulative variance contribution rate of the scale was 73.061% > 60%, and the overall KMO and Cronbach's α values were 0.946 and 0.965, respectively, both greater than 0.7. Meanwhile, the KOM and Cronbach's α values of each dimension also reached the acceptance standard. In the confirmative factor analysis, the standardized factor loading (EFA) was between 0.648 and 0.912, meeting the criteria of greater than 0.5, and the combinatorial reliability (CR) and square extraction variance (AVE) also met the criteria of 0.7 and 0.5. In the goodness of fit test of the model, the chi-square to the degree of freedom ratio (X2/DF) is 2.898, which is less than 3, AGFI and GFI are more significant than 0.8, and NFI, CFI, TLI, and IFI are all greater than 0.9. Based on the above analysis, the measurement scale has good reliability and validity.

The results of model fitting and hypothesis testing are shown in Table 4. The three vectors of clothing brand image that affect purchase intention, self-consistency, and perceived quality are significant at $p < 0.001$. Clothing brand image has a significant impact on purchase intention, self-consistency, and perceived quality. Suppose H1, H2, and H3 Be verified. Moreover, assuming that the vectors of H4 and H9 are also significant at the $p < 0.001$ level, this shows that self-consistency has a significant impact on purchase intention, and perceived quality has a significant impact on self-consistency.

Table 4. Model fitting and hypothesis testing results.

Hypothesis	Estimate	Std. Estimate	CR.	*p*	In Conclusion
H1	0.559	0.055	17.532	***	Valid
H2	0.631	0.076	15.557	***	Valid
H3	0.798	0.045	14.956	***	Valid
H4	0.554	0.060	13.571	***	Valid
H7	0.521	0.052	11.075	***	Valid

Note: *** means *p*-value is less than 0.001.

4.4. Reliability and Validity Test and Factor Analysis

This study uses the Bootstrap method to test self-consistency, the mediating effect of perceived quality (hypothesis H5, H6), and the chain-like mediating effect of self-consistency on perceived quality. Using Model 4 (simple intermediary model) in the SPSS macro written by Hayes, while controlling the demographic factors of gender, education, age, income, and residence, the self-consistency and perceived quality of the clothing brand image, and the mediating effect of the relationship between purchase intentions are tested, and the test results are shown in Tables 5 and 6.

The results show that the direct effect of clothing brand image on purchase intention and the mediating effect of self-consistency and perceived quality do not contain 0 at the upper and lower limits of the Bootstrap 95% confidence interval. This indicates that clothing brand image can directly predict purchase intention and pass the mediating effect of self-consistency and perceived quality predicts purchase intention. The direct effect action value is 0.280, the self-consistency mediation effect action value is 0.217, and the perceived quality action value is 0.180, accounting for 41.33%, 32.04%, and 26.63% of the total effect. Hypothesis H5 and H6 are established.

Table 5. Analysis results of mediation effect.

Path	Effect Size	Boot Standard Error	BootCI Confidence (Lower Bound)	BootCI Confidence (Upper Bound)	Effect Ratio
Indirect effects: mediating effects of perceived quality	0.180	0.0108	0.0034	0.0449	26.63%
Indirect effect: self-consistent mediation effect	0.217	0.0467	0.0997	0.2816	32.04%
Direct effect	0.280	0.0601	0.2634	0.4993	41.33%
Total effect	0.677	0.0410	0.5040	0.6630	—

Note: The interval value of [BootCI, BootCI] does not contain 0, which means it is significant at the 0.05 level.

Table 6. Results of chain mediated effect analysis.

Path	Effect Size	Boot Standard Error	BootCI Confidence (Lower Bound)	BootCI Confidence (Upper Bound)
Total effect	0.773	0.045	0.127	0.306
Direct effect	0.286	0.049	0.290	0.482
Indirect effect: clothing brand image -> self-consistency -> purchase intention	0.224	0.046	0.098	0.277
Indirect effect: clothing brand image -> perceived quality -> purchase intention	0.136	0.012	0.005	0.051
Indirect effect: clothing brand image -> self-consistency -> perceived quality -> purchase intention	0.127	0.0137	0.002	0.005

This research uses clothing brand image as the independent variable X, purchase intention as the dependent variable Y, self-consistency as the first intermediary variable M1, and perception consistency as the second intermediary variable M2 using the SPSS macro written by Hayes. In Model6, a Bootstrap analysis is performed to test the chain mediation effect (i.e., continuous mediation effect) of self-consistency and perceived quality. The results show that the total effect's confidence interval is 0.127~0.306 excluding 0, indicating that the total effect of clothing brand image on purchase intention is significant. The confidence interval of direct effect is 0.290~0.482 excluding 0, indicating that the direct effect of clothing brand image on purchase intention is significant. In the indirect effect, the confidence interval of the action path of clothing brand image -→ self-consistency → purchase intention is 0.098~0.277 excluding 0, indicating that the clothing brand image passes self-consistency but does not pass the perceived quality. The effect on purchase intention is that the mediation effect is significant. The confidence interval of clothing brand image -> perceived quality -> purchasing intention is 0.005~0.051 excluding 0, which means that when clothing brand image passes through perceived quality but not self-consistency, the mediating effect on purchasing intention is significant. The path clothing brand image -> self-consistency -> perceived quality -> purchase intention confidence interval is 0.002~0.005 excluding 0, indicating that when the clothing brand image passes self-consistency and perceives quality, it has a significant effect on purchase intention. In summary, self-consistency and perceived quality play a chain-like intermediary role in the relationship between clothing brand image and purchase intention.

4.5. Moderating Effect Test

This study uses a hierarchical regression model to test the adjustment effect taking purchase intention as a dependent variable, under the condition of controlling the demographic factors of gender, education, age, income, and residence, the cross product of self-consistency and self-motivation, perceived quality and brand familiarity is introduced for regression analysis. The analysis results are shown in Table 7. The results show that the adjustment R^2 gradually increases with the interaction term's introduction, indicating that the model is better fitted. The interaction item of perceived quality and brand familiarity significantly impacts purchase intention ($\beta = -0.153$, $t = -3.239$, $p < 0.001$), which indicates that brand familiarity has a significant moderating effect on the relationship between perceived quality and purchase intention, H8 was established. In addition, the interaction term between self-consistency and self-motivation has a significant positive effect on purchase intention ($\beta = 0.078$, $t = 1.972$, $p < 0.05$), indicating that self-motivation has a moderating effect in the relationship between self-consistency and purchase intention, H7 was established.

Table 7. Test results of adjustment effect.

Variable	Purchase Intention		Purchase Intention		Purchase Intention	
	Standardization Factor	t-Value	Standardization Factor	t-Value	Standardization Factor	t-Value
Gender	0.013	0.251	0.034	0.833	0.041	1.012
Degree	0.039	0.703	0.013	0.278	0.009	0.201
Profession	−0.034	−0.553	−0.003	−0.064	0.006	0.113
Age	0.077	1.358	0.048	1.052	0.04	0.876
Income	−0.023	−0.411	−0.011	−0.251	−0.015	−0.348
Place of residence	−0.027	−0.445	−0.033	−0.657	−0.038	−0.764
Perceived Quality			0.176 ***	4.423	0.255 ***	5.204
Brand Familiarity			0.11 **	2.509	0.14 **	3.063
Self-consistency			0.528 ***	12.527	0.512 ***	11.862
Self-Motivation			−0.052	−1.189	−0.065	−1.485
Perceived Quality × Brand Familiarity					−0.153 ***	−3.239
Self-motivation × self-consistency					0.078 *	1.972
Adjust R^2	−0.008		0.341		0.357	
F Value	0.36		24.861		22.318	

Note: * means significant at 0.005 level, ** means significant at 0.01 level, *** means significant at 0.05 level.

When H7 is established, this study further examines the moderating effects of different self-motivation types (i.e., self-confidence and self-improvement motivation) in the relationship between self-consistency and purchase intention. For self-confirmed motivation, self-consistency has a significant positive effect on purchase intention ($\beta = 0.096$, $t = 2.226$, $p < 0.05$); on the contrary, for self-improvement motivation, self-consistency has no significant effect on purchase intention ($\beta = 0.132$, $t = 1.323$). This shows that self-consistency on purchase intention will only significantly have an effect when consumers have self-confirmed consumer psychology. Hypothesis H7b holds, but H7a does not hold.

Based on the above test results, the summary of the model hypothesis testing in this study is as follows in Table 8.

Table 8. Summary of hypothesis test results.

Numbering	Research Hypothesis	Test Result
H1	Apparel brand image has a significant positive impact on consumers' purchase intention	Valid
H2	Apparel brand image has a significant positive impact on perceived quality	Valid
H3	Clothing brand image has a significant positive impact on self-consistency	Valid
H4	Self-consistency has a significant positive impact on consumers' purchase intention	Valid
H5	Self-consistency plays an intermediary role in the relationship between clothing brand image and consumer purchase intention	Valid (Partial Intermediary)
H6	Perceived quality plays an intermediary role in the relationship between clothing brand image and consumer purchase intention	Valid (Partial Intermediary)
H7	Self-motivation regulates the relationship between self-consistency and purchase intention	Valid
H8	Brand familiarity regulates the relationship between perceived quality and purchase intention	Valid
H9	Perceived quality and self-consistency play a chain-like intermediary role in the process of clothing brand image influencing purchase intention	Valid (Partial Intermediary)

5. Discussion and Marketing Implications

5.1. Discussion

This paper analyzes the influence mechanism of clothing brand image on consumer purchase intention from cognitive consistency theory. It obtains the following research conclusions: First, clothing brand image has a significant positive impact on consumer purchase intention, and part of its influence is the path is realized directly, and the other part is realized through two indirect paths of self-consistency perceived quality. Secondly, self-consistency has a significant impact on the relationship between clothing brand image and consumer purchase intention. Part of its influence is realized through direct paths and two indirect paths, self-improvement motivation and self-recognition motivation, and self-recognition motivation is better than self-improvement while the moderating effect of motivation is substantial. Moreover, perceived quality significantly impacts the relationship between clothing brand image and consumer purchase intentions. Part of its influence is realized through direct paths and partly through indirect paths of brand familiarity. Finally, perceived quality and self-consistency have a chain-like intermediary effect in the relationship between clothing brand image and purchase intention. Consumers form perceived quality after receiving clothing brand information and then affect purchase intention through self-consistency.

The influence of self-consistency and perceived quality on the relationship between clothing brand image and consumer purchase intention provides new ideas for brand marketing. Consumers' perception of clothing brand image affects their willingness and initiative to match their self-concept with the brand image and positively affects self-consistency. In the actual consumption process, people purchase products or brands like their personality and image characteristics to match their accurate self-image or ideal self-image and use the explicitness, display, and brand symbolism of clothing to create their ideas. Moreover, in consumer behavior, individuals will try to avoid choosing brands or products that conflict with their views and values or are different, and they tend to choose brands or products with values and connotations that match them. Otherwise, they will fall into cognitive dissonance. Therefore, compared with self-improvement motivation, self-confirmation motivation has a more substantial moderating effect on the process of self-consistency and purchase intention.

Perceived quality is the perceptual knowledge of the brand and belongs to the category of subjective evaluation. Therefore, an optimistic clothing brand image can prompt consumers to make more positive subjective evaluations, that is, to obtain higher perceived quality. As a subjective judgment produced by comprehensive brand information, perceived quality is also a comprehensive evaluation of brand attitude. A positive attitude guides consumers' thoughts and actions and directly affects their purchase intention. In consumer brand image recognition, the perceived quality is often formed through direct perception of image information, and then this information evaluation clue is used to associate self-image with the brand image; thus, self-consistency is produced. When receiving brand image information to produce positive perception quality, to maintain cognitive consistency, and avoid cognitive dissonance, individuals will more actively retrieve and process brand-related information in memory in the process of forming self-consistency.

5.2. Marketing Enlightenment

The research results on the influence of brand image on consumers' purchase intention improve brand building and marketing. A unilateral effect does not form the brand image but results from enterprise and consumers' joint action. Therefore, in the process of brand image shaping, companies should recognize their cultural advantages and brand and product characteristics, comprehensively sort out and integrate the brand information that needs to be transmitted, and identify the differences from other brands according to the core positioning of the brand and make the brand unique. The brand image is shaped according to sex so that consumers can have deep memories in brand image recognition and quickly review and retrieve brand image information during the consumption process, which triggers a series of positive brand associations.

This research believes that companies should promote clothing brand building and marketing on three levels. First, consumers' perceptions of various clothing brand image dimensions will affect their willingness to buy, and companies should establish a corresponding brand image when formulating brand marketing strategies through the corporate image, product image, brand logo image, publicity image. The shaping and dissemination of service, network, and store images establish a positive brand image. Second, since self-consistency plays a crucial role in the relationship between brand image and consumer purchase intentions, companies should analyze consumer image characteristics, personality preferences, and self-concept in the target market and use this as a basis for targeting. Shaping the brand image and strengthening consumers' enthusiasm to match the brand image's self-concept strengthens the attraction between the brand and consumers invisibly. Finally, because a consistent brand image can achieve more substantial cognitive effects and persistence than a single dimension and enhance consumers' perception of the brand, companies should transmit the same brand core message in different dimensions and communication channels. This creates a consistent brand image to strengthen consumers' clear and complete cognition of the brand image. Delivering the same core brand information through different dimensions and communication channels can help consumers recall relevant brand memories in a competitive market with complex information and numerous choices, promoting purchase intentions.

6. Conclusions

Among other sectors of the textile industry, the clothing and apparel industry heavily rely on consumer purchase intention due to the largest numbers of consumers involved in the sales process. The consumer purchase intention is a complex mixture of several factors that work together to make overall purchase decisions of the buyer. Herein, we aimed to explore several key questions in the clothing industry. It was found that clothing brand image, self-consistency, and perceived quality can significantly affect consumers' purchase intention. Consumers' purchase intention is directly related to clothing brand image, wherein self-consistency and perceived quality play an intermediary role, and it is governed by brand familiarity and self-motivation, including self-improvement and self-confirmation. Therefore, during textile marketing, particularly garment or apparel marketing, these parameters must be analyzed carefully. The results would be of particular interest theoretically for academia exploring textile marketing and practically for textile fashion and clothing brands.

Author Contributions: Conceptualization, L.C. and X.Y.; methodology, L.C.; software, L.C. and N.K.; validation, L.C., X.Y. and C.L.; formal analysis, L.C. and Q.D.; investigation, L.C. and Q.D.; resources, L.C.; data curation, L.C.; writing—original draft preparation, L.C. and H.H.; writing—review and editing, H.H., N.K. and H.M.; visualization, L.C. and H.H.; supervision, C.L.; project administration, X.Y.; funding acquisition, X.Y. All authors have read and agreed to the published version of the manuscript.

Funding: The authors received General Projects of National Social Science Foundation (education) "internationalization development and overseas communication strategy of university brand image driven by education of foreign students in China" (BGA200057) for this research.

Data Availability Statement: The corresponding author can provide the data on request.

Acknowledgments: Authors would like to extend our sincere thanks to anonymous reviewers for providing helpful comments and suggestions on earlier drafts of the manuscript.

Conflicts of Interest: The authors declared no potential conflicts of interest with respect to the research, authorship, and/or publication of this article.

References

1. Ozbal, O.; Duman, T.; Topaloglu, O. A trust-based peer-to-peer digital brand equity (P2P-DBE) model. *J. Mark. Theory Pract.* **2020**, *28*, 497–520. [CrossRef]
2. Kass, J. The symbolic value of fashion brand collaboration. In *Communicating Fashion Brands*; Routledge: Abingdon, UK; New York, NY, USA, 2020; pp. 141–155.
3. Feng, B.; Sun, K.; Chen, M.; Gao, T. The impact of core technological capabilities of high-tech industry on sustainable competitive advantage. *Sustainability* **2020**, *12*, 2980. [CrossRef]
4. Morris, J.; Koep, L.; Damert, M. Labels in the textile and fashion industry: Communicating sustainability to effect sustainable consumption. In *Sustainable Textile and Fashion Value Chains*; Springer: Cham, Switzerland, 2021; pp. 257–274.
5. Zameer, H.; Wang, Y.; Yasmeen, H. Reinforcing green competitive advantage through green production, creativity and green brand image: Implications for cleaner production in China. *J. Clean. Prod.* **2020**, *247*, 119119. [CrossRef]
6. Kaur, H.; Anand, S. Actual versus ideal self: An examination of the impact of fashion self congruence on consumer's fashion consciousness and status consumption tendencies. *J. Glob. Fash. Mark.* **2021**, *12*, 146–160. [CrossRef]
7. Keller, K.L. *Building Customer-Based Brand Equity: A Blueprint for Creating Strong Brands*; Marketing Science Institute: Cambridge, MA, USA, 2001.
8. Kumaravel, V.; Kandasarny, C. To What Extent the Brand Image Influence Consumers' Purchase Decision on Durable Products. *Rom. J. Mark.* **2012**, *1*, 34–38.
9. Kim, H.B.; Kim, W.G.; An, J.A. The effect of consumer-based brand equity on firms' financial performance. *J. Consum. Mark.* **2003**, *20*, 335–351. [CrossRef]
10. Faircloth, J.B.; Capella, L.M.; Alford, B.L. The effect of brand attitude and brand image on brand equity. *J. Mark. Theory Pract.* **2001**, *9*, 61–75. [CrossRef]
11. Jaiprakash, A.T. A conceptual research on the association between celebrity endorsement, brand image and brand equity. *J. Mark. Manag.* **2008**, *7*, 54–64.
12. Nelson, P. Information and consumer behavior. *J. Political Econ.* **1970**, *78*, 311–329. [CrossRef]
13. Keller, K.L. Conceptualizing, measuring, and managing customer-based brand equity. *J. Mark.* **1993**, *57*, 1–22. [CrossRef]
14. Verhoef, P.C.; Nijssen, E.J.; Sloot, L.M. Strategic reactions of national brand manufacturers towards private labels: An empirical study in The Netherlands. *Eur. J. Mark.* **2002**, *36*, 1309–1326. [CrossRef]
15. Anselmsson, J.; Bondesson, N.V.; Johansson, U. Brand image and customers' willingness to pay a price premium for food brands. *J. Prod. Brand Manag.* **2014**, *23*, 90–102. [CrossRef]
16. Tsai, P.-H.; Lin, G.-Y.; Zheng, Y.-L.; Chen, Y.-C.; Chen, P.-Z.; Su, Z.-C. Exploring the effect of Starbucks' green marketing on consumers' purchase decisions from consumers' perspective. *J. Retail. Consum. Serv.* **2020**, *56*, 102162. [CrossRef]
17. Bird, M.; Channon, C.; Ehrenberg, A.S. Brand image and brand usage. *J. Mark. Res.* **1970**, *7*, 307–314. [CrossRef]
18. Jiang, L.; Lu, T. Does image create value? The impact of service brand image on customer value-satisfaction-loyalty relationship. *Manag. World* **2006**, *4*, 106–114.
19. Jin, S.V.; Ryu, E. I'll buy what she's #wearing: The roles of envy toward and parasocial interaction with influencers in Instagram celebrity-based brand endorsement and social commerce. *J. Retail. Consum. Serv.* **2020**, *55*, 102121.
20. Dabbous, A.; Aoun Barakat, K.; Merhej Sayegh, M. Social commerce success: Antecedents of purchase intention and the mediating role of trust. *J. Internet Commer.* **2020**, *19*, 262–297. [CrossRef]
21. Jacoby, J.; Olson, J.C.; Haddock, R.A. Price, brand name, and product composition characteristics as determinants of perceived quality. *J. Appl. Psychol.* **1971**, *55*, 570. [CrossRef]
22. Wang, X.; Zhang, L. The influence of store image on private brand perception and purchase intention. In Proceedings of the British Academy of Management Conference (BAM), Sheffield, UK, 14–16 September 2010.
23. Yulistiana, I.; Dewi, R.V.; Mas'adi, M.; Sunarsi, D.; Erlangga, H. Did Brand Perceived Quality, Image Product and Place Convenience Influence Customer Loyalty Through Unique Value Proposition? *J. Contemp. Issues Bus. Gov.* **2021**, *27*, 2854–2867.
24. Hosany, S.; Martin, D. Self-image congruence in consumer behavior. *J. Bus. Res.* **2012**, *65*, 685–691. [CrossRef]
25. Chon, D.; Sitkin, S.B. Disentangling the Process and Content of Self-Awareness: A Review, Critical Assessment, and Synthesis. *Acad. Manag. Ann.* **2021**, *5*, 607–651. [CrossRef]
26. Jeong, D.; Ko, E. The influence of consumers' self-concept and perceived value on sustainable fashion. *J. Glob. Sch. Mark. Sci.* **2021**, *31*, 511–525. [CrossRef]
27. Sirgy, M.J. Self-concept in consumer behavior: A critical review. *J. Consum. Res.* **1982**, *9*, 287–300. [CrossRef]
28. Rogers, C.R. *Client-Centered Therapy: Its Current Practice, Implications, and Theory*; Houghton Mifflin: Oxford, UK, 1951.
29. Alnawas, I.; Altarifi, S. Exploring the role of brand identification and brand love in generating higher levels of brand loyalty. *J. Vacat. Mark.* **2016**, *22*, 111–128. [CrossRef]
30. He, H.W.; Mukherjee, A. Corporate identity and consumer marketing: A process model and research agenda. *J. Mark. Commun.* **2009**, *15*, 1–16. [CrossRef]
31. Van Gils, S.; Horton, K.E. How can ethical brands respond to service failures? Understanding how moral identity motivates compensation preferences through self-consistency and social approval. *J. Bus. Res.* **2019**, *95*, 455–463. [CrossRef]
32. Ericksen, M.K.; Sirgy, M.J. Employed females' clothing preference, self-image congruence, and career anchorage. *J. Appl. Soc. Psychol.* **1992**, *22*, 408–422. [CrossRef]

33. Tan, Q.; Geng, L.-H. A Study of the Influence of Women's Self-concept On their Garment Consumption Behavior. *J. Southwest Jiaotong Univ.* **2005**, *4*, 511–525.
34. Liu, F.; Li, J.; Mizerski, D.; Soh, H. Self-congruity, brand attitude, and brand loyalty: A study on luxury brands. *Eur. J. Mark.* **2012**, *46*, 922–937. [CrossRef]
35. Lee, J.W. Relationship between consumer personality and brand personality as self-concept: From the case of Korean automobile brands. *Acad. Mark. Stud. J.* **2009**, *13*, 25–44.
36. Aaker, J.L. Dimensions of brand personality. *J. Mark. Res.* **1997**, *34*, 347–356. [CrossRef]
37. Boulding, W.; Kalra, A.; Staelin, R.; Zeithaml, V.A. A dynamic process model of service quality: From expectations to behavioral intentions. *J. Mark. Res.* **1993**, *30*, 7–27. [CrossRef]
38. Lin, C.-Y.; Marshall, D.; Dawson, J. Consumer attitudes towards a European retailer's private brand food products: An integrated model of Taiwanese consumers. *J. Mark. Manag.* **2009**, *25*, 875–891. [CrossRef]
39. Levy, S.; Gendel-Guterman, H. Does advertising matter to store brand purchase intention? A conceptual framework. *J. Prod. Brand Manag.* **2012**, *21*, 89–97. [CrossRef]
40. Dodds, W.B.; Monroe, K.B.; Grewal, D. Effects of price, brand, and store information on buyers' product evaluations. *J. Mark. Res.* **1991**, *28*, 307–319.
41. Kim, J.; Jhang, J.; Kim, S.S.; Chen, S.-C. Effects of concealing vs. displaying prices on consumer perceptions of hospitality products. *Int. J. Hosp. Manag.* **2021**, *92*, 102708. [CrossRef]
42. Kristiyono, Y.R.; Anjani, W. The influence of perceived value, identity, and self-congruity on aqua life purchase intention. *Jurnal Manajemen* **2020**, *17*, 157–175.
43. Ellison, W.D.; Gillespie, M.E.; Trahan, A.C. Individual differences and stability of dynamics among self-concept clarity, impatience, and negative affect. *Self Identity* **2020**, *19*, 324–345. [CrossRef]
44. Aguirre-Rodriguez, A.; Bosnjak, M.; Sirgy, M.J. Moderators of the self-congruity effect on consumer decision-making: A meta-analysis. *J. Bus. Res.* **2012**, *65*, 1179–1188. [CrossRef]
45. Wattanasuwan, K. The self and symbolic consumption. *J. Am. Acad. Bus.* **2005**, *6*, 179–184.
46. Maubisson, L.; Riviere, A. More value for more satisfaction? The moderating role of the consumer's accumulation of experience. *Rech. Appl. Mark.* **2021**, *36*, 5–43. [CrossRef]
47. Szybillo, G.J.; Jacoby, J. Intrinsic versus extrinsic cues as determinants of perceived product quality. *J. Appl. Psychol.* **1974**, *59*, 74. [CrossRef]
48. Karangi, S.W.; Lowe, B. Haptics and brands: The effect of touch on product evaluation of branded products. *J. Consum. Behav.* **2021**, *20*, 1480–1496. [CrossRef]
49. Labeaga, J.M.; Lado, N.; Martos, M. Behavioural loyalty towards store brands. *J. Retail. Consum. Serv.* **2007**, *14*, 347–356. [CrossRef]
50. Grimm, M.S.; Wagner, R. Intra-brand image confusion: Effects of assortment width on brand image perception. *J. Brand Manag.* **2021**, *28*, 446–463. [CrossRef]
51. Das, G. Linkages between self-congruity, brand familiarity, perceived quality and purchase intention: A study of fashion retail brands. *J. Glob. Fash. Mark.* **2015**, *6*, 180–193. [CrossRef]
52. Quester, P.G.; Karunaratna, A.; Goh, L.K. Self-congruity and product evaluation: A cross-cultural study. *J. Consum. Mark.* **2000**, *17*, 525–537. [CrossRef]
53. Graeff, T.R. Image congruence effects on product evaluations: The role of self-monitoring and public/private consumption. *Psychol. Mark.* **1996**, *13*, 481–499. [CrossRef]
54. Biel, A.L. How brand image drives brand equity. *J. Advert. Res.* **1992**, *32*, RC6–RC12.
55. Keller, K.L.; Swaminathan, V. *Strategic Brand Management: Building, Measuring, and Managing Brand Equity*; Pearson Harlow: Harlow, UK, 2020.
56. Fan, X.; Chen, J. Measurement of brand image: A brand identity-based integrated model and empirical study. *Nankai J.* **2002**, *3*, 65–71.
57. Escalas, J.E. Narrative processing: Building consumer connections to brands. *J. Consum. Psychol.* **2004**, *14*, 168–180.
58. Parasuraman, A.; Zeithaml, V.A.; Berry, L. SERVQUAL: A multiple-item scale for measuring consumer perceptions of service quality. *J. Retail.* **1988**, *64*, 12–40.
59. Alexandrov, A.; Lilly, B.; Babakus, E. The effects of social-and self-motives on the intentions to share positive and negative word of mouth. *J. Acad. Mark. Sci.* **2013**, *41*, 531–546. [CrossRef]
60. Napper, L.; Harris, P.R.; Epton, T. Developing and testing a self-affirmation manipulation. *Self Identity* **2009**, *8*, 45–62. [CrossRef]
61. Dursun, I.; Kabadayı, E.T.; Alan, A.K.; Sezen, B. Store brand purchase intention: Effects of risk, quality, familiarity and store brand shelf space. *Procedia-Soc. Behav. Sci.* **2011**, *24*, 1190–1200. [CrossRef]
62. Nepomuceno, M.V.; Laroche, M.; Richard, M.-O. How to reduce perceived risk when buying online: The interactions between intangibility, product knowledge, brand familiarity, privacy and security concerns. *J. Retail. Consum. Serv.* **2014**, *21*, 619–629. [CrossRef]
63. Diallo, M.F. Effects of store image and store brand price-image on store brand purchase intention: Application to an emerging market. *J. Retail. Consum. Serv.* **2012**, *19*, 360–367. [CrossRef]
64. Fishbein, M.; Ajzen, I. Belief, attitude, intention, and behavior: An introduction to theory and research. *Philos. Rhetor.* **1977**, *10*, 130–132.

sustainability MDPI

Review

A Short Review of Recent Progress in Improving the Fracture Toughness of FRP Composites Using Short Fibers

Yongan Wang [1], Zhenxing Wang [1] and Lvtao Zhu [1,2,*]

[1] College of Textile Science and Engineering (International Institute of Silk), Zhejiang Sci-Tech University, Hangzhou 310018, China; wyazstu@163.com (Y.W.); zhenxing@zstu.email.cn (Z.W.)
[2] Shaoxing-Keqiao Institute, Zhejiang Sci-Tech University, Shaoxing 312000, China
* Correspondence: zhult@zstu.edu.cn; Tel.: +86-15900-706-015

Abstract: Fiber-reinforced plastic (FRP) composite laminates have excellent mechanical properties, corrosion resistance, and designability and thus are widely used in various engineering fields. However, their interlayer performance is relatively poor due to no fiber reinforcement between the laminate layers. These layers are adhered through resin bonding only, caused during their usage, which results in composite material delamination damage, thus, limiting its performance. In response, researchers have conducted numerous studies on how to improve the interlaminate properties of laminates through component and structural modifications of the composites and interlaminate toughening. Short fiber toughening is a simple and effective way to solve this problem. This paper reviews the latest research progress in short fiber interlaminate toughening and short fiber modified resin toughening, analyzes the mechanisms of short fiber toughening is fiber bridging, fiber debonding, fracture, and the toughening mechanisms specific to different fibers. This review paper also discusses the current problems encountered in short fiber toughening and provides an outlook on the future development direction for short fiber toughening to provide a reference for subsequent research on short fiber toughened composites.

Keywords: FRP; delamination damage; interlaminar fracture toughness (ILFT); short fiber toughening

Citation: Wang, Y.; Wang, Z.; Zhu, L. A Short Review of Recent Progress in Improving the Fracture Toughness of FRP Composites Using Short Fibers. *Sustainability* **2022**, *14*, 6215. https://doi.org/10.3390/su14106215

Academic Editor: Jorge de Brito

Received: 6 April 2022
Accepted: 30 April 2022
Published: 20 May 2022

1. Introduction

Fiber-reinforced plastic (FRP) composite laminates are widely used in the aerospace field [1], automotive industry [2], wind power generation [3], and civil infrastructure [4] because of their lightweight and relatively high strength characteristics. However, as the laminates are bonded by resin only in the thickness direction, they are prone to cracking and eventually delamination under the action of external forces [5]. Delamination damage limits the development of composite materials as it seriously affects their safety performance and service life. Much work has been carried out to suppress delamination and increase damage tolerance, including Z-directional toughening (3D braiding [6–9], Z-pinning [10], laminate stitching [11], and 3D integrated woven spacer sandwich composites [12,13]), matrix modification [14], and interlaminate toughening [15]. Z-directional toughening, which introduces a separate yarn system in the Z-direction of the composite, has been shown to significantly improve interlaminar fracture toughness [16], but this may come at the cost of compromising the in-plane properties and is difficult to operate. Interlayer toughening and resin modification requires only minor changes to conventional processes for toughening purposes but use materials such as nanofilms and nanoparticles, which require high costs.

Short fibers are a variety of sources, excellent performance, and inexpensive materials that can be added to laminate plies alone for interlaminar toughening or mixed in resin to modify the resin matrix, which can simply and effectively improve the interlaminar strength of composites. This paper presents the research in recent years on two aspects of short fiber interlaminate toughening and short fiber resin modification.

2. Interlaminar Fracture Toughness Characterization and Testing of Composite Laminates

According to the different forms of load, the composite laminate delamination mainly has the types: opening mode (mode I), sliding mode (mode II), and tearing mode (mode III). The three forms of delamination, as shown in Figure 1, of which mode III delamination is less studied, account for a negligible fraction [17,18]. Interlaminar fracture toughness (ILFT) is a quantitative measure of the ability of composite laminates to resist crack extension and an estimate of material toughness. The stress-based stress intensity factor K [19] and the strain-based strain energy release rate G [20] are often used to characterize delamination extension. Since composites are materials with anisotropic mechanical properties, the calculation of the stress intensity factor K for interlaminar cracks is particularly complicated, so most of the studies are conducted using the strain energy release rate G as the control parameter for delamination expansion. The strain energy release rate is the strain energy released per unit area of crack growth of the deformed object, and its critical value is called the critical strain energy release rate, expressed by the symbol GC, which indicates the energy required for crack expansion per unit area; generally, there is the mode I (G_{IC}) and mode II (G_{IIC}) [21]. From the ASTM standard, it is known that the mode I and mode II ILFT of composites is mainly obtained by the double cantilever beam (DCB) test [22] and end notched flexure (ENF) test [23], with the test setup diagram in Figure 2. The ILSS testing [24] is a popular method for characterizing the interfacial strength of composite materials. Composites are very sensitive to low-velocity impact loading in practical applications, so the low-velocity impact test [25] and compression after impact (CAI) test [26] are also often used together to characterize the resistance of composites to delamination when subjected to low-velocity impacts.

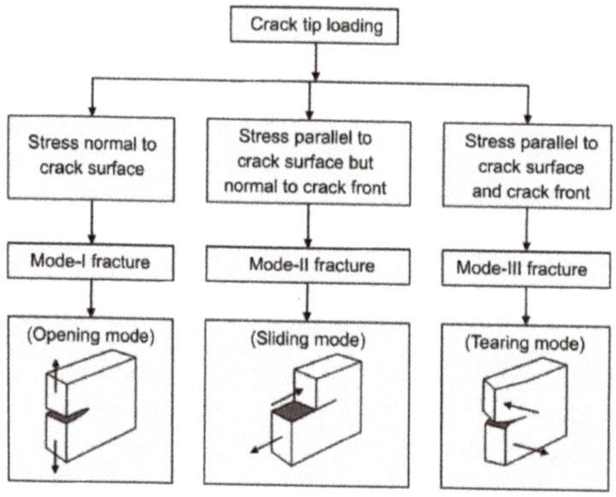

Figure 1. Modes of fracture [18].

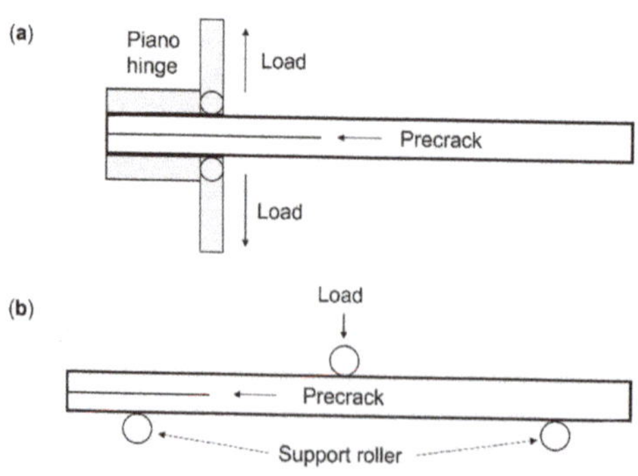

Figure 2. Schematic diagrams of (**a**) double cantilever beam (DCB) test and (**b**) end notch flexure (ENF) test.

3. Short Fiber Toughening

3.1. Short Fiber Interlayer Toughening

In the 1990s, Sohn and Hu covered the surface of carbon fiber prepreg with chopped aramid fibers that accounted for 0.4% to 0.8% of the weight of the laminate and then cured it according to the conventional process. It was found that the addition of short fibers could produce fiber bridging during delamination extension to improve the mode I (G_{IC}) and mode II (G_{IIC}) interlaminar fracture toughness of the laminate by up to more than two times [27]. Since then, more scholars have begun to study short fibers as interlaminar toughening materials to improve the interlaminar fracture toughness of composites [28–32]. Liu et al. investigated the effect of three basic factors on the interlaminar fracture toughness of CF/EP composites using three-factor four-level orthogonal tests, as in Table 1, which are short fiber type, length, and density [33]. The results showed that the effects of each factor on the ILFT were in the order of fiber type, fiber density, and fiber length. The addition of Kevlar fiber most significantly enhanced G_{IC} and G_{IIC}; with the increase of short fiber density, G_{IC} showed a positive correlation, and G_{IIC} showed a decreasing and then increasing trend. Both G_{IC} and G_{IIC} tend to increase as the length of the staple fiber increases and tend to decrease when the length is too long. This provides some reference for future studies in terms of material selection and dosage, but the authors do not further analyze how length and density affect the interlaminar fracture properties.

Table 1. Orthogonal test table (Data from [33]).

Number	Fiber Length (mm)	Fiber Density (g/m^2)	Fiber Type
1	3	5	Kevlar
2	3	10	carbon fiber
3	3	15	glass fiber
4	3	20	basalt fiber
5	7	5	Kevlar
6	7	10	carbon fiber
7	7	15	glass fiber
8	7	20	basalt fiber
9	11	5	Kevlar
10	11	10	carbon fiber
11	11	15	glass fiber
12	11	20	basalt fiber
13	15	5	Kevlar
14	15	10	carbon fiber
15	15	15	glass fiber
16	15	20	basalt fiber

Aramid fibers (AF) have the characteristics of high toughness, high elongation at break, and easy fibrillation on the surface [34]. Under the action of fiber bridging, the deformation and longitudinal tear of the fiber itself can absorb a lot of energy, which plays a very good role in toughening effect [35]. Wang et al. studied the interfacial toughness of CFRP reinforced by short aramid fibers with different lengths and areal densities based on the double cantilever beam test [36]. The results showed that the short aramid fibers with a length of 3 mm and an areal density of 24 g/m^2 had the best toughening effect, the ultimate load, and the mode I interlaminar fracture toughness can be increased by 292.87% and 168.59%, respectively. To ensure the feasibility and effectiveness of aramid fiber interlaminar toughening CFRP, they adopted wettability treatment technology with resin pre-coating to treat short aramid fiber tissues (6 mm, 24 g/m^2), and the peak load and G_{IC} of CFRP were increased by 112%, and 160%, respectively. The pre-coating contributed 37% and 56% improvement, respectively [37]. This explains that the toughening ability of short Kevlar fibers of 3 mm is stronger than 6 mm at the same surface density. This provides a simple way to improve the interfacial compatibility of fibers and resins and also demonstrates that modification of short fibers can further improve the toughening effect.

To enable the short fibers to be evenly dispersed between the layers, Hu et al. impregnated chopped Kevlar fibers in the resin and uniformly coated them on the mid-plane of the laminate. The laminate structure is shown in Figure 3. The effect of chopped Kevlar fiber toughening on the low-speed impact performance of CFRP through drop-weight impact test and compression after impact (CAI) test was studied [38]. The test results showed that the damage resistance of the toughened laminate did not show significant advantages at lower impact energies, and the effect of short fiber toughening was shown only at higher energy impacts. However, the CAI strength and damage strain values increased at all energy levels because the pullout of the chopped fibers consumed part of the energy in delamination expansion under compressive loading, slowing down the yielding process and increasing the CAI strength. The authors considered that short Kevlar fiber toughening is very effective in increasing the damage tolerance when the damage is dominated by mode I fracture toughness.

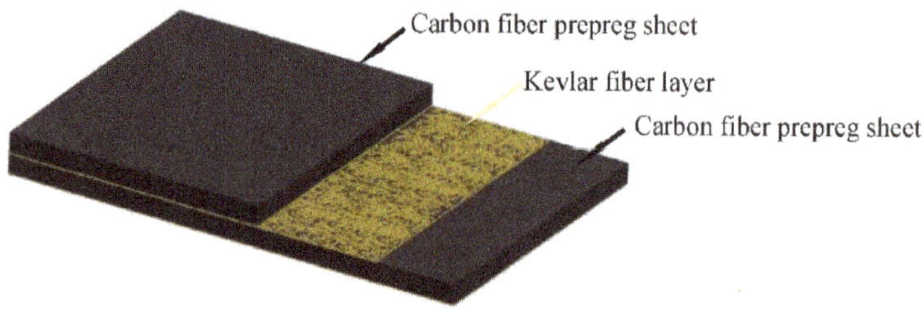

Figure 3. Structure of composite laminate with Kevlar-fiber toughening. The sandwich structure was adopted, with a carbon fiber prepreg sheet at the top and bottom and a Kevlar-fiber layer in the middle [38].

It is believed that adding short fibers to the interlayer is only beneficial to post-impact properties. Still, the interfacial staggering leads to a decrease in the flexural properties of CFRP. Yuan et al. placed an ultra-thin nonwoven short aramid fiber veil (UTNWSAFV) at the interface of all layers of CFRP laminates and found that not only the impact resistance of CFRP could be improved, but also the flexural strength and modulus were improved accordingly [39]. They prepared a toughened prepreg by interweaving the unbonded UTNWSAFV to the surface of the carbon fiber layer in the prepreg production process [40]. Impact tests and compression after impact tests showed that adding UTNWSAFV can effectively suppress out-of-plane deformation and layered expansion. The CAI failure pattern of the laminate with SAFV was changed from delamination damage to shear damage, as seen in Figure 4. The addition of short fibers to the prepreg production process to create short fiber toughened prepreg is a promising technology that focuses on how the short fibers are added to the prepreg in large quantities and uniformly, as this has a major impact on the stability of the composite properties.

Carbon fibers (CF) have high strength and modulus and are often used to make composite materials. Due to their excellent properties, chopped carbon fibers are also used as interlaminar toughening materials to improve the fracture toughness of composites. Xu et al. compared the interlaminar fracture toughness and bending properties of CF/EP composites toughened with two lengths of chopped carbon fibers [41]. It showed that the shorter (0.8 mm) carbon fiber performed better than the longer (4.3 mm) carbon fiber in improving mode I fracture toughness. The longer carbon fiber contributed more to mode II fracture toughness, which was achieved without compromising the bending mechanical properties. Short carbon fibers easily form a three-dimensional interwoven network structure in the epoxy resin of the middle layer of the laminate, and this structure can play a better role in obstructing cracks, as shown in Figure 5. Moreover, due to the addition of short carbon fibers, the electrical conductivity in the thickness direction of the carbon fiber laminates has been improved. According to this conclusion, different lengths of staple fibers can be selected for selective toughening, depending on the form of loading the laminate is subjected to.

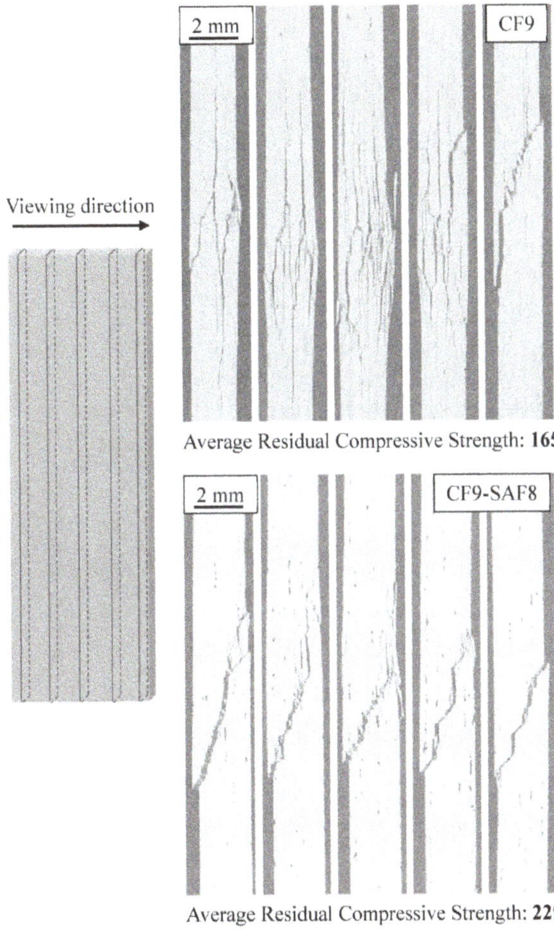

Figure 4. CAI test failure morphology of CF9 and CF9-SAF8 (X-ray µCT) [40].

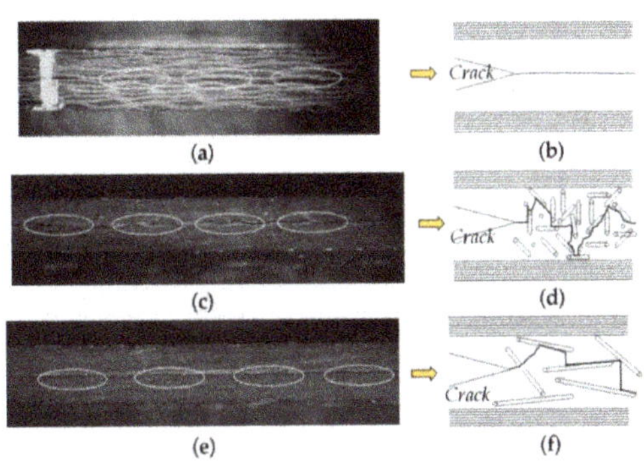

Figure 5. The crack propagation path in the mid-layer of laminates during Mode I DCB tests: (**a**,**b**) pure; (**c**,**d**) SCFs (7.8 mg/cm^2) and (**e**,**f**) LCFs (7.8 mg/cm^2) [41].

To improve the lightning strike resistance of composite materials, Liu et al. bonded chopped nickel-coat carbon fibers (NiCF) together with adhesive to make a NiCF veil, which was used as a multifunctional toughening layer for CFRP [42]. It is shown that the multifunctional NiCF veil can not only improve the interlayer fracture toughness of CFRP, which can increase the G_{IC} by 74.75% and G_{IIC} by 36.22%, but also improve the electrical conductivity in the thickness direction by 220.49%. Its toughening mechanisms include debonding and fracture of NiCF, as well as flaking and splitting of the nickel coating. The above study shows that the rupture and debonding of the coating absorbs additional energy to improve the interlaminar fracture toughness of the composite without losing the interfacial compatibility of the short fibers and the resin. Du et al. used CNT grafted short carbon fibers as interlaminar toughening materials for CF/EP composites, and the type I interlaminar fracture energy of CFRP could be increased by 125% when the density of CNT-SCF was 10 g/m^2 [43]. Of particular note, he investigated the crack damage sensing ability of toughened CFRP composites using a simple electrical response method using an experimental setupand the results showed that the change in resistance increased monotonically with cracking, which provides a new method for measuring crack extension.

Ma et al. produced five different structures of short glass fiber interlayer toughened GF/EP composites by the VARTM process, as shown in Figure 6 [44]. The experimental results showed that the interlaminar tearing performance of the short glass fiber toughened composites was significantly enhanced, and the toughening effect was most evident for structures B and C, and the shorter the glass fibers, the better the toughening result. This research indicates that in addition to fiber type, length, and content, the arrangement of short fibers between the layers should also be considered, which is an important factor affecting the mechanical properties of toughened composites.

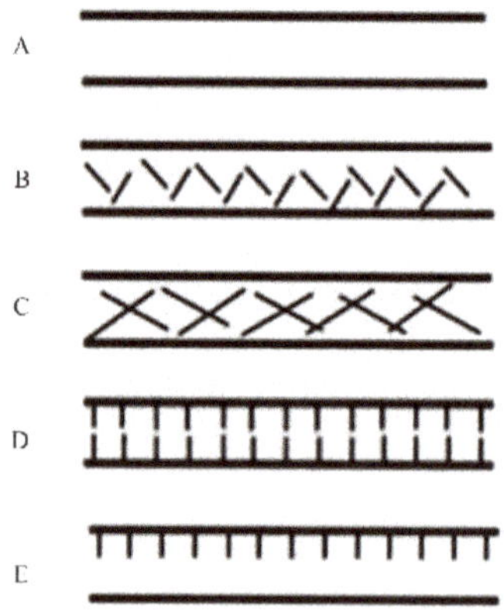

Figure 6. Five kinds of reinforcement structure (**A**): two layers of plain glass fiber fabric without short fibers; (**B**,**C**): short fibers are randomly distributed between the layers; (**D**): two layers of fabric containing short fibers; (**E**): one layer of fabric containing short fibers and one layer of plain glass fiber fabric) [44].

While environmental problems are a growing concern in today's society, natural fibers have the advantages of sustainability, biodegradability, and lower cost, which allows them to replace some synthetic fibers in the manufacture of composite materials [45–54]. It is worth establishing that short natural fibers are a potential interlayer toughening material. Jeong et al. covered short silk fiber within two resin films, placed them in the middle of CFRP prepreg for normal curing, and evaluated mode I and mode II fracture toughness by DCB and ENF tests [55]. The test results showed that the addition of silk was useful in improving the interlaminar fracture toughness of the laminate. The mode I interlaminar fracture toughness was 59% higher compared with the untoughened laminate, the main factors for the increase being the adhesion force on the silk fibers to the epoxy matrix and the relatively high tensile strength of the silk fibers; mode II interlaminar fracture toughness increased by 44%, mainly because the bridging and fracture of the silk fibers inhibited the crack expansion. If spun silk can be processed and appropriately used, it will not only improve the mechanical properties of the composite material but also avoid the waste of raw materials.

The feasibility of interlaminar toughening of CF/EP composites with short ramie fiber (SRF) was demonstrated by Mo et al. [56]. The SRF layer effectively hindered the crack expansion in the matrix and significantly improved the interlaminar fracture toughness of the composite laminates. The toughening mechanism of SRF can be analyzed from the type II interlaminar fracture microscopic morphology shown in Figure 7: firstly, the fracture surface becomes rougher, which increases the surface area of the fracture surface of the composite material and requires more energy to be consumed during crack propagation; secondly, the SRF bridging in the composite absorbs the energy of the crack tip, which effectively hinders the crack propagation in the matrix; furthermore, the SRF pulling and splitting from the matrix also needs to absorb a certain amount of energy. At the same time, the tensile and bending properties of the composites have been raised to a certain extent. They also treated SRF with surface coupling agents, which proved to be effective in improving the fracture toughness of the composites.

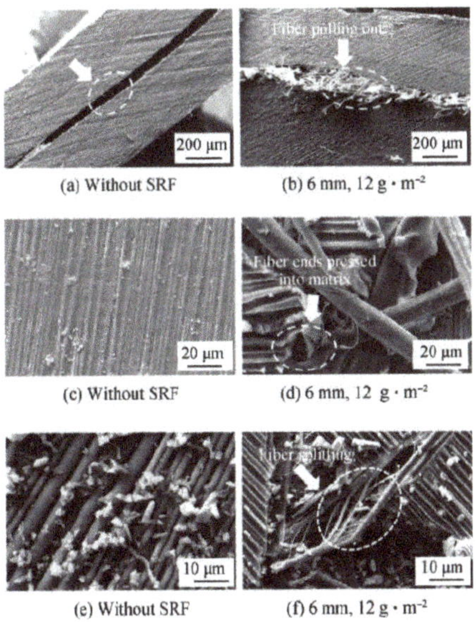

Figure 7. Mode II interlaminar fracture morphology of CF/EP composite laminate specimens without and with SRF [56].

Zheng et al. compared the toughening effect of short flax fiber, short carbon fiber, and short Kevlar fiber interlayer toughened CF/EP composites, and the test results showed the toughening effect of short flax fiber was between short carbon fiber and short Kevlar fiber [35]. The reason is that the multi-layer level damage of the fiber itself occurs during the crack extension, which absorbs additional energy and further improves the toughening effect. Li et al. investigated the impact of chopped flax yarns (CFY) on the mode I interlaminar fracture toughness of flax fiber/epoxy composites. The experimental results showed that the mode I fracture toughness of the composites increased with the addition of CFY [57]. The CFY layer hindered the crack expansion and led to more energy consumption in the delamination process, but excessive yarn length and content direct unstable crack extension, weakening the toughening effect. When the yarn length and content are moderate (10 mm, 20 g/m^2), the toughening effect is the best, and the G_{IC} is increased by 31%. Zhang et al. proposed a multiscale cellulose fiber interlayer consisting of chopped flax fibers (CFF) and cellulose nanofibers (CNF) to improve the interlayer fracture toughness of CF/EP composites [58]. The results showed that the addition of cellulose fiber interlayer could significantly improve the mode II interlaminar fracture toughness due to the synergistic effect of CFF and CNF, and the toughening mechanism can be seen in Figure 8, which includes fiber bridging, fiber pullout, fiber fracture, and the fibrillation of flax fiber. The existence of CNF can improve the interfacial compatibility between CFF and epoxy resin matrix, which is conducive to the fibrillation of flax fibers and hinders the propagation of interlaminar cracks. Obviously, flax fibers, due to their multi-layered structure, can absorb more energy, which is unique to toughening unlike other fibers mechanism. The price of flax fiber compared to aramid, carbon fiber, and others is lower, but there is a very good toughening effect, which is a cost-effective toughening material.

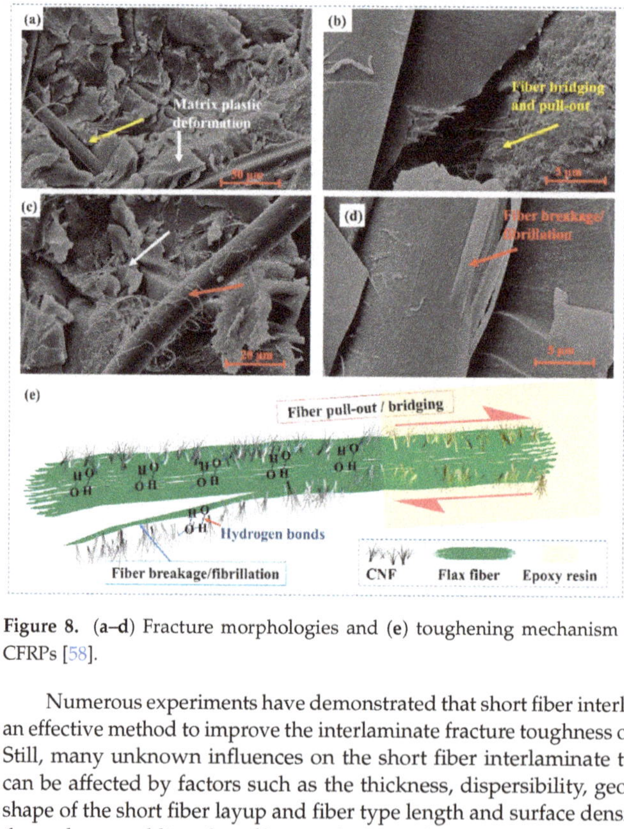

Figure 8. (**a–d**) Fracture morphologies and (**e**) toughening mechanism of CNF/FF-interleaved CFRPs [58].

Numerous experiments have demonstrated that short fiber interlaminate toughening is an effective method to improve the interlaminate fracture toughness of composite laminates. Still, many unknown influences on the short fiber interlaminate toughening technique can be affected by factors such as the thickness, dispersibility, geometry, and final cure shape of the short fiber layup and fiber type length and surface density. Instead of limiting themselves to adding short fibers to the layers by simple hand-spreading, researchers have developed methods such as making nonwovens, muslins, films, and blending in resins, all of which are designed to eliminate the effects of uneven distribution of short fibers between the layers [59]. If the effects of these factors can be balanced and allow for the control and manipulation of delamination damage, this will achieve the goal of preparing higher performance, more damage-resistant composites.

Natural fibers are also starting to be used in composite interlayer toughening because of their high-cost performance, but the natural fibers are not as good as the synthetic fibers, which limits their large-scale promotion. The hybrid short fiber interlayer toughening is a method that should be considered for interlayer toughening, which can reduce the cost and make up for the shortage of single-type short fiber toughening.

3.2. Short Fiber Modified Resin Toughening

Another way to improve the toughness of FRP is to modify the resin matrix, and it has been shown that nanoparticles are very good toughening agents, such as nano titanium dioxide [60], nano clay [61], multi-walled carbon nanotubes [62], etc. Nanoparticles have strong chemical activity and can fully bond with epoxy resin to form an ideal interface to blunt cracks and hinder crack expansion from achieving toughening while improving modulus and heat resistance. However, nanomaterials are easy to agglomerate in the resin matrix, unfavorable to operation and processing, and nanoparticles are usually relatively high in cost, which limits their application to some extent [63].

Short-fiber toughened resins offer a viable, low-cost method for resin modification. Chawla et al. prepared unidirectional glass fiber/epoxy laminates from two different resins, one pure and the other containing 2% volume fraction short glass fiber reinforcement, and performed DCB tests [64]. The experimental results showed that fiber bridging was observed in the laminates with the presence of short fibers and that the toughened laminates increased G_{IC} by about 19.6% compared to the untoughened laminates. Imagawa et al. added 0.3 wt% weight content of micro-glass fibers to thermoplastic epoxy resin and thermoset epoxy resin to make CFRP and conducted ENF tests [65]. The results showed that the addition of glass fibers increased fiber bridging to prevent crack extension and improved the G_{IIC} of composite regardless of the resin. Fiber bridging was observed in all of these studies, and the toughness of the composites was improved, demonstrating the feasibility of short fiber-modified epoxy resin toughening.

Ravindran et al. prepared CFRP composite laminates by adding CNF, CSF, and CNF/SCF fillers to an epoxy resin matrix, and the experimental results showed that the addition of CNF or/and SCF fillers improved the delamination toughness, impact damage tolerance, and CAI strength of CFRP laminates without reducing their in-plane bending and interlaminar shear strength properties compared to CFRP laminates made with the unmodified epoxy resin matrix [66]. In terms of mode II fracture toughness, the multiscale short fiber toughening effect is better than the sum of CNF and CSF alone. Future research on multiscale toughening needs to focus on the synergistic effect of multiscale toughening, which is similar to hybrid fiber toughening. Dasari et al. investigated the role of short carbon fibers (SCFs) as secondary reinforcement in glass fiber/epoxy composites. They evaluated the effect of SCF concentration on the mechanical properties of GFRP by tensile, flexural, and interlaminar fracture toughness tests [67]. The results showed that when the SCF content was 0.1 wt%, the tensile and flexural strengths of the composites were increased by 29.02% and 16.08%, respectively. In mode I and mode II interlaminar fracture toughness tests, the samples with 0.1 wt% exhibited the highest strain energy release rate, with an increase of 13.49% and 20.45% in G_{IC} and G_{IIC}, respectively. They prepared glass fiber composites (SGF-GE) by modifying epoxy resin with short glass fibers of 2–5 mm at different concentrations, and the results of three-point bending tests showed that the short fibers as secondary reinforcement could improve the flexural strength and flexural modulus of the composites [68]. Yu et al. mixed short basalt fibers (SBF) in unsaturated polyester (UP) resin and then coated them in the middle of basalt fabric curing molding to make BF/UP composites. The results showed that SBF had an excellent toughening effect on BF/UP, including mode II interlaminar fracture toughness, tensile strength, elongation at break, bending strength, and energy absorption were significantly enhanced [69]. From these studies, it can be seen that the addition of very small amounts of short fibers not only improves the interlaminar fracture toughness of the composite but also the in-plane properties to a greater or lesser extent. This is further evidence of the feasibility and superiority of short fiber-modified resin toughening.

The use of recycled fibers to manufacture useful engineered composite components is a current research trend and requirement to protect the environment and save costs. Cholake et al. modified epoxy resin by recycled short-milled carbon fiber, 5 wt% and 10 wt% of SMCF could increase the G_{IC} of epoxy matrix by 261% and 692%, respectively. The G_{IC} of ground laminate using modified epoxy resin was increased by 50% and 64%, respectively, which proved to be a low-cost and effective method to toughen CFRP [70]. Saravana Kumar et al. investigated the effect of milled glass fibers filler on the interlaminar fracture toughness of GF/EP composites. They showed that the fracture toughness of mode I and mode II could be maximized by adding 5 wt% milled glass fibers by 102% and 175%, respectively, while the bending strength was also enhanced [71]. The interlaminar fracture toughness of composites can be effectively improved by adding recycled short fiber fillers. Still, fiber recycling is a very big topic, and it is of great concern to ensure that the recycled fibers have less property damage that may affect its toughening effect.

Graphene oxide (GO) is commonly used to treat fibers to improve the interfacial properties of composites [72]. Dang et al. used GO-modified short glass fibers to improve the ILSS of GF/EP composites. The experimental results showed that GO significantly increased the surface roughness of SGFs and improved the interfacial adhesion between SGFs and EP, while GO-SGFs could improve the ILSS of GF/EP composites by 18.4% [73]. Nie et al. added short carbon fibers treated with ethanol and concentrated nitric acid to the epoxy matrix for the ILSS improvement of CF/EP composites, and the preparation process of the composites is shown in Figure 9a [74]. The ILSS testing showed that adding oxidized short carbon fibers equivalent to a very low epoxy resin content could effectively improve the ILSS. They also prepared GO-SCF by a simple aqueous solution method, as shown in Figure 9b, and studied the effect of GO-SCF toughening resin on the ILSS of CF/EP composites, which showed that the ILSS of composites containing GO-SCF was higher than that of SCF at the same SCF content because GO increased the bonding force between SCF and epoxy resin, and GO-SCF could increase the ILSS by up to 14.7% [75]. The authors concluded that this method has practical application in improving the interlaminar shear properties of CF/EP composites due to the low cost and easy processability of SCF.

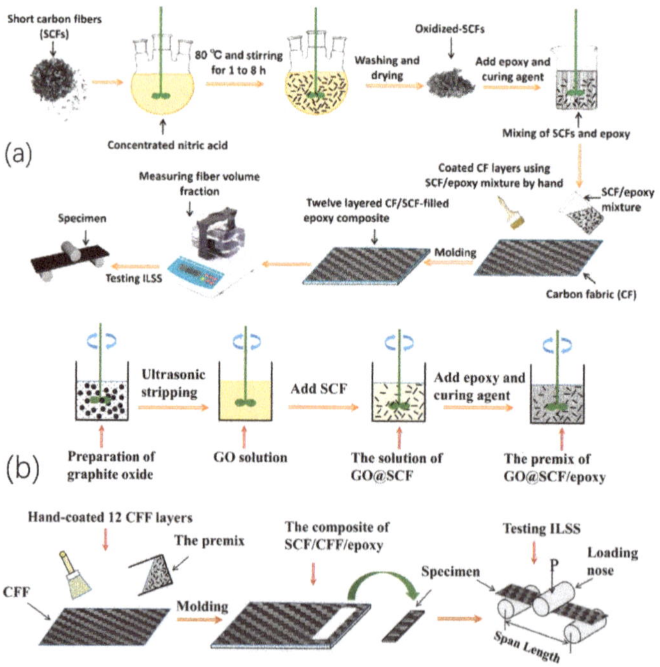

Figure 9. Schematic diagrams of preparation and mechanical testing of (**a**) oxidized SCF/CFF/EP composites; (**b**) GO@SCF/CFF/EP composites [74,75].

As stated in the literature [71], until the problem of agglomeration of short fibers in the resin is solved, only a limited amount of short fibers can be added to the resin, and this method cannot be used on a large scale yet.

4. Conclusions

This paper reviews the enhancement of composite interlaminar fracture toughness and other mechanical properties by incorporating short fibers in the composite preparation. In summary, short fiber interlaminar toughening techniques and short fiber modified resins can effectively improve the interlaminar fracture toughness of composites. Short fiber interlaminate toughening is specifically designed to improve the interlaminate properties of laminates. It is highly targeted to improve the interlaminate properties of laminates; short fiber modified resin toughening can improve the interlaminate properties and improve the in-plane properties. By adding a small number of short fibers, improving the interlaminar fracture toughness of composites without damaging other properties can be achieved. For different toughening methods, the toughening mechanism of short fibers is discussed. The toughening mechanisms are fiber bridging to hinder crack expansion, fiber debonding, and fracture from consuming additional energy and specific to different fibers and methods.

5. Prospect

It is not difficult to find that the short fibers are randomly distributed in the composite during the experiment, which brings many uncertainties to the short fiber toughening. If the short fibers can be oriented and arranged to eliminate the uncertainties of random distribution, it may make short fiber toughening stable and applicable.

Moreover, the issue of the source of staple fibers is also an issue worth considering; with the development of a low-carbon economy, natural fibers and recycled fibers have gradually become the hotspot of composite materials. Reasonable use of these fibers as toughening materials can reduce costs on the one hand and reduce environmental damage and waste of resources on the other hand. However, these fibers may have deficiencies in performance compared to synthetic fibers, so hybridization with synthetic fibers should be considered.

Hybrid fiber composites can make up for the shortcomings of single fiber composites, especially the hybrid of synthetic fibers and natural fibers, which is the trend of composite material development [76,77]. There are few research reports on hybrid short fiber toughened composites. Further research is needed to verify the feasibility and effectiveness of hybrid short fiber toughening and explore the mechanism of hybrid toughening.

Author Contributions: Conceptualization, L.Z.; supervision of the study, L.Z.; Funding acquisition, L.Z.; investigation, Y.W.; data curation, Y.W.; writing—original draft preparation, Y.W.; writing—review and editing, L.Z. and Z.W. All authors have read and agreed to the published version of the manuscript.

Funding: This research was funded by the Zhejiang Provincial Natural Science Foundation of China under Grant No LGG21E050025, the Fundamental Research Funds of Zhejiang Sci-Tech University (Project Number: 20202113-Y), as well as the Fundamental Research Funds of Shaoxing Keqiao Research Institute of Zhejiang Sci-Tech University (Project Number: KYY2021001G).

Institutional Review Board Statement: Not applicable.

Informed Consent Statement: Not applicable.

Data Availability Statement: Not applicable.

Conflicts of Interest: The authors declare no conflict of interest.

References

1. Aamir, M.; Tolouei-Rad, M.; Giasin, K.; Nosrati, A. Recent Advances in Drilling of Carbon Fiber-Reinforced Polymers for Aerospace Applications: A Review. *Int. J. Adv. Manuf. Technol.* **2019**, *105*, 2289–2308. [CrossRef]

2. Alshahrani, H.; Ahmed, A. Enhancing Impact Energy Absorption, Flexural and Crash Performance Properties of Automotive Composite Laminates by Adjusting the Stacking Sequences Layup. *Polymers* **2021**, *13*, 3404. [CrossRef] [PubMed]
3. Zangenberg, J.; Brondsted, P.; Koefoed, M. Design of a Fibrous Composite Preform for Wind Turbine Rotor Blades. *Mater. Des.* **2014**, *56*, 635–641. [CrossRef]
4. Ullah, H.; Harland, A.R.; Silberschmidt, V.V. Dynamic Bending Behaviour of Woven Composites for Sports Products: Experiments and Damage Analysis. *Mater. Des.* **2015**, *88*, 149–156. [CrossRef]
5. Ouyang, T.; Sun, W.; Bao, R.; Tan, R. Effects of Matrix Cracks on Delamination of Composite Laminates Subjected to Low-Velocity Impact. *Compos. Struct.* **2021**, *262*, 113354. [CrossRef]
6. Wang, J. Research on Properties of Three-dimensional UHMWPE Fiber Composite with Difference Structures. *Adv. Text. Technol.* **2021**, *29*, 12–17. [CrossRef]
7. Tao, N.; Gao, X. Research on Interface Modification and Bend of 3D Orthogonal Woven Glass Fiber Composites. *Adv. Text. Technol.* **2019**, *27*, 12–18. [CrossRef]
8. Shen, X.; Liu, X.; Tian, W.; Zhu, C. Study on Mechanical Properties of Orthogonal and Quasi-Orthogonal 3D Woven Composite with Combined Structure. *Adv. Text. Technol.* **2019**, *27*, 6–11. [CrossRef]
9. Hu, Q.; Memon, H.; Qiu, Y.; Liu, W.; Wei, Y. A Comprehensive Study on the Mechanical Properties of Different 3D Woven Carbon Fiber-Epoxy Composites. *Materials* **2020**, *13*, 2765. [CrossRef]
10. Liao, B.; Zhou, J.; Zheng, J.; Tao, R.; Xi, L.; Zhao, T.; Li, Y.; Fang, D. Effect of Z-Pin Inclined Angle on the Impact Damage Suppression Effectiveness for Cross-Ply Composite Laminates. *Compos. Pt. A-Appl. Sci. Manuf.* **2021**, *142*, 106264. [CrossRef]
11. Drake, D.A.; Sullivan, R.W.; Lovejoy, A.E.; Clay, S.B.; Jegley, D.C. Influence of Stitching on the Out-of-Plane Behavior of Composite Materials—A Mechanistic Review. *J. Compos Mater.* **2021**, *55*, 3307–3321. [CrossRef]
12. Zhu, L.; Bin Rahman, M.; Wang, Z. Effect of Structural Differences on the Mechanical Properties of 3D Integrated Woven Spacer Sandwich Composites. *Materials* **2021**, *14*, 4284. [CrossRef] [PubMed]
13. Bin Rahman, M.; Zhu, L. Low-Velocity Impact Response on Glass Fiber Reinforced 3D Integrated Woven Spacer Sandwich Composites. *Materials* **2022**, *15*, 2311. [CrossRef] [PubMed]
14. Ning, N.; Wang, M.; Zhou, G.; Qiu, Y.; Wei, Y. Effect of Polymer Nanoparticle Morphology on Fracture Toughness Enhancement of Carbon Fiber Reinforced Epoxy Composites. *Compos. Pt. B-Eng.* **2022**, *234*, 109749. [CrossRef]
15. Saghafi, H.; Palazzetti, R.; Heidary, H.; Brugo, T.M.; Zucchelli, A.; Minak, G. Toughening Behavior of Carbon/Epoxy Laminates Interleaved by PSF/PVDF Composite Nanofibers. *Appl. Sci.-Basel* **2020**, *10*, 5618. [CrossRef]
16. Hu, Q.; Memon, H.; Qiu, Y.; Wei, Y. The Failure Mechanism of Composite Stiffener Components Reinforced with 3D Woven Fabrics. *Materials* **2019**, *12*, 2221. [CrossRef]
17. Shrivastava, R.; Singh, K.K. Interlaminar Fracture Toughness Characterization of Laminated Composites: A Review. *Polym. Rev.* **2020**, *60*, 542–593. [CrossRef]
18. Sharma, P.; Mali, H.S.; Dixit, A. Mechanical Behavior and Fracture Toughness Characterization of High Strength Fiber Reinforced Polymer Textile Composites. *Iran. Polym. J.* **2021**, *30*, 193–233. [CrossRef]
19. Irwin, G.R. Structural Aspects of Brittle Fracture. *Appl. Mater. Res.* **1964**, *3*, 65–81.
20. Griffith, A.A. The Phenomena of Rupture and Flow in Solids. *Philos. Trans. R. Soc. A Math. Phys. Eng. Sci.* **1920**, *A221*, 163–198. [CrossRef]
21. Dikshit, V.; Bhudolia, S.; Joshi, S. Multiscale Polymer Composites: A Review of the Interlaminar Fracture Toughness Improvement. *Fibers* **2017**, *5*, 38. [CrossRef]
22. Standard Test Method for Mode I Interlaminar Fracture Toughness of Unidirectional Fiber-Reinforced Polymer Matrix Composites. Available online: https://www.astm.org/d5528_d5528m-21.html (accessed on 3 May 2022).
23. Standard Test Method for Determination of the Mode II Interlaminar Fracture Toughness of Unidirectional Fiber-Reinforced Polymer Matrix Composites. Available online: https://www.astm.org/d7905_d7905m-19e01.html (accessed on 3 May 2022).
24. Standard Test Method for Short-Beam Strength of Polymer Matrix Composite Materials and Their Laminates. Available online: https://www.astm.org/d2344_d2344m-16.html (accessed on 3 May 2022).
25. Standard Test Method for Measuring the Damage Resistance of a Fiber-Reinforced Polymer Matrix Composite to a Drop-Weight Impact Event. Available online: https://www.astm.org/d7136_d7136m-20.html (accessed on 3 May 2022).
26. Standard Test Method for Compressive Residual Strength Properties of Damaged Polymer Matrix Composite Plates. Available online: https://www.astm.org/d7137_d7137m-17.html (accessed on 3 May 2022).
27. Sohn, M.-S.; Hu, X.-Z. Comparative Study of Dynamic and Static Delamination Behaviour of Carbon Fibre/Epoxy Composite Laminates. *Composites* **1995**, *26*, 849–858. [CrossRef]
28. Park, B.Y.; Kim, S.C. A Study of the Interlaminar Fracture Toughness of a Carbon-Fiber/Epoxy Composite Containing Surface-Modified Short Kevlar Fibers. *Compos. Sci. Technol.* **1998**, *58*, 1599–1606. [CrossRef]
29. Park, B.Y.; Kim, S.C.; Jung, B. Interlaminar Fracture Toughness of Carbon Fiber/Epoxy Composites Using Short Kevlar Fiber and/or Nylon-6 Powder Reinforcement. *Polym. Adv. Technol.* **1997**, *8*, 371–377. [CrossRef]
30. Yadav, S.N.; Kumar, V.; Verma, S.K. Fracture Toughness Behaviour of Carbon Fibre Epoxy Composite with Kevlar Reinforced Interleave. *Mater. Sci. Eng. B* **2006**, *132*, 108–112. [CrossRef]
31. Yasaee, M.; Bond, I.P.; Trask, R.S.; Greenhalgh, E.S. Mode I Interfacial Toughening through Discontinuous Interleaves for Damage Suppression and Control. *Compos. Part A Appl. Sci. Manuf.* **2012**, *43*, 198–207. [CrossRef]

32. Yasaee, M.; Bond, I.P.; Trask, R.S.; Greenhalgh, E.S. Mode II Interfacial Toughening through Discontinuous Interleaves for Damage Suppression and Control. *Compos. Part A Appl. Sci. Manuf.* **2012**, *43*, 121–128. [CrossRef]
33. Liu, W.; Li, W.; Ma, C.; Wang, S. Study on Properties of Interlayer Short Fiber Reinforced Carbon Fiber/Epoxy Composite Laminates. *J. Phys. Conf. Ser.* **2019**, *1215*, 012040. [CrossRef]
34. Qi, G.; Zhang, B.; Du, S.; Yu, Y. Estimation of Aramid Fiber/Epoxy Interfacial Properties by Fiber Bundle Tests and Multiscale Modeling Considering the Fiber Skin/Core Structure. *Compos. Struct.* **2017**, *167*, 1–10. [CrossRef]
35. Zheng, H.; Li, Y.; Tu, H. Research on Interlayer Properties of Short Fiber Intercalated Carbon Fiber/Epoxy Composites. *Acta Mater. Compos. Sin.* **2021**, 1–13. [CrossRef]
36. Wang, B.; Dong, N.; Ding, G. Mode I Interlaminar Fracture Toughness of CFRP Laminates Reinforced with Short Aramid Fibers. *J. Adhes. Sci. Technol.* **2020**, *34*, 2522–2536. [CrossRef]
37. Wang, B.; Ding, G.; Wang, G.; Kang, S. Effects of Resin Pre-Coating on Interfacial Bond Strength and Toughness of Laminar CFRP with and without Short Aramid Fibre Toughening. *J. Compos Mater.* **2020**, *54*, 3883–3893. [CrossRef]
38. Hu, Y.; Liu, W.; Shi, Y. Low-Velocity Impact Damage Research on CFRPs with Kevlar-Fiber Toughening. *Compos. Struct.* **2019**, *216*, 127–141. [CrossRef]
39. Yuan, B.; Ye, M.; Hu, Y.; Cheng, F.; Hu, X. Flexure and Flexure-after-Impact Properties of Carbon Fibre Composites Interleaved with Ultra-Thin Non-Woven Aramid Fibre Veils. *Compos. Pt. A-Appl. Sci. Manuf.* **2020**, *131*, 105813. [CrossRef]
40. Yuan, B.; Tan, B.; Hu, Y.; Shaw, J.; Hu, X. Improving Impact Resistance and Residual Compressive Strength of Carbon Fibre Composites Using Un-Bonded Non-Woven Short Aramid Fibre Veil. *Compos. Pt. A-Appl. Sci. Manuf.* **2019**, *121*, 439–448. [CrossRef]
41. Xu, F.; Yang, B.; Feng, L.; Huang, D.; Xia, M. Improved Interlaminar Fracture Toughness and Electrical Conductivity of CFRPs with Non-Woven Carbon Tissue Interleaves Composed of Fibers with Different Lengths. *Polymers* **2020**, *12*, 803. [CrossRef]
42. Liu, H.; Guo, Y.; Zhou, Y.; Wan, G.; Chen, Z.; Jia, Y. Multifunctional Nickel-Coated Carbon Fiber Veil for Improving Both Fracture Toughness and Electrical Performance of Carbon Fiber/Epoxy Composite Laminates. *Polym. Compos.* **2021**, *42*, 5335–5347. [CrossRef]
43. Du, X.; Zhou, H.; Liu, H.; Zhou, H.; Mai, Y. Surface Functionalized CNTs Grafted Short Carbon Fibers in CFRP Composites as Interleaving Toughener and Damage Sensor. In Proceedings of the 21st International Conference on Composite Materials, Xi'an, China, 20–25 August 2017.
44. Ma, P.; Nie, X. Interface Improvement of Multi Axial Warp-Knitted Layer Composite with Short Glass Fiber. *Fiber. Polym.* **2017**, *18*, 1413–1419. [CrossRef]
45. Alkbir, M.F.M.; Sapuan, S.M.; Nuraini, A.A.; Ishak, M.R. Fibre Properties and Crashworthiness Parameters of Natural Fibre-Reinforced Composite Structure: A Literature Review. *Compos. Struct.* **2016**, *148*, 59–73. [CrossRef]
46. Huang, X.; Yu, H.; Li, F.; Wang, F. Study on Preparation Process of Unsaturated Polyester Resin Composite Reninforced by Banana Fiber. *Adv. Text. Technol.* **2019**, *27*, 5–9. [CrossRef]
47. Qian, W.; Xu, P.; Wang, L. Review on Polyester Fiber Recycling and Progress of Its Environmental Impact Assessment. *Adv. Text. Technol.* **2021**, *29*, 22–26. [CrossRef]
48. Zhao, X.; Shao, X.; Jiang, G.; Pang, J. Application of Modefied Lignin in Carbon Fiber Reinforced Composite Materials. *Packag. Eng.* **2022**, *43*, 103–110.
49. Memon, H.; Wei, Y.; Zhu, C. Recyclable and Reformable Epoxy Resins Based on Dynamic Covalent Bonds Present, Past, and Future. *Polym. Test* **2022**, *105*, 107420. [CrossRef]
50. Memon, H.; Wei, Y.; Zhang, L.; Jiang, Q.; Liu, W. An Imine-Containing Epoxy Vitrimer with Versatile Recyclability and Its Application in Fully Recyclable Carbon Fiber Reinforced Composites. *Compos. Sci. Technol.* **2020**, *199*, 108314. [CrossRef]
51. Yue, D. Current Situation and Development Prospects of the Resource Utilization of Coir Fiber. *Packag. Eng.* **2020**, *41*, 37–43. [CrossRef]
52. Wang, H.; Memon, H.; AM Hassan, E.; Miah, M.; Ali, M. Effect of Jute Fiber Modification on Mechanical Properties of Jute Fiber Composite. *Materials* **2019**, *12*, 1226. [CrossRef]
53. Wang, H.; Hassan, E.; Memon, H.; Elagib, T.; Abad AllaIdris, F. Characterization of Natural Composites Fabricated from Abutilon-Fiber-Reinforced Poly (Lactic Acid). *Processes* **2019**, *7*, 583. [CrossRef]
54. Wang, H.; Memon, H.; Hassan, E.A.; Elagib, T.H.; Hassan, F.E.A.; Yu, M. Rheological and Dynamic Mechanical Properties of Abutilon Natural Straw and Polylactic Acid Biocomposites. *Int. J. Polym. Sci.* **2019**, *2019*, 8732520. [CrossRef]
55. Jeong, J.-S.; Cheong, S.-K. Interlaminar Fracture Toughness of CFRP Laminates with Silk Fibers Interleave. *J. Mech. Sci. Technol.* **2013**, *27*, 3651–3656. [CrossRef]
56. Mo, Z.; Hu, C.; Huo, Y.; Ye, D. Interlayer-Toughening Carbon Fiber/Epoxy Composites with Short Ramie Fiber. *Acta Mater. Compos. Sin.* **2017**, *34*, 1237–1244. [CrossRef]
57. Yan, L.; Di, W.; Hao, M. Improving Interlaminar Fracture Toughness of Flax Fiber/Epoxy Composites with Chopped Flax Yarn Interleaving. *Sci. China-Technol. Sci.* **2015**, *58*, 1745–1752. [CrossRef]
58. Zhang, Z.; Fu, K.; Li, Y. Improved Interlaminar Fracture Toughness of Carbon Fiber/Epoxy Composites with a Multiscale Cellulose Fiber Interlayer. *Compos. Commun.* **2021**, *27*, 100898. [CrossRef]
59. Hu, C.; Zhou, Y.; Zhou, X.; Zhao, X.; Hong, J. Research Progress of Electrospun Oriented Nanofiber Yarns. *Adv. Text. Technol.* **2021**, *29*, 27–33. [CrossRef]

60. Prasad, V.; Sekar, K.; Varghese, S.; Joseph, M.A. Enhancing Mode I and Mode II Interlaminar Fracture Toughness of Flax Fibre Reinforced Epoxy Composites with Nano TiO_2. *Compos. Pt. A-Appl. Sci. Manuf.* **2019**, *124*, 105505. [CrossRef]
61. Gabr, M.H.; Okumura, W.; Ueda, H.; Kuriyama, W.; Uzawa, K.; Kimpara, I. Mechanical and Thermal Properties of Carbon Fiber/Polypropylene Composite Filled with Nano-Clay. *Compos. Pt. B-Eng.* **2015**, *69*, 94–100. [CrossRef]
62. Borowski, E.; Soliman, E.; Kandil, U.F.; Taha, M.R. Interlaminar Fracture Toughness of CFRP Laminates Incorporating Multi-Walled Carbon Nanotubes. *Polymers* **2015**, *7*, 1020–1045. [CrossRef]
63. Vallack, N.; Sampson, W.W. Materials Systems for Interleave Toughening in Polymer Composites. *J. Mater. Sci.* **2022**, *57*, 6129–6156. [CrossRef]
64. Chawla, K.; Ray-Chaudhuri, S.; Kitey, R. Interlaminar Fracture Toughness of Short Fibre Reinforced GFRP Laminates. *Procedia Struct. Integr.* **2019**, *14*, 571–576. [CrossRef]
65. Imagawa, S.; Nishida, H.; Okubo, K.; Fujii, T. Improvement of Mode-II Interlaminar Fracture Toughness of Carbon Textile Composites with Modified Matrix of Thermoplastic and Thermoset Epoxy-Addition of Glass Fibers. In *IOP Conference Series: Materials Science and Engineering*; IOP Publishing: Bristol, UK, 2018; Volume 406, p. 012045. [CrossRef]
66. Ravindran, A.R.; Ladani, R.B.; Kinloch, A.J.; Wang, C.-H.; Mouritz, A.P. Improving the Delamination Resistance and Impact Damage Tolerance of Carbon Fibre-Epoxy Composites Using Multiscale Fibre Toughening. *Compos. Pt. A-Appl. Sci. Manuf.* **2021**, *150*, 106624. [CrossRef]
67. Dasari, S.; Lohani, S.; Prusty, R.K. An Assessment of Mechanical Behavior of Glass Fiber/Epoxy Composites with Secondary Short Carbon Fiber Reinforcements. *J. Appl. Polym. Sci.* **2022**, *139*, e51841. [CrossRef]
68. Dasari, S.; Lohani, S.; Sumit Dash, S.; Omprakash Fulmali, A.; Kumar Prusty, R.; Chandra Ray, B. A Novel Study of Flexural Behavior of Short Glass Fibers as Secondary Reinforcements in GFRP Composite. *Mater. Today: Proc.* **2021**, *47*, 3370–3374. [CrossRef]
69. Yu, K.; He, X.; Li, J.; Liang, C. Toughening Effect of Basalt Fiber on Unsaturated Polyester Resin Composites. *J. Jilin Univ.* **2021**, 1–7. [CrossRef]
70. Cholake, S.T.; Moran, G.; Joe, B.; Bai, Y.; Raman, R.K.S.; Zhao, X.L.; Rizkalla, S.; Bandyopadhyay, S. Improved Mode I Fracture Resistance of CFRP Composites by Reinforcing Epoxy Matrix with Recycled Short Milled Carbon Fibre. *Constr. Build. Mater.* **2016**, *111*, 399–407. [CrossRef]
71. Saravanakumar, K.; Arumugam, V.; Souhith, R.; Santulli, C. Influence of Milled Glass Fiber Fillers on Mode I & Mode II Interlaminar Fracture Toughness of Epoxy Resin for Fabrication of Glass/Epoxy Composites. *Fibers* **2020**, *8*, 36. [CrossRef]
72. Chen, J.; Zhang, H.; Gao, X. Compressive Properties of Graphene Oxide Modified Carbon/Glass Hybrid Reinforced Composite. *Adv. Text. Technol.* **2022**, *30*, 75–84. [CrossRef]
73. Dang, C.-Y.; Tang, B.-L.; Zeng, X.-L.; Xu, J.; Feng, M.-J.; Jiang, Y.; Shen, X.-J. Improved Interlaminar Shear Strength of Glass Fiber/Epoxy Composites by Graphene Oxide Modified Short Glass Fiber. *Mater. Res. Express* **2019**, *6*, 085324. [CrossRef]
74. Nie, H.-J.; Shen, X.-J.; Tang, B.-L.; Dang, C.-Y.; Yang, S.; Fu, S.-Y. Effectively Enhanced Interlaminar Shear Strength of Carbon Fiber Fabric/Epoxy Composites by Oxidized Short Carbon Fibers at an Extremely Low Content. *Compos. Sci. Technol.* **2019**, *183*, 107803. [CrossRef]
75. Nie, H.-J.; Xu, Z.; Tang, B.-L.; Dang, C.-Y.; Yang, Y.-R.; Zeng, X.-L.; Lin, B.-C.; Shen, X.-J. The Effect of Graphene Oxide Modified Short Carbon Fiber on the Interlaminar Shear Strength of Carbon Fiber Fabric/Epoxy Composites. *J. Mater. Sci.* **2021**, *56*, 488–496. [CrossRef]
76. Lin, J.; Qin, X. Design and Performance of Hybrid Fiber Composite for Packaging Box. *Packag. Eng.* **2020**, *41*, 103–108.
77. Li, W.; Wang, J.; Gai, Y.; Deng, C.; Li, C. Investigation on Tensile Properties of Three-Dimensional Orthogonal Basalt/Carbon Fiber Hybrid Composite. *Adv. Text. Technol.* **2019**, *27*, 1–5. [CrossRef]

sustainability MDPI

Systematic Review

Slow Fashion Consumer Behavior: A Literature Review

Mariana Domingos [1], Vera Teixeira Vale [2,*] and Silvia Faria [3]

1 DEGEIT—Department of Economics, Management, Industrial Engineering and Tourism, University of Aveiro, 3810-193 Aveiro, Portugal; marianadomingos@ua.pt
2 Research Unit on Governance, Competitiveness and Public Policies (GOVCOPP), CEOS.PP, DEGEIT-Department of Economics, Management, Industrial Engineering and Tourism, University of Aveiro, 3810-193 Aveiro, Portugal
3 Research on Economics, Management and Information Technologies (REMIT), Universidade Portucalense, 4200-072 Porto, Portugal; sfaria@upt.pt
* Correspondence: v.c.vale@ua.pt

Abstract: This study aims to understand the dimensions associated with the behavior of the Slow Fashion consumer, their values, attitudes, and motivations, as well as to know in depth the literature related to Slow Fashion. The present article is a literature review related to the concept of "Slow Fashion", and follows a qualitative methodology based on an in-depth literature review of 25 papers from the Scopus and Web of Science research databases. For this literature review, a content analysis was initially performed through a bibliometric analysis. Then, a mind map was developed where the five major dimensions related to Slow Fashion consumer behavior were identified: "ethical values", "sustainable consumption", "consumer motivations", "consumer attitudes", and "sustainability awareness". We than related the mind map with the main conclusions of the literature review. The main limitation of this research results from the low number of published papers approaching the research concept. We were, however, able to identify the main concepts associated with the movement of Slow Fashion, thereby contributing to the available information on the variables that impact consumers' purchasing decisions.

Keywords: slow fashion; slow fashion consumer; consumer behavior; sustainability

Citation: Domingos, M.; Vale, V.T.; Faria, S. Slow Fashion Consumer Behavior: A Literature Review. *Sustainability* **2022**, *14*, 2860. https://doi.org/10.3390/su14052860

Academic Editors: Hafeezullah Memon, Xiaoke Jin, Tian Wei, Chengyan Zhu and Stefan Hoffmann

Received: 15 December 2021
Accepted: 21 February 2022
Published: 1 March 2022

1. Introduction

The social and ecological impact of the fashion industry has been the subject of much research and considerable documentation in recent years. It is recognized as one of the most polluting industries at all stages of its life cycle [1–3]. At an early stage and until approximately twenty years ago, two collections were produced per year (Spring/Summer and Autumn/Winter); currently, trend cycles have sped up and between 50 and 100 mini-collections are placed on the market per year [2]. An example is the Inditex Group, especially the Zara brand, which changes its in-store collection every fifteen days [4].

Nowadays, some companies are increasingly adopting green initiatives such as cultural innovation and social responsibility as a part of their business philosophy and core values. This is mainly due to three reasons: being socially and environmental responsible (one of the main concerns of all markets), gaining competitive advantages, and following consumer trends [5].

In 2017, Europeans bought 6.4 million tons of new clothes; between 1996 and 2012, studies estimate that each person increased by 40% the amount of new clothes they purchased [2], increasing the ecological negative impact of this industry. Actual well-informed clients are concerned and focused on the long-term well-being of people and the planet itself, and thus a new movement has appeared called Slow Fashion [6]. It represents the need to adopt sustainable performance and a change in core values in the fashion industry [1,6]; the challenge is to focus on more durable products and traditional production techniques or design concepts that have no season [6–8], emphasizing quality in order to achieve sustainability [9,10]. In other words, the movement encourages brands to embrace a quality-based rather than a time-based business philosophy based around slower production, ethical attitudes, and well-made and long-lasting products.

There are data on the values that drive consumers to invest in Slow Fashion, namely increased awareness of sustainability, the key explanation for the acceptance of this movement. Concern for preserving the environment and guaranteeing the well-being of individuals is leading to a consequent change in consumers' value perceptions and purchasing behavior. Slow Fashion is seeing an increasing support over the last several years as consumers begin to demand higher standards in sustainability and corporate ethics [7,9,10].

Despite several studies proving that social and ecological awareness influence purchasing decisions, it has been discovered that other variables such as style are often related to fashion buying decisions [1,5,11–13]. This leads us to the following question: in which way and how do consumers engage in this movement by adopting more responsible consumption habits and demands?

A study developed in Finland in 2009 by Niinimaki [13] on eco-fashion consumption and consumer buying decisions concluded that during a purchase journey, consumers think about aspects related to ethics and the environmental impact of the products. However, at the moment of buying these thoughts tend to decrease. This can be related to individual and personality traits and consequent ethical commitment; it seems that the greater consumer ethical commitment is, the higher the importance of sustainability and the lower the attention towards physical appearance [13]. This leads to more sustainable consumption; in other words, consumers with higher ethical commitment and stronger values place more importance on their ideals than on aspects related to their own identity and/or beauty [13].

The main reasons that seem to lead consumers to engage in sustainable consumption are related to the capacity to find alternatives that minimize the damage caused to the planet without giving up on style, that is, "sustainability as a facilitator of style and sustainable fashion as a source of pleasure and well-being" [12] (p. 12). This complements Szmigin and Carrigan's study [14] that associates sustainable consumption to feelings of personal growth, well-being, and experiential pleasure.

Previous studies on Slow Fashion consumer behaviour have addressed specific issues such as the sales journey and the fashion industry's evolution over time. The consumer and consumption habits are addressed as contexts and often analysed as secondary data. Furthermore, the already extant literature is dispersed due to lack of systematization, and there are no studies that allow for tracing consumer profiles in order to understand how they behave. Thus, in order to be able to profile them in the future, it is necessary to understand in depth the related literature and the dimensions associated with such consumers.

The present study has a two primary aims: (1) to permit understanding of the main dimensions associated with Slow Fashion consumer behavior (e.g., values, attitudes, and motivations); and (2) to better understand in depth the main research pertaining to Slow Fashion.

2. Materials and Methods

A systematic literature review was carried out on Slow Fashion following a qualitative methodology based on an in-depth literature review.

On 2 March 2021, research was carried out on all the papers published to date on recognized databases, namely, Scopus and Web of Science.

On Scopus, we searched using the keyword "Slow Fashion", limiting the research areas to "business", "social science" and "environmental" and obtaining a total of 49 research documents.

In the Web of Science database, we replicated the survey using "Slow Fashion" as the keyword and defining the following research areas: "business", "green sustainable science technology", "environmental", "management", "environmental studies" and "interdisciplinary social sciences". We found a total of 24 research documents. The chosen areas were intended to limit the results to slow fashion consumer behavior and its impact on sustainability both as a whole and from a management perspective.

The decision to set a limit on the search was due to our intention to focus as much as possible on the industry under analysis.

A total of 73 articles were obtained, of which 16 were excluded because they appeared in duplicate in both databases; in all, 57 documents were considered (see also supplementary material).

The collected documents were placed in an Excel sheet and sorted by year of publication. The data collected included author(s), title, type of document, the journal where the publication was made, keywords, abstract, objectives, research questions, variables, hypotheses, context, theories addressed, sample, methodology, achieved results, and recommendations for future studies. In several articles it was not possible to complete all fields for reasons of consistency, document context, or access. Through this choice and compilation of data it was possible to create a comprehensive screening of all the considered research papers.

After synthesizing all the information collected from the final 57 papers in Excel, one was excluded for not being in Portuguese or English and 13 were excluded for not allowing full access. We than read the 43 documents that could be fully accessed; it was possible to separate them into "+/− relevant" (13 articles), "relevant" (12 articles) or "not relevant" (18 articles). The 18 articles that were not relevant were excluded for not being related to consumer behavior in the context of Slow Fashion. We carried out a thorough analysis of the 25 documents considered relevant for the study in order to better understand their respective alignment with the topic under analysis. PRISMA Article was used for the analysis and exclusion of the articles [15], as shown in Figure 1.

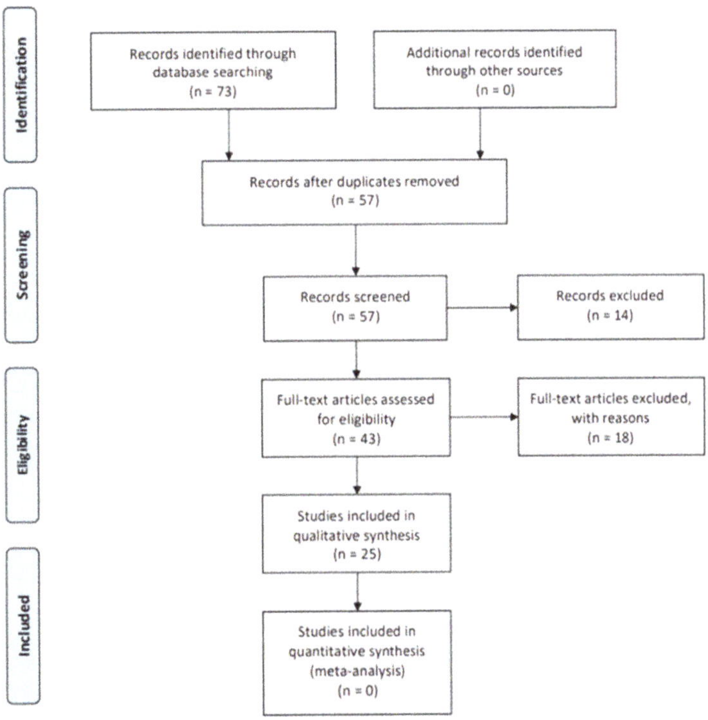

Figure 1. Slow Fashion PRISMA (Source [15]).

Data collected from the papers were descriptively analyzed by year of publication, number of papers per journal, and content. In the content analysis the theories presented in the papers, research methodologies used, recommendations for future research, keywords, and relevance were studied. A bibliometric analysis in VowViewer allowed for keyword analysis and the creation of a word cloud.

3. Results

3.1. *Brief Descriptive Analysis*

2017 was the year with the highest number of documents published in databases (11). Of the eleven documents, four were excluded for not allowing full access; 2019 turned out to be the year with the highest number of publications (7).

Slow Fashion is a relatively recent topic in academic studies, and in fact the first article to address the concept dates back to 2008, only twelve years ago. Figure 2 describes the number of articles per year.

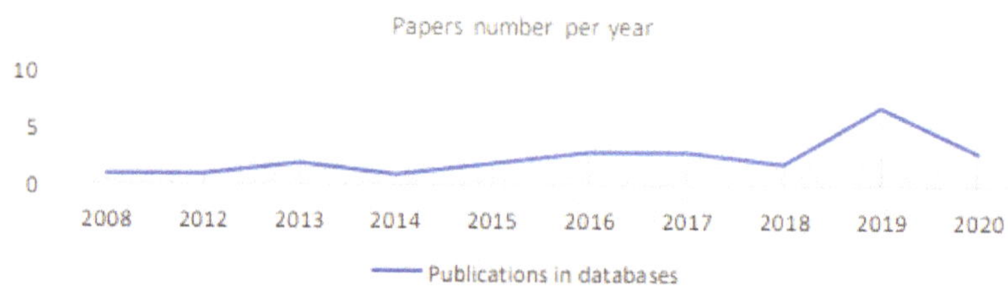

Figure 2. Papers: number per year.

Sustainability (Switzerland) was the journal with the highest number of papers (4). Two of the papers found did not clearly identify the journal. Figure 3 summarizes the number of papers per journal that we were able to find and use for content analysis.

Figure 3. Number of papers per journal.

3.2. Content Analysis

The content analysis of the final 25 documents used in this study provides an overview of the literature supporting the concept of "Slow Fashion". The theories presented throughout the papers were divided into seven groups, as they are associated with the same theoretical environment. Table 1 presents the groups of theories and corresponding authors.

Table 1. Theory groups.

Theory Groups	Theories
Sustainability	Approach to contribute to sustainable development [1]; Eco design [16]; Life cycle assessment [16]; Green and sustainability marketing [17]; Sustainable design theory [18]; Five dimensions of Slow Fashion [7,10]; Consumer Orientation to Slow Fashion [9,10]; Slow theory [8]; Rubbish theory [8].
Marketing	Marketing Theory [11,19]; Green and sustainability marketing [17]; Churchill's paradigm [9].
Culture	Fashion Cultures [1]; Cultural theory [20]; Attributes of classic design [21].
Fashion	Fashion theory [7,8,19,22–28]; Differences in patterns of dressing in modern and postmodern fashion [23]; Five dimensions of Slow Fashion [7,10]; The fashion business theory [8]; Attributes of classic design [21]; Eco design ([16]; Consumer Orientation to Slow Fashion [9,10]; Slow theory [8].
Human behavior	Theory Perspective on Consumer Quests for Greater Choice in Mainstream Markets [19]; Attitude-behavior relation [29,30]; Theory of planned behavior [29]; Consumer Orientation to Slow Fashion [9,10]; Consumption Theory [19]; Modern Social Theory [23]; Age ordering [23]; Sociological theory [18]; Feminist theory [8]; Balance theory [31]; Theory of Reasoned Action [29,32]; Schwartz values [9]; Institutional theory [18].
Methods	Grounding grounded theory [30]; Grounded theory [28].
Business	Theory of the relational view [33]; The fashion business theory [8]; Trickle-down theory [18]; Churchill's paradigm [9]; Rubbish theory [8].

The seven groups were assigned the following designations: sustainability, marketing, culture, fashion, human behavior, methods, and business. The name of each group is related to the main topic analysed. However, there are theories that are present in more than one group and that approach other topics as well.

The groups with the highest representation of theories were Human Behavior and Sustainability. Fashion and Business bring together a considerable number of theories as well.

The published papers considered in this research used different methodologies, which are summarized in Table 2.

Two types of methodological analysis were used: qualitative research and quantitative research. In quantitative analyses, surveys were the most used methodology. As for qualitative studies, the most used methodologies were mainly semi-structured interviews along with literature reviews and case studies.

For suggestions for future research, Table 3 summarizes the main recommendations.

People make up organizations, whether customers in general or other stakeholders. Therefore, research needs to be linked to consumer behavior, attitudes, ethics, and demographic characteristics.

When the Slow Fashion movement emerged, a main concern was to understand what should be done from a business perspective to embrace the concept and rethink business models. Initial papers focused on the production of Slow Fashion clothes such as yarn, yarn producers, and craftsmen. These studies attempted to help craftsmen/producers to develop their work, increase their activity, rethink their business models, gain recognition from consumers, and at the same time to understand and properly communicate the growing importance of adopting sustainable fashion practices.

Table 2. Methodologies.

Methodologies	Authors of the Articles
Content analysis of promotional materials used in sustainable fashion	Hammond, C. [22].
Interviews	Hammond, C. [22]; Henninger, C.E. [17]; Barnes, L., Lea-Greenwood, G., Zarley Watson, M. and Yan, R.-N. [27].
Literature review	Freudenreich, B. and Schaltegger, S. [1]; McNeill, L.S. and Snowdon, J. [34]; Da Costa, A.G., Soares, I.M., Pinto, B.F., Au-Yong-Oliveira, M., Szczygiel, N.; Amorim M.P.C., Costa C., Au-Yong-Oliveira M. and Amorim M.P.C. [35]; Štefko, R. and Steffek, V. [24]; Ozdamar Ertekin, Z. and Atik, D. [18]; Gwilt, A. and Rissanen, T. [36]; Clark, H. [8].
Survey	Pencarelli, T., Taha, V.A., Škerháková, V., Valentiny, T. and Fedorko, R. [31]; Casto, M.A. and DeLong, M. [21]; Delong, M.R., Bang, H. and Gibson, L. [23]; Gupta, S., Gwozdz, W. and Gentry, J. [19]; Şener, T., Bişkin, F. and Kılınç, N. [7]; Sung, J. and Woo, H. [29]; Magnuson, B., Reimers, V. and Chao, F. [25]; Jung, S. and Jin, B. [10]; Reimers, V., Magnuson, B. and Chao, F. [32]; Henninger, C.E. [17].
Case studies and case profiles	Clark, H. [8]; McNeill, L.S. and Snowdon, J. [34]; Cimatti, B., Campana, G. and Carluccio, L. [16]; Matheny, R. and Hernandez, A. [37].
Lab testing	Henninger, C.E. [17].
Use of social networks for research and study (YouTube, Facebook)	Da Costa, A.G., Soares, I.M., Pinto, B.F., Au-Yong-Oliveira, M., Szczygiel, N.; Amorim M.P.C., Costa C., Au-Yong-Oliveira M. and Amorim M.P.C. [35].
Focus groups	Pookulangara, S. and Shephard, A. [28].

Figure 4 shows a word cloud concerning the first keyword of each paper used in this study. The size of the words varies according to the relative frequency of their appearance, which allows us to identify fields directly related to Slow Fashion such as Ethics, Sustainability, and Design.

Considering the 25 documents used, 23 keywords were identified with more than one mention: "Slow Fashion" (7), "sustainability" (3), "design" (2), and "ethics" (2). All other keywords appeared only once. Researchers usually place the central themes of their studies in descending order according to their relevance to the paper.

A bibliometric analysis was performed to map the relationships between keywords. First, the keyword network was examined to identify all possible research fields and associated concepts (Figure 5). A co-occurrence analysis was performed with a minimum of one occurrence per word for a total of 79 keywords from the 25 papers. The size of the circles represents the frequency of keyword appearance; the larger the circle, the greater the occurrence, while the color of an item is related to the cluster where that keyword belongs. The lines between the circles show which keywords were mentioned together in a paper. In Figure 5, the keywords with most prominence are "Slow Fashion", "sustainability", and "fast fashion". The most prominent term, "Slow Fashion", is at the center of this paper as well and is linked to almost all of the other words.

Table 3. Future research field.

Authors of the Articles	Future Research Field
Freudenreich, B. and Schaltegger, S. [1]	Consumer's acceptance of the proposals.
Pencarelli, T., Taha, V.A., Škerháková, V., Valentiny, T. and Fedorko, R. [31]	Larger sample; do a comparative analysis between consumers.
Casto, M.A. and DeLong, M. [21]	Potential for extended wear; comparison of attributes with classic design; influence over *Slow Fashion* design.
Delong, M.R., Bang, H. and Gibson, L. [23]	Implications of dress patterns on sustainable practices; motivation to dress every day; ways of dressing from other populations.
Gupta, S., Gwozdz, W. and Gentry, J. [19]	Style versus fashion.
McNeill, L.S. and Snowdon, J. [34]	Larger sample; *Slow Fashion* salling strategies.
Şener, T., Bişkin, F. and Kılınç, N. [7]	Different dimensions.
Sung, J. and Woo, H. [29]	Test the model in various consumer markets; consumer behavior against excessive consumption.
Cimatti, B., Campana, G. and Carluccio, L. [16]	Sustainable fashion production.
Magnuson, B., Reimers, V. and Chao, F. [25]	Different attitude contexts.
Jung, S. and Jin, B. [10]	Cross-cultural research; attributes of *Slow Fashion*; consumer profile.
Reimers, V., Magnuson, B. and Chao, F. [32]	Consumer perceptions.
Henninger, C.E. [17]	Larger sample.
Ozdamar Ertekin, Z. and Atik, D. [18]	Different factors in the fashion field; market change.
Barnes, L., Lea-Greenwood, G., Zarley Watson, M. and Yan, R.-N. [27]	*Slow Fashion* consumers; fast fashion consumers.
Pookulangara, S. and Shephard, A. [28]	The power and impact from the *Slow Fashion* movement.
Clark, H. [8]	Redefinition of fashion.

The VOSviewer software allows for joining keywords into groups called "clusters" based on affinity and/or proximity. The VOSviewer program presented 18 clusters for the 79 items.

The two clusters with the higher number of items are shown in red and green. The red cluster presents eight items: clothing, clothing industry, ethic, social responsibility, sufficiency, sustainable consumption, sustainable management, and sustainable production. The green cluster has eight items as well: classic design, consumer perception, eco design, esthetic, green fashion, recycling, Slow Fashion, and sustainability. All other unnamed items were divided into clusters grouping from six to two items each. This analysis allowed us to understand how the keywords in different papers are interconnected. VOSviewer made it possible to identify which keywords were most prominent as well as to create clusters of related items.

The red cluster shows a higher interconnectedness between words; however, the expressions are not the ones that appear most often, such as "Slow Fashion" or "sustainability". These results seem to reveal that at the time of the database research there were still few articles that named these items as keywords, and/or that these two concepts have not yet been related to Slow Fashion.

Figure 4. Word Cloud.

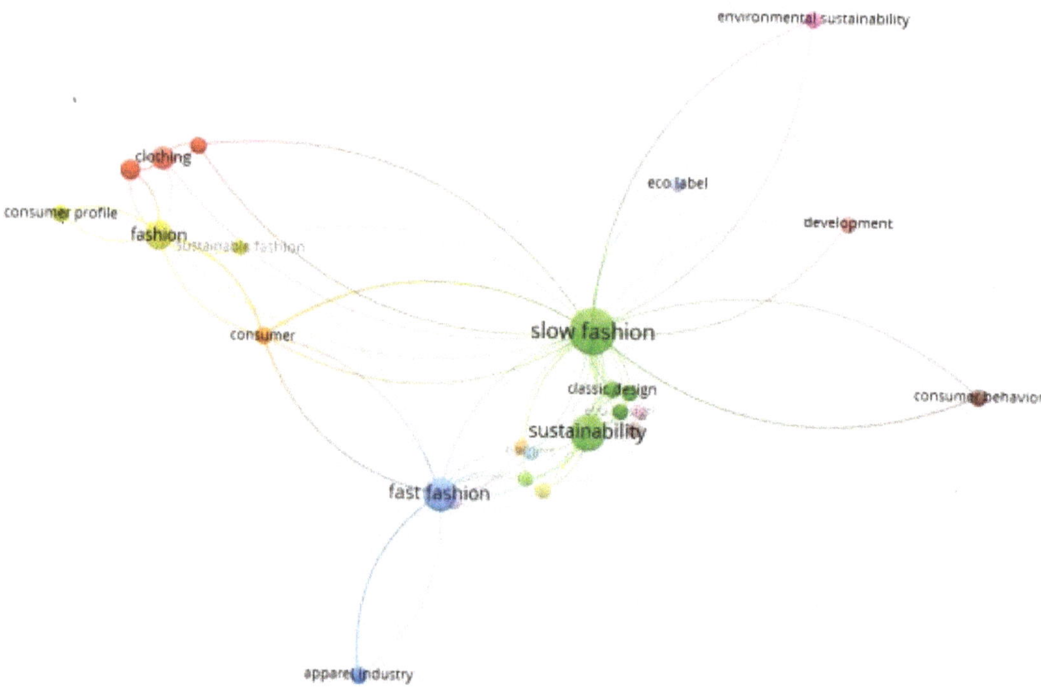

Figure 5. Bibliometric analysis.

This analysis made it possible to understand how the keywords of different papers are interconnected.

4. Discussion

Nowadays, consumers are increasingly looking for unique products and designs and willing to pay more if they feel that their needs and preferences are met [24]. This search for exclusivity is related to any human being's need to be perceived as different from others, enhance self-esteem and social image, mark a position, and consequently to avoid mass consumption [38–40]. In fashion, this search for uniqueness seems to be related to the style aspect; people prefer to reorient the quality of the products they use bearing in mind certain integral aspects, not only appearance. Clothes that are customized and made of recycled materials reveal a high concern for sustainability. In other words, sustainable consumption is positively associated with environment and social responsibility [19].

Several studies have revealed that the more classic a piece of clothing is, the lesser its temporality, and therefore its alignment with a sustainable perspective. A simple and discreet classical design is usually considered an expression of maturity and sophistication [21], revealing a specific social image in which self-image seems to impact an individual's clothing choice. Self-image may act as a variable leading to the choice of slow or fast fashion [27,41], as slow fashion consumers have a non-modern self-image. They follow their individual style and bet on more classic and simple clothes [24,27]. Promoting the idea that a sustainable consumption leads to a very positive and effective image could be a potential solution for achieving an increase in Slow Fashion.

The Slow Fashion movement shows the importance of sustainability to the fashion industry players along with a change in peoples' values [1,6]; it is important to understand which values have changed, the reasons for this, and which reasons currently guide consumer purchasing decisions.

According to the papers used in this research, two specific generations have been studied, namely, Generation Y and Generation Z. Data show that both generations are currently looking for more mature and timeless clothing, especially females. In different contexts, both generations have similarities in their dressing practice and are moving in the direction of timeless clothing; however, they feel the need to be up to date as well. This arises from an external influence resulting from a location with an eco-conscious culture that leads to the involvement in more sustainable shopping [23].

Environmental sustainability has increasingly emerged as an important issue in the fashion industry. On one hand, the chemicals used in the production and finishing of the goods are recognized as capable of harming humans and the environment; on the other hand, there are a lot of products that end up not being consumed, therefore becoming wasted and or being destroyed by brands. Slow Fashion promotes a change in the consumer mindset that encourages a preference for quality over quantity [42].

The Slow Fashion consumer cares about all the participants in the value chain, bearing in mind both producers and society itself. They care about sustainable companies in sustainable markets [28,42].

Based on the existing literature it is possible to state that culture is crucial to consider when developing a marketing strategy. Cultural values influence how consumers evaluate and respond to marketing strategies and how they decide to buy products and services [43]. In other words, human beings are influenced by the community where they were born and grew up. This impacts their personality traits, beliefs, values, thoughts, and concerns, which result in motivations and later actions. In order to better understand how consumers perceive slow fashion, a mind map was developed (Figure 6).

This mind-map tries to present the top five expressions related to Slow Fashion consumer behavior. Five groups, in pink, reveal the key concepts associated with Slow Fashion: "ethical values", "sustainable consumption", "consumer motivations", "consumer attitudes", and "awareness of sustainability". Each group highlights certain questions that need to be properly answered in order to understand who the Slow Fashion consumer is and create a possible profile.

Figure 6. Slow Fashion Mind-Map.

Thus far, in considering consumer attitudes studies have looked at various attributes impacting buying decisions: environmental responsibility, employee welfare, animal welfare, Slow Fashion, costs, and physical and extrinsic attributes. Four of these were found to be important dimensions for ethical fashion perception: fashion employee welfare, Slow Fashion attributes, animal welfare, and environmental responsibility, with only the first two appearing to influence consumer attitudes [25,32].

Employee wellbeing has become a fundamental aspect for society, mainly due to news reporting on unfair earnings and working conditions, child labor, and social inequalities. Consumers are increasingly paying attention to these issues, enjoying brands that adopt working policies that value their employees and respect their rights. Being more concerned with socially responsible practices, consumers express it by preferring to make ethical purchases [42,44–46].

However, consumers do not need to associate Slow Fashion with environmental and employee welfare good practices to perceive ethical clothing; the durability of the goods and the materials used in their production is enough for this consumer perception [32], as well as for animal welfare and environmental responsibility.

A study that looked at 221 consumers in the US revealed that delivering an exclusive product affects the Slow Fashion customer's perception of value creation. Slow fashion and customized products increase perceived value, which in turn positively affects Slow Fashion purchasing intentions and the acceptance of premium prices [10].

When adopting sustainable consumption habits, the consumer tends to pay attention to transparency in terms of the way brands communicate and support local communities [31].

Regarding Slow Fashion consumer values, several studies have been found. A study developed in Turkey showed that main perceived values in Turkish Slow Fashion consumers are authenticity, locality, and exclusivity, while in Kazakhstan the main values are equity, functionality, locality, and exclusivity. Perceived values in both groups positively affect the intention to buy Slow Fashion products and predispose consumers to pay higher prices [7,24]. Based on US consumer orientation, four target groups of consumers have been identified: (1) a group highly involved in the movement; (2) a conventional group; (3) an exclusivity-oriented group; and (4) a group little-involved in the movement. These groups have been created based on values, purchasing behaviors and demographic aspects [9]. For each group, a distinct marketing strategy should be implemented.

In other research, certain values have been clearly identified: (1) quality versus emotionality; (2) price; and (3) social values. Consumers of Slow Fashion products perceive them as having a higher quality standard, which encourages individuals to consume these types of products, feel good about doing it, and invest in more durable goods [29].

Table 4 shows the information presented in this discussion regarding consumer behavior around Slow Fashion and synthesizes and relates it to the five groups identified in the mind map. These five key concepts seem to be somehow related to the information collected from the articles analyzed. However, further data are needed in order to assertively define the Slow Fashion consumer profile.

Table 4. Five categories and outcomes.

5 Categories	Slow Fashion Consumer Behaviour	Outcomes of Articles Analysed
Consumer motivations	Unique and classic designs	– Feeling of sophistication and maturity; – A taste for exclusivity that leads to a willingness to pay extra for it.
Ethical values	Values	– Functionality, exclusivity, equity, locality and authenticity.
Awareness of sustainability, consumer motivation and sustainable consumption	Concerns	– Impact of fashion industry on workers and society; – Where the product is produced; – Be less and less driven by fashions and wear your clothes for longer; – Fair trade; – Organic products; – Sustainable production.
Awareness of sustainability, consumer attitudes	Style orientation	– Style orientation seems to be interconnected with the purchase decision, buyers think more and more about recycling and reuse, they try to make more sustainable purchases, they worry about corporate responsibility, about the transparency of brands' communication and about how they support local communities.
Consumer motivations	Self-image and social Image	– This consumer likes to distinguish himself from others and pays attention to his social image and self-image. Self-image refers to the search for their own style and non-trendy, but always paying attention to the style factor, which increases their self-esteem.

5. Conclusions

According to Freudenreich et al. [1], one way to reduce clothing consumption is to create a structure to offer potential businesses. By developing these offerings, the clothing sector could modernize itself and move away from high-speed and low-price production towards a slower, more responsible, more sustainable business approach [1].

The consequences of the environmental and ethical impacts of this industry are now a concern for brands that encourage their consumers to prioritize longevity and ethical consumption over price and novelty [34].

This paper, after a careful analysis of the current existing bibliography on the topic and considering the gaps that still exist, highlights possible paths for future and complementary work. The focus must be on gathering data that allow a deep understanding of who the Slow Fashion consumer is and which factors influence their adoption of this new global tendency.

Highlighting specific research results, Štefko, R. and Steffek, V. [24], complemented by the studies of Santos [38], Giglio, E.M. [39], Vigneron, F., and Johnson, L.W. [40] show that Slow Fashion consumers look for unique designs which are associated with a taste for exclusivity, which in turn influences the probability of being willing to pay more for a Slow Fashion product and a future purchase.

Several authors, such as Štefko, R. and Steffek, V. [24], Barnes, L. et al. [27], and Yoo, B. and Lee, S.H. [41] argue that such consumers seek to distinguish themselves from others, and pay attention to their social and self-image. Self-image is related to creating a unique style instead of merely being trendy; a different and very-own style seems to increase self-esteem. Classic clothes with a simple and discreet design and longer durability due to the materials used expresses maturity and sophistication. Style orientation seems to be interconnected with purchasing decisions and the individual's main concerns and values. Buyers think more and more about recycling and reuse; they try to make more sustainable purchases, as they are worried about corporate responsibility, the transparency of brands' communication, good environmental practices, and how they support local communities. When buying, the Slow Fashion consumer considers the place where the product is produced and whether it is fair trade, sustainably produced, and organic. Increasingly, Slow Fashion consumers are concerned with buying items that are less driven by fashion trends, and with wearing them for longer. In addition, they care about the impact of the fashion industry on workers and society [19].

The current literature allows us to conclude that there are already identified values; these values plus individual ethics have a positive influence on sustainable consumption. The five values identified by Şener, T. et al. [7], and Stefko, R. and Steffek, V. [24] allow a first characterization of this consumer: (1) authenticity; (2) locality; (3) exclusivity; (4) equity; and (5) functionality. It should be noted that these authors pointed out that their samples should be extended and more research topics should be added.

We recommend that future research use a qualitative and quantitative approach gathering data through questionnaires applied to final consumers, focus groups, or Delphi interviews with well-recognized fashion experts.

The main limitations of this paper are the existence of few published papers on Slow Fashion and even fewer on Slow Fashion consumer behavior.

This paper presents a theoretical contribution, being the first literature review focusing on Slow Fashion Consumer behavior. In the context of sustainability, this paper relates Slow Fashion to companies' need to adopt a sustainable internal and external policy. It provides insights into the associations, values, and key motivations that drive sustainability-conscious consumers.

From a management perspective, it is important to highlight that understanding Slow Fashion consumer behavior is important for any brand; therefore, being able to understand the motivations, attitudes, beliefs, and values of these consumers, which have already been presented in this study, becomes a high-value asset and should be a priority for companies acting in the fashion industry.

Slow Fashion is a concept that consumers and brands are now beginning to pay attention to. Brands want to understand whether consumers are willing to pay more if they adopt a new business model moving from fast to slow fashion. Brands such as H&M and Mango have created new brands with timeless products, higher quality raw materials, and more responsible supply chains, and Zara is now implementing classical and durable products at their stores, which carry a higher price. How this will be accepted by customers and how to properly communicate demands for more detailed information about Slow Fashion adopters is as yet unknown. It demands further market studies to support such strategic decisions as whether to prioritize new brands, new product lines, prices levels, and policies. The number of consumers who are concerned about sustainability and consider Slow Fashion as a way to achieve sustainable businesses is increasing, especially among young individuals, and the challenge is for companies to answer promptly.

Supplementary Materials: The following are available online at https://www.mdpi.com/article/10.3390/su14052860/s1, CONSORT 2010 Checklist. Table S1—Theory groups. Table S2—Methodologies. Table S3—Future research field. Table S4—5 categories and outcomes.

Author Contributions: M.D. developed research ideas based on an extensive literature review, collected, and analyzed the data, wrote the paper; V.T.V. supervised this investigation; V.T.V. and S.F. provided a thorough review to enhance the overall quality of this paper. All authors have read and agreed to the published version of the manuscript.

Funding: This research received no external funding.

Institutional Review Board Statement: Not applicable.

Conflicts of Interest: The authors declare no conflict of interest.

References

1. Freudenreich, B.; Schaltegger, S. Developing sufficiency-oriented offerings for clothing users: Business approaches to support consumption reduction. *J. Clean. Prod.* **2020**, *247*, 119589. [CrossRef]
2. A Pegada da Nossa Roupa. Available online: https://www.publico.pt/2019/11/29/infografia/pegada-roupa-391 (accessed on 10 March 2020).
3. Caniato, F.; Caridi, M.; Crippa, L.; Moretto, A. Environmental sustainability in fashion supply chains: An exploratory case based research. *Int. J. Prod. Econ.* **2012**, *135*, 659–670. [CrossRef]
4. Zara: Staying Fast and Fresh. Available online: https://www.thecasecentre.org/main/ (accessed on 10 March 2020).
5. Choi, D.; Han, T.-I. Green Practices among Fashion Manufacturers: Relationship with Cultural Innovativeness and Perceived Benefits. *Soc. Sci.* **2019**, *8*, 138. [CrossRef]
6. Fletcher, K. Slow Fashion: An Invitation for Systems Change. *J. Des. Creat. Process Fash. Ind.* **2010**, *2*, 259–265. [CrossRef]
7. Şener, T.; Bişkin, F.; Kılınç, N. Sustainable dressing: Consumers' value perceptions towards Slow Fashion. *Bus. Strategy Environ.* **2019**, *28*, 1548–1557. [CrossRef]
8. Clark, H. Slow + Fashion—An oxymoron—Or a promise for the future ... ? *Fash. Theory J. Dress Body Cult.* **2008**, *12*, 427–446. [CrossRef]
9. Jung, S.; Jin, B. From quantity to quality: Understanding Slow Fashion consumers for sustainability and consumer education. *Int. J. Consum. Stud.* **2016**, *40*, 410–421. [CrossRef]
10. Jung, S.; Jin, B. Sustainable development of Slow Fashion businesses: Customer value approach. *Sustainability* **2016**, *8*, 540. [CrossRef]
11. Gwozdz, W.; Nielsen, K.S.; Gupta, S. *The Relationship between Fashion and Style Orientation and Well-Being*; Mistra Future Fashion Report. 2017. Available online: http://mistrafuturefashion.com/wp-content/uploads/2017/11/D3.1.1.2-Style-versus-fashion-and-wellbeing.pdf (accessed on 10 March 2020).
12. Bly, S.; Gwozdz, W.; Reisch, L.A. Exit from the high street: An exploratory study of sustainable fashion consumption pioneers. *Int. J. Consum. Stud.* **2015**, *39*, 125–135. [CrossRef]
13. Niinimaki, K. Eco-Clothing, consumer identity and ideology. *Sustain. Dev.* **2010**, *18*, 150–162. [CrossRef]
14. Szmigin, I.; Carrigan, M. Exploring the dimensions of ethical consumption. *Adv. Consum. Res.* **2005**, *7*, 608–613. Available online: https://www.researchgate.net/publication/42797313_Exploring_the_dimensions_of_ethical_consumption (accessed on 21 March 2020).
15. Liberati, A.; Altman, D.G.; Tetzlaff, J.; Mulrow, C.; Gotzsche, P.C.; Ioannidis, J.P.; Clarke, M.; Devereaux, P.J.; Kleijnen, J.; Moher, D. The PRISMA Statement for Reporting Systematic Reviews and MetaAnalyses of Studies That Evaluate Health Care Interventions: Explanation and Elaboration. *PLoS Med.* **2009**, *6*, 1–28. [CrossRef] [PubMed]

16. Cimatti, B.; Campana, G.; Carluccio, L. Eco Design and Sustainable Manufacturing in Fashion: A Case Study in the Luxury Personal Accessories Industry. *Procedia Manuf.* **2017**, *8*, 393–400. [CrossRef]
17. Henninger, C.E. Traceability the new eco-label in the slow-fashion industry? Consumer perceptions and micro-organisations responses. *Sustainability* **2015**, *7*, 6011–6032. [CrossRef]
18. Ozdamar Ertekin, Z.; Atik, D. Sustainable Markets: Motivating Factors, Barriers, and Remedies for Mobilization of Slow Fashion. *J. Macromarketing* **2015**, *35*, 53–69. [CrossRef]
19. Gupta, S.; Gwozdz, W.; Gentry, J. The Role of Style Versus Fashion Orientation on Sustainable Apparel Consumption. *J. Macromarketing* **2019**, *39*, 188–207. [CrossRef]
20. Clark, H. Slow + Fashion–Women's Wisdom. *Fash. Pract.* **2019**, *11*, 309–327. [CrossRef]
21. Casto, M.A.; DeLong, M. Exploring Esthetic Response to Classic as a Means to Slow Fashion. *Fash. Pract.* **2019**, *11*, 105–131. [CrossRef]
22. Hammond, C. Stitching Time: Artisanal Collaboration and Slow Fashion in Post-disaster Haiti. *Fash. Theory J. Dress Body Cult.* **2020**, *24*, 33–57. [CrossRef]
23. Delong, M.R.; Bang, H.; Gibson, L. Comparison of patterns of dressing for two generations within a local context. *Fash. Style Pop. Cult.* **2019**, *6*, 99–117. [CrossRef]
24. Štefko, R.; Steffek, V. Key issues in Slow Fashion: Current challenges and future perspectives. *Sustainability* **2018**, *10*, 2270. [CrossRef]
25. Magnuson, B.; Reimers, V.; Chao, F. Re-visiting an old topic with a new approach: The case of ethical clothing. *J. Fash. Mark. Manag.* **2017**, *21*, 400–418. [CrossRef]
26. Keith, S.; Silies, M. New life luxury: Upcycled Scottish heritage textiles. *Int. J. Retail. Distrib. Manag.* **2015**, *43*, 1051–1064. [CrossRef]
27. Barnes, L.; Lea-Greenwood, G.; Zarley Watson, M.; Yan, R.-N. An exploratory study of the decision processes of fast versus Slow Fashion consumers. *J. Fash. Mark. Manag.* **2013**, *17*, 141–159. [CrossRef]
28. Pookulangara, S.; Shephard, A. Slow Fashion movement: Understanding consumer perceptions-An exploratory study. *J. Retail. Consum. Serv.* **2013**, *20*, 200–206. [CrossRef]
29. Sung, J.; Woo, H. Investigating male consumers' lifestyle of health and sustainability (LOHAS) and perception toward Slow Fashion. *J. Retail. Consum. Serv.* **2019**, *49*, 120–128. [CrossRef]
30. Stannard, C.R.; Mullet, K. Consumption of Raw Materials by Crafters: Desired Characteristics of Yarn and Retailers. *Cloth. Text. Res. J.* **2018**, *36*, 17–32. [CrossRef]
31. Pencarelli, T.; Taha, V.A.; Škerháková, V.; Valentiny, T.; Fedorko, R. Luxury products and sustainability issues from the perspective of young Italian consumers. *Sustainability* **2020**, *12*, 245. [CrossRef]
32. Reimers, V.; Magnuson, B.; Chao, F. The academic conceptualisation of ethical clothing: Could it account for the attitude behaviour gap? *J. Fash. Mark. Manag.* **2016**, *20*, 383–399. [CrossRef]
33. Shih, W.Y.C.; Agrafiotis, K. The Sustainable Luxury Craft of Bespoke Tailoring and Its' Enduring Competitive Advantage. *Environ. Footpr. Eco-Des. Prod. Processes* **2016**, *2*, 137–161. [CrossRef]
34. McNeill, L.S.; Snowdon, J. Slow Fashion—Balancing the conscious retail model within the fashion marketplace. *Australas. Mark. J.* **2019**, *27*, 215–223. [CrossRef]
35. Da Costa, A.G.; Soares, I.M.; Pinto, B.F.; Au-Yong-Oliveira, M.; Szczygiel, N.; Amorim, M.P.C.; Costa, C.; Au-Yong-Oliveira, M.; Amorim, M.P.C. Innovating in the fashion industry for a more sustainable production and consumption. In Proceedings of the European Conference on Innovation and Entrepreneurship, ECIE, Aveiro, Portugal, 20–21 September 2018.
36. Gwilt, A.; Rissanen, T. *Shaping Sustainable Fashion: Changing the Way We Make and Use Clothes*; Routledge: Abingdon, UK, 2012. [CrossRef]
37. Matheny, R.; Hernandez, A. Slow Fashion in retail environments: Why storytelling is critical for product longevity. In *PLATE: Product Lifetimes and The Environment*; IOS Press: Amsterdam, The Netherlands; Delft University Press: Delft, The Netherlands, 2017; pp. 250–255. [CrossRef]
38. Santos, A. Estudo da Estratégia de Design e do Comportamento do Consumidor da Marca de Luxo Just Cavalli. Master's Thesis, Departamento de Ciência e Tecnologia Têxteis da Universidade da Beira Interior, Covilhã, Portugal, 2009.
39. Comportamneto do Consumidor. Available online: https://pt.scribd.com/doc/269575045/GIGLIO-Ernesto-Michelangelo-O-Comportamento-Do-Consumidor (accessed on 24 April 2020).
40. Vigneron, F.; Johnson, L.W. Measuring Perceptions of Brand Luxury. *J. Brand Manag.* **2004**, *11*, 484–506. [CrossRef]
41. Yoo, B.; Lee, S.H. Buy genuine fashion products or counterfeits? *Adv. Consum. Res.* **2009**, *36*, 280–286.
42. Jung, S.; Jin, B. A theoretical investigation of Slow Fashion: Sustainable future of the apparel industry. *Int. J. Consum. Stud.* **2014**, *38*, 510–519. [CrossRef]
43. Sobol, K.; Cleveland, M.; Laroche, M. Globalization, national identity, biculturalism and consumer behavior: A longitudinal study of Dutch consumers. *J. Bus. Res.* **2018**, *82*, 340–353. [CrossRef]
44. Sneddon, J.; Soutar, G.; Lee, J. Exploring wool apparel consumers' ethical concerns and preferences. *J. Fash. Mark. Manag.* **2014**, *18*, 169–186.

45. Ogle, J.; Hyllegard, K.; Yan, R.-N. An investigation of mothers' and tween daughters' clothing preferences and purchase intentions toward a prosocial clothing company. *J. Fash. Mark. Manag.* **2014**, *18*, 70–84.
46. Carrington, M.; Neville, B.; Whitwell, G. Why ethical consumers don't walk their talk: Towards a framework for understanding the gap between the ethical purchase intentions and actual buying behaviour of ethically minded consumers. *J. Bus. Ethics* **2010**, *97*, 139–158. [CrossRef]

MDPI
St. Alban-Anlage 66
4052 Basel
Switzerland
Tel. +41 61 683 77 34
Fax +41 61 302 89 18
www.mdpi.com

Sustainability Editorial Office
E-mail: sustainability@mdpi.com
www.mdpi.com/journal/sustainability

www.ingramcontent.com/pod-product-compliance
Lightning Source LLC
Chambersburg PA
CBHW040747250926
51850CB00007BB/1434
9783036555256